Security, Privacy, and Trust in WBANs and E-Healthcare

Wireless Body Area Networks (WBANs) are vulnerable to cyberattacks and security breaches that could unlock the door for cybercriminals to penetrate hospital networks. This book covers the fundamental concepts of security and privacy in WBANs including security requirements, issues, and challenges.

Security, Privacy, and Trust in WBANs and E-Healthcare highlights the taxonomy of threats and attacks in WBANs and Internet of Medical Things (IoMT) and presents all technical aspects related to the security and privacy of WBANs. In addition to outlining viable solutions that take into account constrained resources at WBAN end-devices, hybrid network architecture, application characteristics, and communication protocols, the book covers the core concepts of WBAN security, privacy, and trust. It describes both theoretical and practical aspects for those working in security in the WBAN and IoMT, emphasizing the most significant potential WBAN security issues and challenges. The book also covers intrusion detection and security risk assessments in WBANs as well as lightweight security solutions for WBANs, blockchain-based solutions for WBANs, and authentication and access control in WBANs through various applications and case studies.

This book is highly relevant to the graduate/postgraduate students, academicians, security system designers, security analysts, computer scientists, engineers, researchers, digital forensic experts, and other personnel working in information security, IoMT, and WBAN.

Security, Privacy, and Trust in WBANs and E-Healthcare

Edited by Anuj Kumar Singh and Sachin Kumar

CRC Press
Taylor & Francis Group
Boca Raton London New York

CRC Press is an imprint of the
Taylor & Francis Group, an **informa** business

Designed cover image: © Shutterstock

First edition published 2025
by CRC Press
2385 NW Executive Center Drive, Suite 320, Boca Raton FL 33431

and by CRC Press
4 Park Square, Milton Park, Abingdon, Oxon, OX14 4RN

CRC Press is an imprint of Taylor & Francis Group, LLC

ISBN: 978-1-032-62236-1 (hbk)
ISBN: 978-1-032-63375-6 (pbk)
ISBN: 978-1-032-63510-1 (ebk)

DOI: 10.1201/9781032635101

Contents

Preface

Wireless Body Area Network, popularly known as WBAN, is a network of interconnected wearable technology, sensors, medical devices, and clinical services. It makes it easier to implement a variety of healthcare solutions that reduce healthcare costs, offer immediate medical aid, and raise hospital treatment standards. Through the provision of individualized and focused care as well as the smooth exchange of medical data, E-Health Systems coupled with WBAN are transforming the medical industry. It is primarily propelled by developments in big data analysis, cloud services, portable devices, wireless communication networks, and sensor technologies. WBAN is unquestionably beneficial to patients in the healthcare industry since it enables better diagnosis and treatment through technical advancements and a mix of analytics. E-Health Systems and WBAN devices are more trustworthy in monitoring and tracking due to higher accuracy levels, which also assist minimizing human errors and inaccurate reporting.

These ubiquitous medical devices offer numerous benefits, but they also present serious privacy and security dangers. Medical systems use this sensitive health information to gather, process, and make decisions. Cybercriminals may very well be able to breach the hospital network and obtain unauthorized access to confidential patient data if they take advantage of the holes and weaknesses in these WBAN devices. Attacks on these WBAN devices may put the patients' lives in jeopardy and result in significant bodily injury.

The security and privacy concerns that patients and medical practitioners confront in E-Health Systems and WBANs are being covered in this book. The book identifies a topical subject since E-Health Systems and WBANs are vulnerable to cyberattacks, and a security breach there could unlock the door for cybercriminals to penetrate hospital networks. In addition to outlining viable solutions that consider constrained resources at WBAN end-devices, hybrid network architecture, application characteristics, and communication protocols, the book covers the core concepts of WBAN security, privacy, and trust. The book provides a solid foundation for those working in the domain of security of WBAN and E-Health Systems, emphasizing the most significant potential WBAN security issues and challenges.

The book has been structured into three assorted segments. The first section, which includes four chapters presenting the foundations of WBAN and E-Health System's security, is titled Part 1. Four chapters in Part 2 highlight latest developments in security and privacy of wearable and implantable devices. Five chapters have been grouped in Part 3, which focuses on security frameworks and security solutions for WBANs and E-Health Systems.

In **Chapter 1** of the book, various security needs, risks, potential security attacks, major issues with WBAN communication technology, main obstacles, and security solutions in E-Health Systems are analyzed with the goal of assisting system designers in creating complex systems including preventive measures. The importance of WBANs in E-Health Systems, their applications, the security, and privacy issues they raise, and solutions are going to be discussed in **Chapter 2**. The chapter also discussed patients' resistance to adopt WBANs and E-Health Systems, as well as their limitations.

Wireless body area networks are briefly introduced in **Chapter 3**, with an emphasis on security-related issues. The security shortcomings in WBAN, its security standards, and risk analysis are highlighted in the conclusion. Moreover, the technologies that are used frequently in WBANs are the ones that were given the greatest consideration during the investigation. WBANs, security concerns, physical sensor network architectures, and communication technologies are all covered in **Chapter 4** of the book. To pinpoint problems and enhance protection and monitoring precisely, we carried out a risk assessment (Practical Impact Assessment) in a real-world scenario to talk about potential WBAN security weaknesses and threats. As a result, we can design a safe network architecture for WBAN systems that are ready for forensic analysis.

The purpose of **Chapter 5** is to give a comprehensive overview of wearable medical technology and a current study of the situation of the market. Lastly, the survey presented in this chapter identifies important issues as well as current and novel solutions. In **Chapter 6**, real-world breaches and vulnerabilities are used to explore the security issues related to clinical equipment. Concerns about patient confidentiality, data gathering, storage, sharing, and ethical implications are all well covered. In addition, it examines new technologies that have the potential to improve security and privacy standards, such as developments in blockchain applications, authentication methods, and encryption techniques, while also analyzing current security protocols, frameworks, and approaches.

The need of preserving and improving WBAN dependability is emphasized in **Chapter 7**. With their clear relevance and significance, privacy and security in WBAN technologies are set to play critical roles in the future. This chapter aims to increase awareness and promote cooperation among stakeholders to protect lives and data in the continuously changing field of healthcare technology by providing an overview of privacy and security in wearable and implantable medical devices. **Chapter 8** covers the many security risks and problems related to wearable Internet of Medical Things (IoMT) devices, the privacy solutions and architectures that can be used to prevent assaults, and the requirements for guaranteeing WBAN security and safeguarding user privacy and well-being. Not to mention, the main shortcomings and inconsistencies between the proposed alternatives and the existing model are emphasized.

In **Chapter 9**, different frameworks related to security and privacy are examined that can be applied in the development of secure WBANs and E-Health Systems. Here, regulatory, ethical, and legal frameworks as well as access control, authentication, security management, intrusion detection/prevention, and communication security frameworks have all been discussed in detail. The important ideas and problems to consider while putting these into practice for WBANs and E-Health Systems have been emphasized in this chapter. A comprehensive security architecture to address the specific issues WBANs in e-healthcare is presented in **Chapter 10**. The framework incorporates technologies for intrusion detection, access control, encryption, and authentication to ensure the privacy, accuracy, and availability of critical patient data. Using modern security measures would improve patient outcomes and provide a more stable and safe hospital environment, encouraging trust among patients, medical professionals, and technology.

The application of machine learning (ML) in WBANs is examined in **Chapter 11**, with a focus on its functions in anomaly detection, intrusion detection, and privacy protection. The efficacy of ML-based intrusion detection systems in smart healthcare wearable and implanted medical devices is illustrated through case studies in this chapter. To guarantee secure transmission and encryption of sensitive medical data, the chapter also examines cutting-edge cryptographic techniques and privacy-enhancing technology. Adversarial assaults, resource limitations, and data privacy issues are only a few of the difficulties that highlight the necessity of ongoing study and innovation in WBAN security. Tiny Machine Learning, which is utilized to make decisions in emergency scenarios in the WBAN, is the focus of **Chapter 12** and is implemented in microcontrollers (Edge Devices) that are interfaced with sensors. Less memory space needed, reduced latency, reduced bandwidth, and low power, which translates to longer battery life, are the primary benefits of edge computing. They can be used in wearable applications because of their small size.

A two-layer communication architecture for wireless body area networks is covered in the book's last **Chapter 13**, with the first layer focusing on personal digital assistant (PDA) and sensor authentication and the second layer on access point and PDA authentication. Data protection was achieved by implementing a multi-authentication framework, and tier 1 and tier 2 security methods were built through key management systems. Tiers 1 and 2 saw the development of the security measures. Lightweight cryptography algorithms were explored in order to secure the networks, and symmetric key encryption and decryption were employed to increase security.

The book covers every important facet of WBANs and E-Health Systems and gives readers an overview of current research trends in the field of security and privacy. This book has a unique approach that addresses nearly all aspects of WBANs and E-Health Systems security.

About the Editors

Dr. Anuj Kumar Singh is working as Associate Professor in the Department of Computer Science and Engineering at Amity University, Madhya Pradesh, India. He has more than 20 years of teaching experience in technical education. He holds a PhD degree in the field of Computer Science and Engineering from Dr. A.P.J. Abdul Kalam Technical University, Lucknow. He passed MTech degree with First Distinction from Panjab University, Chandigarh, and BTech degree with First Honours from Uttar Pradesh Technical University, Lucknow, in Computer Science and Engineering. In addition to these, he has also qualified UGC NET. Having published more than 40 research papers in journals and conferences including SCIE and Scopus, he has also authored one book and edited five. He has also filed five patents. He has also supervised more than ten dissertations at postgraduation level. His areas of specialization include Intelligent Systems, Cryptography, Network and Information Security, Blockchain Technology, and Algorithm Design.

Dr. Sachin Kumar is an Associate Professor in the Akian College of Science and Engineering at American University of Armenia, Yerevan. Previously, he was a leading research fellow in the laboratory of Big Data and Machine Learning at South Ural State University, Chelyabinsk, Russia. He has around 16 years of teaching and research experience. He is actively engaged in teaching and research in the field of machine learning, natural language processing, smart healthcare etc. He is a senior member of IEEE, and member of ACM, IAENG etc. Additionally, he is an associate editor and reviewer in various journals. He has also delivered various keynote speeches in various conferences, workshops and events. He has also edited various books, and contributing in authored books. He has worked on various research grants as project leader and participant.

Contributors

Ashwini A.
Vel Tech Rangarajan Dr. Sagunthala R&D Institute of Science and Technology, Avadi
Chennai, Tamil Nadu, India

R. Abirami
School of Computing, SRM Institute of Science and Technology
Kattankulathur, Tamil Nadu, India

Pabitha B.
National Institute of Technology Puducherry
Karaikal, Union Territory of Puducherry, India

Rhythma Badola
Ayurvedic Advisor
Patanjali India Limited, Haridwar
Uttarakhand, India

R.M. Bommi
Chennai Institute of Technology
Chennai, Tamil Nadu, India

P. Chandrasekar
Siddartha Institute of Science and Technology, Puttur
Andhra Pradesh, India

Gagandeep
Chitkara Business School
Chitkara University
Punjab, India

Ankit Garg
Apex Institute of Technology-CSE and University Center of Research Development Chandigarh University
Punjab, India

K. Juliana Gnanaselvi
Rathinam College of Arts and Science
Coimbatore, Tamil Nadu, India

Meena Gupta
Amity Institute of Health Allied Science, Amity University, Noida
Uttar Pradesh, India

Rahul Joshi
School of Media Studies & Humanities
Manav Rachna International Institute of Research & Studies, Faridabad
Haryana, India

S. Karthika
Rathinam College of Arts and Science
Coimbatore, Tamil Nadu, India

A. Mary Joy Kinol
Saveetha School of Engineering, SIMATS
Chennai, Tamil Nadu, India

P. Muthu Krishnammal
School of Electronics Engineering, VIT-AP University
Andhra Pradesh, India

Ajay Kumar
KIET Group of Institutions, Delhi-NCR
Ghaziabad, Uttar Pradesh, India

Sachin Kumar
Akian College of Science and Engineering
American University of Armenia
Yerevan, 0019, Armenia

Suman Kumari
School of Media Studies & Humanities, Manav Rachna International Institute of Research & Studies
Faridabad, Haryana, India

C. Malathy
School of Computing
SRM Institute of Science and Technology, Kattankulathur
Tamil Nadu, India

H. Mangalam
Electronics and Communication Engineering, Sri Ramakrishna Engineering College
Coimbatore, Tamil Nadu, India

Tanvi Mishra
Amity Institute of Health Allied Science
Amity University, Noida
Uttar Pradesh, India

Karthik N.
National Institute of Technology Puducherry
Karaikal, Union Territory of Puducherry, India

Banu Priya Prathaban
School of Computing, College of Engineering and Technology
SRM Institute of Science and Technology, SRM Nagar, Kattankulathur
Chennai, Tamil Nadu, India

P. Rajasekar
Electronics and Communication Engineering, Sri Krishna College of Technology
Coimbatore, Tamil Nadu, India

ED Kanmani Ruby
Vel Tech Rangarajan Dr. Sagunthala R&D Institute of Science and Technology
Chennai, Tamil Nadu, India

Shridhar Sanshi
National Institute of Technology Karnataka
Surathkal, Karnataka, India

M. Sathesh
Vel Tech Rangarajan Dr. Sagunthala R&D Institute of Science and Technology, Avadi
Chennai, Tamil Nadu, India

D. Ferlin Deva Shahila
Vel Tech Rangarajan Dr. Sagunthala R&D Institute of Science and Technology, Avadi
Chennai, Tamil Nadu, India

Anjali Sharma
Mittal School of Business
Lovely Professional University
Punjab, India

Anuj Kumar Singh
Amity University Madhya Pradesh
Gwalior, India

R. Subash
School of Computing
SRM Institute of Science and Technology, SRM Nagar, Kattankulathur
Chennai, Tamil Nadu, India

G. Uganya
Vel Tech Rangarajan Dr. Sagunthala R&D Institute of Science and Technology
Chennai, Tamil Nadu, India

Vinod K. S.
Vel Tech Rangarajan Dr. Sagunthala R&D Institute of Science and Technology
Chennai, Tamil Nadu, India

Part 1

Introduction to Security of WBANs and E-Health Systems

1 Security Requirements and Challenges in WBANs and E-Health Systems

R. Abirami and C. Malathy

1.1 INTRODUCTION

Wireless Body Area Networks (WBANs) are crucial for real-time patient monitoring, requiring simple devices, low power consumption, and energy efficiency. During COVID-19, enabling secure patient information sharing and requiring the American Health Insurance Portability and Accountability Act (HIPAA) safeguards for healthcare professionals became crucial to ensure patient privacy and data protection amidst heightened digital interactions and telemedicine practices. Table 1.1 describes various medical sensors that are deployed on the human body.

TABLE 1.1
Various Medical Sensors Deployed on the Human Body

Sensor	Description
ECG	The electrical activity of the heart.
Blood circulation	Measuring forces that accelerate the body in three-dimensional space.
Blood pressure	Blood circulation exerts pressure on the walls of blood vessels.
Temperature of the body	The measure of the body's capability to produce and emit heat.
Muscle signal	The electrical activity of skeletal muscles is related to the nervous and muscular systems.
Electroencephalography	Measures automatic brain activity and other cognitive capabilities.
Blood sugar	Measures the sugar levels, including its type, source, and energy, in the bloodstream.
Respiration rate	Rate of inhalation and exhalation movements over a given period.
Heart rate	The frequency of the cardiac cycle.
Oxygen level	Indicates the level of oxygen present in the patient's bloodstream.

DOI: 10.1201/9781032635101-2

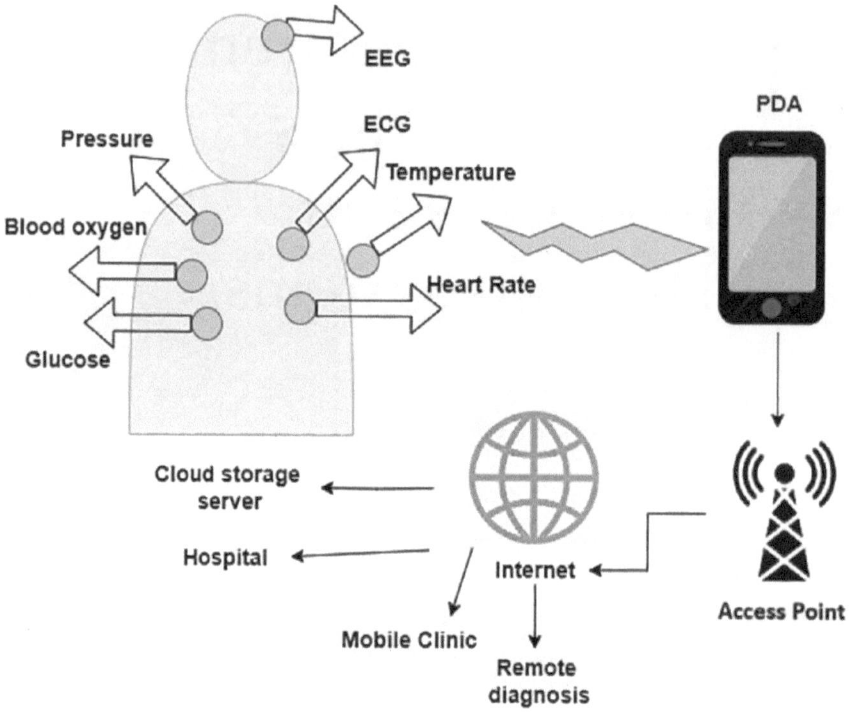

FIGURE 1.1 Generic WBAN architecture.

Body sensor networks use in-body communication for implanted sensor nodes and on-body communication for wearable devices. In-body communication uses a Medical Implant Communication System, while on-body communication uses Industrial, Scientific, and Medical Band and Ultra-wideband Communication. The tiny sensors can be implanted or worn to capture body-centered information and interface with a smarter portable device (hub). The hub processes data, generates alerts, and distributes it to medical professionals for long-term storage and study. This allows for earlier patient discharge at hospitals and home surveillance of problems. A PDA, or personal data assistant, is a device that gathers information from sensors and sends it to users through a Bluetooth connection. The communication network comprises the base station/access point to a cloud server, hospital, mobile clinic, and remote diagnosis through the internet. Body sensors and a gateway include WBANs, which monitor a user's physiological and environmental conditions. The sensory data from the body sensors is transmitted via the internet or Bluetooth to the gateway, which then sends it to an online Remote Health Monitoring server in the service domain. This is explained in Figure 1.1, the generic WBAN architecture.

1.2 BACKGROUND

E-health systems must be secure in digital ecosystems like the cloud, Internet of Things (IoT), and smart cities. Some medical WBAN applications include electrocardiogram

(ECG), electromyography (EMG), blood saturation, glucose monitoring, temperature, audio, and motion sensors. Elderly individuals receive home-based patient information monitoring for urgent situations. The safety and privacy of sensitive health information must be ensured through urgent and diligent measures. Neglecting security could seriously affect individuals, healthcare providers, and society. However, maintaining the security of patient data is still complicated because interfering with their medical information can have undesirable consequences. Proactive steps must be taken to protect the integrity of e-health systems. Ensuring security at all levels is crucial when it comes to e-health systems. With the potential for passive and active attacks, all necessary measures must be taken to prevent breaches. According to a survey by the Identity Theft Resource Center, data breaches in the healthcare sector topped that of business, government, and other sectors. Traditional healthcare systems struggle to meet the needs of the global population. In 2001, more than 25 million patients in the United States utilized medical implants for ongoing health monitoring, while remote healthcare was advancing to include biomedical nano-networks. The modern system comprises the WBAN, which consists of networked sensor nodes monitoring physical parameters like heart rate, blood pressure, and sugar levels and is widely used in healthcare. It also allows patients to measure biological functioning, enhancing their quality of life and lowering healthcare expenditures. Wireless transmission of the data to the cloud for analytics and medical guidance facilitates real-time monitoring, enables predictive insights, and enhances collaborative decision-making in healthcare settings. When developing WBANs, it is essential to prioritize using compact sensors and actuators, secure connections, timely emergency response, constant patient monitoring, and short-range communication. WBAN use sensors on the body to monitor patients' health. There are two types of sensors: implantable and wearable. Implantable sensors are surgically inserted into the human body and do not require removal, offering a reliable and long-lasting monitoring solution. Diabetic and trauma care people and soldiers in action (military) benefit greatly from implantable biosensors. Wearable sensors are easily removable and monitor vital signs like blood glucose, temperature, heart function, and blood pressure. Wearable sensor devices monitor functions and the surroundings, delivering feedback to ensure the best possible state. GPS watches the distance, heat, and motion, while electroencephalogram (EEG) and blood pressure devices monitor critical patients. Four tiers can be identified in a WBAN and remote healthcare system ecosystem: tier 1, which includes senders and receivers that are both located inside the human body; tier 2, which provides for at least one device inside the body; tier 3, which provides for at least one device off the body; and tier 4, which includes communications that extend beyond the gateway [1]. Federated Body Area Network (FBAN) enhances the security, scalability, and speed of data transmission in WBAN [2]. The chapter examines potential security risks and suggests solutions for e-health systems connected with recent advances like cloud computing, IoT, and smart cities. It also identifies upcoming difficulties in achieving complete security and privacy for electronic health records. Cyber attackers are drawn to healthcare data breaches because of the high value of stolen personal medical records. These records contain valuable personal, insurance, and financial information that can be exploited [3]. The analysis identifies advanced solutions for future Smart BANs. Figure 1.2 explains the various types of WBAN communication.

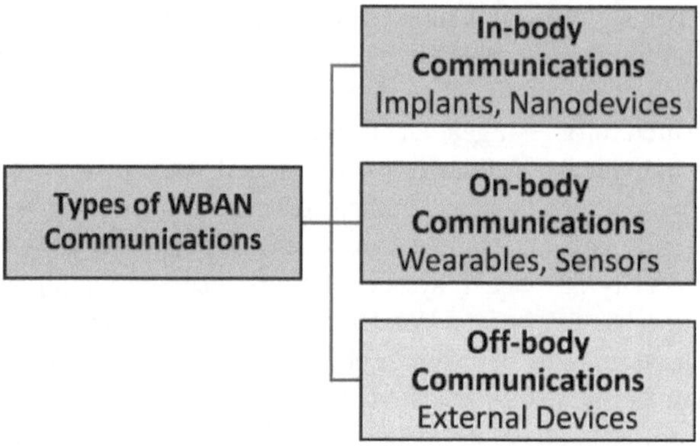

FIGURE 1.2 Types of WBAN communications.

TABLE 1.2
Four Tiers of Various Communication Methods in WBAN

Four Tiers of Communication	Communication Name	Communication Methods
Tier 1 (Intra-WBAN)	In body communications	Classical communications, biological communications
Tier 2 (Inter-WBAN)	On-body and in-body communications	Wireless radio frequency communications, optical and ultrasound communications
Tier 3 (Inter-WBAN)	On-body and off-body communications	Near Field Communications, IEEE 802.15.6
Tier 4 (Beyond-WBAN)	Off-body communications	WiFi links, LPWAN

The WBAN stores health data in digital healthcare record systems for improved service. The ability of patients to roam freely and leave the hospital can improve their quality of life while also lowering hospital costs. Long-term data obtained in the patient's natural environment allows for more precise and, in some cases, speedier diagnosis. Table 1.2 depicts the various four tiers of communication and its methods in WBAN. This chapter systematically reviews the literature on possible breaches and security requirements for WBAN applications to develop effective data protection and regulatory security solutions.

1.3 SECURITY REQUIREMENTS OF WBANS

Information and communication technology (ICT) advancements have increased the delivery of E-health services, influencing society by making high-quality services accessible in remote regions. IoT solutions for e-health use data gathered from WBAN to track patient health, which is subject to analysis, and the recommendations can be provided as medical advice. Connectivity, efficiency,

TABLE 1.3
WBAN Security Measures

Security Measures	Descriptions
Data originality	Services for licensing and verification
Data confidentiality	Encryption to avoid unauthorized access and robust to user collusion
Data integrity	Data not to be changed by an attacker
Data availability	Maintaining medical services accessibility even under denial-of-service (DoS) attacks
Data freshness	Old data are not recycled
Data authentication	Ensures information is sent by using public and private keys
Secure management	Manage key distribution between network nodes independently
Flexibility and accountability	Medical facilities must record and maintain patient information for timely and accurate care. Emergency information can be shared with others
Privacy and compliance requirements	International rules and regulations protect patient data and vital information

accessibility, and trustworthy communication are essential for advancing e-health. Due to limited resources, body communication security remains underexplored. Failure to save and utilize sensitive information can result in discomfort, incorrect treatment, relationship problems, and job loss. However, security and privacy are put at risk by electronic medical data sharing and remote consultation. Healthcare organizations are proactively safeguarding patient data and preventing costly data breaches. These measures include increasing cyber security budgets and implementing critical measures such as risk management, security controls, and monitoring. It's important to note that the expense of implementing preventive cyber security measures is only \$8. In contrast, the expense of dealing with a major breach after it has occurred is \$200 per patient record [4]. The physiological parameters, sensor design, and transmission technology of disease monitoring systems vary, with WBAN technology being critical for heterogeneous systems that demand various sensors and transmission protocols. In WBAN applications, ensuring data security involves maintaining confidentiality, integrity, and availability (CIA). However, achieving CIA compliance requires implementing additional security measures described in Table 1.3.

1.3.1 Data Originality

To ensure the accuracy of sensor data in WBAN medical care applications, it is necessary to verify and license each sensor and base station.

1.3.2 Data Integrity

To ensure the legitimacy of the information received, it's important to confirm its genuineness by using both public and private keys to verify the sender's authenticity.

This process is crucial in maintaining the integrity of data. Data integrity ensures data is not modified during transmission or storage and protects sensitive patient health information (PHR) in WBAN applications. Performing data integrity checks prevents attackers from manipulating data, which can cause system failure or patient injury and lead to illegal goals being achieved. To maintain data integrity, a Message Authentication Code (MAC) or a hashed MAC is utilized, and its validity is confirmed by reversing the code generation process for authentication.

1.3.3 Data Confidentiality

In medical scenarios, WBAN nodes transmit highly sensitive information. WBANs rely on sensor nodes to store and share PHR data, but protecting confidentiality is crucial due to security challenges and potential unauthorized access. If this data is overheard, it can cause harm to the patient, provider, or the system. Maintaining data confidentiality is crucial, and encryption is vital in achieving this. It ensures that data remains secure and protected from unauthorized access.

1.3.4 Data Availability

Unauthorized access or capture can result in the loss of valuable medical data. The essential physical system of WBAN relies on data and resource availability, which makes it vulnerable to DoS attacks. Consistent availability of medical sensor nodes is critical in healthcare applications to ensure uninterrupted access to health data. Because the unavailability of patient health data can result in patient loss in medical crises, WBAN applications need data availability to authorized workplaces.

1.3.5 Data Freshness

Data freshness techniques are crucial to maintain the safety and privacy of data transmitted through a WBAN. Unwanted parties attempting to replay old data can confuse the WBAN coordinator [5]. To prevent this, employ two types of data freshness—strong freshness with delay and frame ordering guarantees and weak freshness with frame ordering guarantees only. Strong freshness is mandatory for beacon synchronization, while soft freshness suffices for low-duty WBAN nodes. The freshness technique effectively prevents any attempt by an attacker node to reuse, record, replay, or publish old data. This is a crucial measure to ensure the security and integrity of the system.

1.3.6 Data Authentication

Data authentication and authorization are essential measures to enable secure monitoring, and developments in data security are being examined. WBAN applications use authentication to confirm the user's identity. Man-in-the-middle attacks are decreased by mutual authentication mechanisms that allow senders and receivers to authenticate one another. To guarantee security, the authentication system must force each node to identify and establish the sources from which it receives information. This can be accomplished through the use of techniques like symmetric authentication. Authentication and encryption are used in WBAN security schemes. Identity authentication ensures that nodes have permission to connect to the network's resources and prevents unauthorized intruders. Authentication schemes must be improved as WBAN applications become increasingly diverse.

1.3.7 Secure Management

Firewalls, authentication, and routers are examples of security measures. Users must have permission to get to and keep their health information. To ensure secure communication in WBAN, it is essential to implement a high-performing encryption technique and a highly effective key management plan. The role of the coordinator is to maintain secure control over the encryption and decryption operations involved in distributing keys to a WBAN. This ensures that nodes can be safely added or removed from the network during integration and disintegration. Communication and computer capacity, asymmetric cryptography systems, and symmetric cryptographic methods should be examined when selecting security strategies. It is intended to function effectively for both ad hoc and wired networks, considering critical issues such as data security, battery life, capacity, computing capability, and sensor network dynamic topological limits [6].

1.3.8 Flexibility and Accountability

It is crucial to follow the proper protocol when sharing a patient's vital information during an emergency. Prior consent from the hospital and the secondary party must be obtained before sharing occurs. Failure to adhere to this protocol will result in significant consequences. Staff in medical institutions are responsible for maintaining patient information, recording and retaining it, and are accountable for any misuse of patient's personal information.

1.3.9 Privacy and Compliance Requirement

Various international regulations, such as HIPAA, exist to protect sensitive patient information. Violations of these regulations can lead to severe consequences, including civil and criminal penalties of up to USD 250,000 or imprisonment for as long as 10 years [7]. For WBAN devices to work seamlessly together, they must be normalized. Failure to do so can result in similarity issues that could delay their functionality.

1.4 CHALLENGES IN WBANS

Sharing information through e-health systems is convenient, but it also raises valid concerns about security and privacy. WBANs aid in monitoring and responding to emergencies by guaranteeing safe application usage for patient healthcare at every stage. Every level of security must be handled, with simple defenses against active attacks like DoS as well as passive ones like those on the confidentiality of patient data. These concerns involve a range of challenges, such as accurately identifying users, maintaining precise data, safeguarding patient confidentiality, and guaranteeing the safety of each patient. The difficulty is in making sure that body sensors are hidden and not noticeable while they are being put in place on a person's body. The use of ICT in e-health systems plays a vital role in improving patient outcomes and reducing costs. However, these technologies also present vulnerabilities that can be exploited by attackers, putting medical data security and privacy at risk. As e-health systems become more integrated into digital ecosystems, researchers must work to address these challenges. WBAN data traffic is managed by algorithms for scheduling and routing protocols, assuring high-quality service levels. Power optimization techniques are used to limit energy use and data loss. The creation of a protocol that simultaneously provides energy competence and quality of service (QoS) support is also the most difficult task in WBANs. Various challenges in WBAN are mentioned in Figure 1.3.

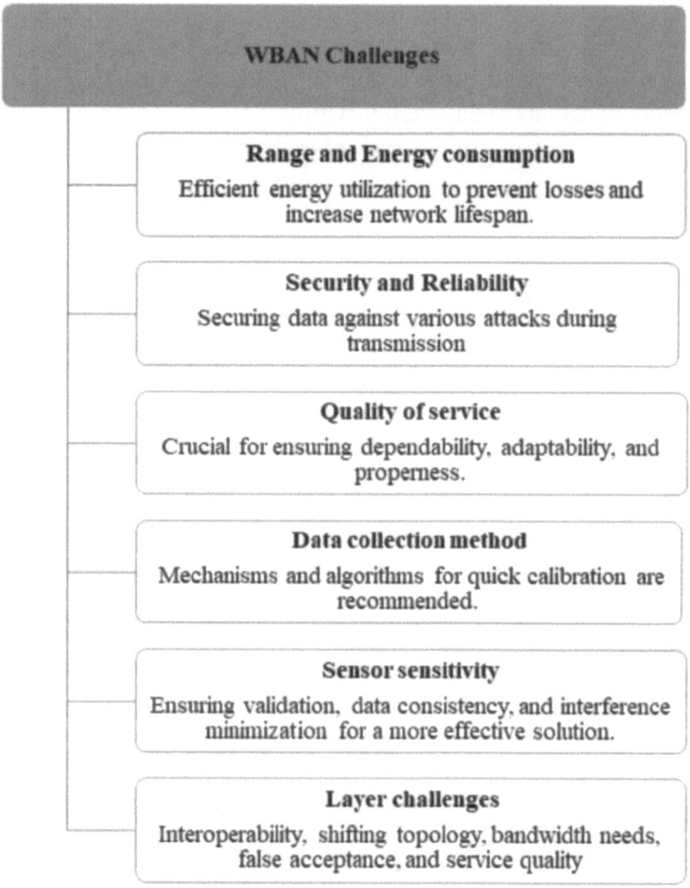

FIGURE 1.3 Various WBAN challenges.

1.4.1 Range and Energy Consumption

Wireless body sensor networks play a vital role in medical treatments. It can considerably lower patient care expenses by tracking vital sign data over time. However, optimizing these networks is necessary to reduce energy consumption and improve service and security. The body sensor nodes' construction is influenced by their weight, dimensions, and energy consumption limitations since they are placed on the body and rely on energy supply. Body sensor networks are built on a star topology and have restricted energy resources. Efficient and effective e-health systems are essential for society's welfare because compromised nodes might result in losses. Energy utilization is crucial in such networks since decreasing energy sources can drastically reduce the network's lifespan. Battery life is a concern for WBAN, and the routing standards for data transfer may change depending on the network design. Energy conservation can help increase the lifespan of a network. Clustering and routing protocols reduce energy consumption in WBANs by controlling cluster members' energy levels. In certain cases, such as in medical clinics, enclosed spaces, or when dealing with patients who have Alzheimer's or forgetfulness, charging sensors can be difficult. One possible solution is to harness the natural heat of the patient's body or to charge batteries through movement. The long lifetime of battery nodes in better-performing nodes allows for continuous communication between nodes. These methods can address potential issues and provide a solution.

1.4.2 Security and Reliability

The goal of network security is to secure data during transmission against attacks. To detect malicious data, both active and passive attacks are used. Security weaknesses are used at numerous levels, with attacks generated at each layer of the Transmission Control Protocol/Internet Protocol (TCP/IP) architecture. Small antennas on wireless sensor devices make it difficult to transmit large amounts of data over long distances, leading to reliability issues. To improve reliability and prevent inaccurate data, coding and multiplexing methods are recommended as solutions [8]. DoS attacks limit access to information and prevent authorized individuals from gaining access to resources. Distributed denial-of-service (DDoS) assaults target numerous hosts at the same time, causing network performance to suffer.

1.4.3 Quality of Service

QoS measures are critical for assuring dependability, adaptability, and properness. Identifying the right record for each of these metrics is a complex, specific job that necessitates back-to-back measurements such as redundant storage, message repetitions, and delays [9]. When designing QoS for Wireless Sensor Networks (WSNs), it is important to take into account the importance of the information and the requirement for a consistent supply. The amount of complexity of QoS needs in sensor network applications necessitates tests at every network layer, similar to the Open System Interconnection (OSI) seven-layer approach for sensor applications. These requirements are simply listed in Table 1.4.

TABLE 1.4
Quality of Service Requirements in Different Layers of WBANs

Layers	Quality of Service Requirements
Application layer	System lifespan, time to response, data novelty, data resolution, and dependability.
Transport layer	Cost, availability, delay, and throughput.
Network layer	Route latency, congestion likelihood, resilience of the routing, and energy consumption.
Connectivity maintenance layer	Network size, capability, standard path cost, connection, resilience, and maintenance of connectivity.
Coverage maintenance layer	Proportion, dependability, robustness, and maintenance of coverage.
Medium access control layer	Range, speed, dependability, and energy efficiency of communications.
Physical layer	The system can physically restrict resources on higher-layer QoS needs, affecting all other layers.

1.4.4 Data Collection Method

WBANs use intelligent micro- or nano-sensors to process and transfer data to base stations. This technology enables continuous medical treatment while allowing patients to continue going about everyday activities without restriction. WBAN helps save energy in tiny sensors and actuators by securely storing patient information without data loss. Sensors detect changes in the environment and relay data to electronic systems. WSNs improve data-collecting efficiency while being limited by memory capacity, computer power, and energy. In industrial settings, sensors with high sensitivity for response to inflammable compounds need to be sensitive to body temperature changes. Excessive sweat and steam can damage sensors, causing inaccurate physiological data recording. A robust sensor can jam the sensor node and stop it from regularly collecting patient data. There are concerns about the accuracy of decisions made after processing human body information collected by sensors. To solve this problem, it is recommended to utilize mechanisms and algorithms that can calibrate on time and quickly. IoT technology uses sensors in intelligent objects to sense internal or external conditions, perform actions based on collected data, and make decisions based on this data. Hierarchical data collection entails processing locally at sensor nodes, with gathering unit collaboration data for better categorization, extraction of features, and sensor coordination at higher levels. Sensor-based data gathering of essential human bodily data raises questions since it could result in poor judgment. To record sensitively with precision, sensors can be integrated.

1.4.5 Sensor Sensitivity

Implantable sensors are placed in intimate contact with the skin, and occasionally even inside the human body, to detect health data. They are an important class of biosensors because they can continually assess metabolite levels without human intervention and regardless of the person's physiological condition (sleep, rest, etc.). Sensors may not always accurately record and interpret sensitive human body information. It is imperative to address the accuracy concerns surrounding decisions made on the basis of human body information collected by sensors. Relying solely on sensors may not yield the necessary level of precision and sensitivity required for accurate recording. Thus, it is crucial to integrate sensors for both collection and processing to ensure a more effective solution. Sensor validation, data consistency, and interference minimization enhance coexistence.

WBAN sensors must be capable of sensing vital health indications that are necessary for healthcare practitioners to get timely and reliable monitoring data.

1.4.6 Layer Challenges

The challenges associated with the physical layer are interoperability, shifting topology, bandwidth needs, false acceptance, fluctuating data rates, continual signaling, safety, and service quality. The dynamic channels assignment, control messages and protocols overhead, control of errors, coordination, throughput, verification, packet scheduling, and over-emitting are some of the issues associated with the MAC layer of the WBAN. Transportation, location network, real-time streaming wellness services, multiple path routing, temperature management, and acceptance of faults are merely some of the different network layer-related problems.

1.5 SECURITY AND PRIVACY ISSUES

Implementing e-health systems poses significant security and privacy challenges. It's imperative to verify the user's identity, maintain data accuracy and confidentiality, and safeguard patient privacy both locally and remotely. E-health faces challenges with cloud computing, including data accuracy and safety, secure server architecture, energy efficiency, and managing dispersed clouds. IoT is transforming e-health. It should address patient mobility, technical security, and customized computing. Healthcare IoT hardware requires strict requirements, such as low power consumption and real-time operating systems. Frameworks and libraries can assist developers in efficiently using data and resources. Smart Health (S-Health) is crucial in smart cities, driven by e-health, which supports electronic and mobile health services. Challenges include lightweight, compact sensors, collaboration among stakeholders, and proper auditing procedures for pervasive healthcare in smart cities.

IEEE 802.15.4 executes security suites that are classified as secured or unsecured. It defines eight distinct suites, the first of which is the Null suite. The IEEE 802.15.6 standard, which was approved in 2012, establishes an international standard for low-power, short-term wireless transmission of information within and beyond the human body. A WBAN is susceptible to many different key attacks. These attacks are carried out in a variety of ways, including physical attacks, privacy violations, and DoS attacks. Protection against these attacks is a difficult task due to restrictions on the power used by the sensor nodes. Network security protects data from potential threats during transmission. Passive attacks modify data without degrading

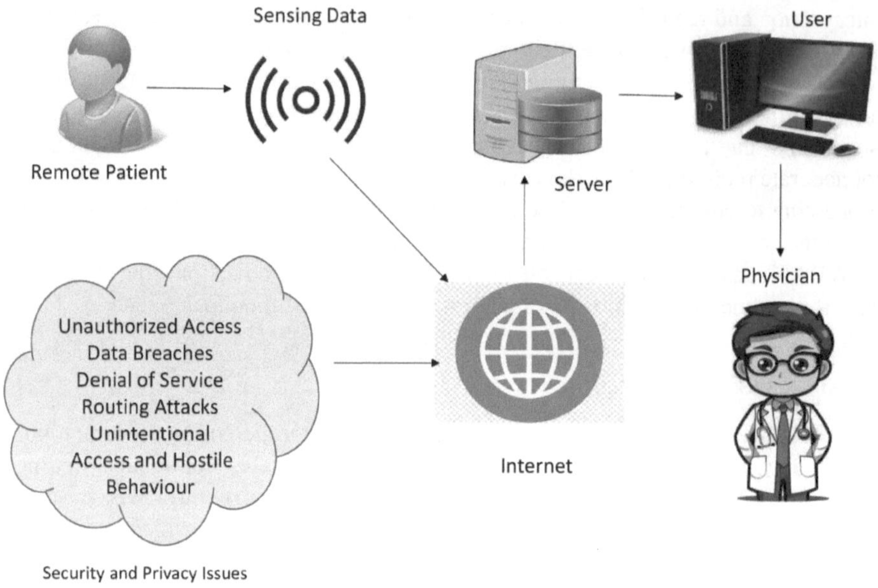

FIGURE 1.4 Security and privacy issues in WBAN.

resources, while active attacks impact system operation. The various issues related to security and privacy are pictured in Figure 1.4. Exploring the diverse range of factors that influence WBANs are as follows:

- Security measures address privacy, authentication, and security concerns.
- Depending on their usage, WBAN devices rely on batteries that may need frequent replacement or re-energization.
- The key management scheme incorporates cryptographic techniques, mutual authentication, and session keys.
- Attacker detection detects DoS attacks based on resemblance.
- Avoid intruders to detect attackers and communicate effectively by using alternate routes.

1.5.1 Unauthorized Access

Authentication methods aid in determining the origin of data. A centrally managed device in WBAN systems can help with data transmission. Sensitive information can be intercepted or changed by unauthorized access, which compromises patient safety. Ensuring data authenticity means that the data comes from its intended source, which helps prevent malicious actors from adding false information to the WBAN. This is achieved through the use of public key cryptography systems. Personal health information must be encrypted to avoid unauthorized data modification and to ensure that only trustworthy devices can insert or create data. Although many authentication methods are available to secure data transmission, there are still security and performance concerns, such as counterfeiting and disruption attacks. Biometrics could be a viable solution, but patient data security is still a problem.

1.5.2 Data Breaches

A data breach occurs when unauthorized parties gain access to personal information, financial records, or medical records. WBAN security breaches may occur at the physical, interaction, or data processing and storage levels. Physical layer breaches are uncommon and simple to spot, but they can be prevented with the right access control procedures. Jamming or radio frequency (RF) interference is frequent hazards that interrupt the WBAN spectrum and prevent the utilization of tier 1 and tier 2 devices.

1.5.3 Denial of Service

Wireless networks' current routing protocols are inadequate in preventing irregularities caused by DoS attacks and malicious data when sensor nodes transmit data to non-radio nodes. DoS attacks happen when resources are not provided promptly, whether actively or passively. Attacks that cause a DoS block a particular sensor by jamming or exhausting it with interference. These attacks are extremely concerning for WBAN systems, and more advanced measures are required to address passive attacks such as lazy node behavior. Improving solutions for dealing with DoS attacks is an ongoing area of research in e-health systems.

1.5.4 Routing Attacks

Data snooping, intercepting, and data modification are common security threats in WBANs. Passive attacks involve receiving a copy of transmitted data between sensors or nodes, making it hard to rectify without a key security management scheme. These threats may originate from internal or external parties, which complicate identification. A network layer can be attacked by a malicious user who steals or alters packets and sends them to a remote wireless center while changing the address information before passing them on to the next node [10]. This can lead to a deviated path or an infinite routing loop. Attacks that can occur at different levels within the TCP/IP architecture are described in Table 1.5.

TABLE 1.5
Attacks at Different Levels of TCP/IP Architecture

Layer	Attacks	Result
Logic link control layer	IP attacks	Address spoofing
Media access control layer	Internet Control Message Protocol (ICMP) attacks	Sniffing and a man in the middle
Network layer	Routing attacks	Black hole and eavesdropping
Transport layer	TCP attacks	High synchronization flooding
Application layer	Application layer attacks	Authentication issues

1.5.5 Unintentional Access and Hostile Behavior

Modifications in user conduct, information usage, security flaws, and regulatory difficulties are some unintended effects of wearable sensors. These problems can be reduced by using intelligent design methods. Sensors can have unintended consequences if they give patients or healthy people feedback on their physiological parameters and encourage them to alter their behavior. Changes in the behavior of sensors are brought about by feedback for better health. Complex human behavior, however, may have unintended consequences that have an impact on preventative measures [11]. Unintentional consequences from the following could occur in one or more cases:

- People exhibit worsening behavioral changes.
- People might experience increased health anxiety.
- The risk of device addiction among people.
- People may unintentionally self-diagnose a condition.
- Participants or caretakers put too much faith in the system's effectiveness.

1.6 SECURITY SOLUTIONS IN WBANS AND E-HEALTH SYSTEMS

WBANs implement safety measures such as intrusion detection and cryptography to enhance data authentication. Interoperability enhances WBANs' scalability, allowing priority transfer of patient information to doctors. Table 1.6 explains various methods and solutions obtained for WBAN security-related problems.

TABLE 1.6
Various Methods and Results for WBAN Security-Related Problems

Author[s]	Problems Identified	Methods Applied	Result Obtained
Hussain et al. [12]	Security issues	Key hash message authentication	Implement wearable devices with less cost
Fotouhi et al. [13]	Information attacks and intruders	Cryptographic hash function using real-or-random (ROR) model	Secured wireless network
Haider et al. [14]	DoS attack	Cooperative model using efficient resource utilization	Safe and reliable WBAN communication
Kasyoka et al. [15]	Eavesdropping attack and integrating attack	Efficient pairing-free certificate-less authentication scheme	Use on resource-constrained environment with efficient communication cost
Achour et al. [16]	Malicious attacks, selective jamming	Time-division multiplexing scheme for guaranteed time slots	Improve WBAN packet delivery ratio

TABLE 1.6 ***(Continued)***
Various Methods and Results for WBAN Security-Related Problems

Author[s]	Problems Identified	Methods Applied	Result Obtained
Arya et al. [17]	Data security attack	IoT technology in monitoring the health of critical patients	Ensuring data authentication and authorization
Thamilarasu et al. [18]	Intrusion attacks at the network level	Mobile agent-based intrusion detection systems (IDS) using machine learning and regression algorithms	Ensuring exact results while minimizing resource consumption
Dharshini et al. [19]	Vulnerable attacks	DoS Resistant Movement-Aided Authentication and Secret Key Extraction (DMASK-BAN) enables secret key extraction from DoS attacks	Optimize power consumption while maintaining high QoS

Prioritizing emergency patients ensures faster, secure communication by reducing complexity during network installments. Cooperation enhances system throughput and communication through reliable nodes while reducing data rate and channel quality due to distance and mobility. It can be a useful strategy for addressing issues related to network performance, such as adaptability, service quality, energy consumption, information dissemination, and distance.

However, it also introduces new challenges, such as potential obstacles, slowdowns, and additional workload. For solutions to be effective, they must strike a balance between safety, effectiveness, mobility, and usability.

1.6.1 Biometrics

WBAN sensors, which are small devices that measure physiological parameters, are subject to energy, memory, and computation constraints. Because of their high costs, conventional safety mechanisms are insufficient. Key establishing methods based on ECG enable WBAN nodes to establish trustworthy security keys while consuming the least amount of energy.

Biometric technology can provide accountable security solutions in e-health systems. These include processing speed, patient privacy connected to biometric features, biometric database protection, and trust management approaches to help authenticate an entity when the requester and the receiver are not previously known. Biometric technologies can improve security in e-health, but concerns about processing time, patient privacy, database protection, and trust management need further exploration.

1.6.2 Bluetooth and Wireless Security Protocols

Bluetooth is a standardized protocol that uses low-energy radio waves for short-range communication. It is an ideal choice for body sensor networks as it enables the observation of human body information. A Bluetooth chip, small in size, can create a personal wireless network that allows communication with nearby devices [20]. To improve energy efficiency in wireless body sensor networks, it's important to assess the energy usage of various components such as the sensor unit, signal processor, power supply, and communication unit. This can help to reduce the frequency of battery charging and overall energy consumption [21]. Wired Equivalent Privacy (WEP), Wi-Fi Protected Access (WPA), and Wi-Fi Protected Access version 2 (WPA-2) are wireless network security technologies. WEP, which had faults when it was first developed, was eventually substituted by WPA and WPA-2. For data encryption, WPA uses a pre-shared key and Temporal Key Integrity Protocol (TKIP), but WPA-2 utilizes Advanced Encryption Standard (AES) for safer and dependable encryption [22].

1.6.3 Encryption and Watermarking

To ensure better confidentiality, encryption can be used. A shared key is established on a secure communication channel between WBAN nodes and coordinators. Medical WBANs require strong encryption functions to ensure patient privacy and security. However, selecting the correct cryptography for medical sensor nodes can be difficult due to restricted resources. To maintain the security of the decryption and authentication process among multiple nodes, the coordinator must be responsible for distributing keys to all network nodes. The system creates watermarks from biometric information, embeds them at the sensor, and sends them via WBAN to the base station. It extracts made watermarks for integrity verification and recovers original data for medical applications. In [23] generation of watermarks in biometric data, the embedding of those watermarks, and the restoration of original data for healthcare applications are proposed and implemented in a novel integrity authentication scheme for WBAN based on reversible watermarks.

1.6.4 Key Management

Key management is the process of overseeing the generation, sharing, storage, and use of encryption keys. This is one of the most difficult parts of cryptography in practice. Key safety is critical for security since intruders and decryption analyzers frequently target encryption with public key systems and symmetric techniques via key management. As a result, a trustworthy and durable key management design is critical to ensure secure information flow. The Trust Key Management Scheme is a security solution for WBANs that utilizes ECGs. This scheme handles the creation and distribution of symmetric cryptographic keys to WBAN sensors, ensuring the privacy and security of the network.

1.6.5 Secure Routing

WBANs encounter energy consumption problems because of their small node size, which demands higher energy for transmitting data. This poses a significant threat to medical care applications that require mobility and dynamism, as existing protocols may not meet real-time security requirements. In addition, security concerns arise due to the possibility of patient information being hacked. A routing protocol is a technique for categorizing and safeguarding paths in a structured system with a focus on energy use. Routing methods are mostly designed for static wireless networks. By utilizing efficient routing protocols and clustering techniques, it is possible to minimize energy usage and mitigate security risks. DoS attacks, malicious data, and conflicts can all compromise routing protocols. Medical care applications require mobility and flexibility, impacting protocol security.

1.6.6 Trust Management

The WBAN design ensures that data is transmitted to medical staff or database administrators by ensuring authentic interaction among nodes in the network and the local processing unit (LPU). Trust management is an approach for ensuring the reliability, safety, and integrity of all nodes and LPUs. Trust is the relationship between two trusted nodes sharing data, determining the degree of secure reliability. Medical programs rely on trust evaluation in sensor node behavior, making trust management systems crucial for identifying the degree of trust in a node. WBAN faces challenges such as adaptive behavior, treating new users fairly, context perception, modifying identity, reliability, assurance score, and attacks on trust management.

1.6.7 Blockchain

The combination of WSNs and IoT is seen as a solution to power supply issues. However, security and privacy issues in data and commerce constitute significant barriers to smart grid adoption. To overcome these issues, Blockchain technology is being developed, which is a decentralized system that uses cryptographic techniques to record activities from the beginning of the chain. Blockchain technology has great promise for secure extra-WBAN communications in healthcare, allowing for data patient management, identity protection, drug integrity tracking, payment, and planning. It necessitates widespread acceptance, scalability, access security, and data privacy. It improves clinical care coordination by improving managing patients, identity security, drug integrity monitoring, and patient care coordination. Blockchain data formats enhance treatment and patient outcome analyses by enabling continuous availability and access to real-time data. Blockchain technology is being considered for smart grid applications such as WBAN, a distributed network that uses cryptographic techniques to record transactions.

Blockchain keys are private and public keys encrypted using mathematical algorithms, rendering the produced digital code irreversible and unchangeable. With a

TABLE 1.7
Problems and Recommended Solutions for Various Attacks

Attacks	Problems	Solutions
Masquerade attacks	Fake identity to access and modify the system, denial of service	Mutual authentication before access to the data and network
Medical device attacks	Eavesdrop, control, and spoofing of medical equipment	Rolling code cryptography
Data injection attack	Sends false data, exhausting resources	The Prediction-based Secure and Reliable (PSR) routing framework for WBANs enhances reliability
Clone attacks on Blind Carbon Copy (BCC)	Type of insider attack, replicate the legitimate nodes, and launch various attacks	Identifying user memberships and distinguishing between normal and clone users
Accountability attack	Unauthorized users exploit key abuse to access a secret key for decrypting e-health data	Detects and reveals the key abuser's identity in an n-bit space, ensuring no two users share a common identity
Privacy attack	Obtain the private identity and location information of a mobile patient while forwarding the protected health information (PHI) packets	Profile matching and privacy protection in mobile health networks, attribute-oriented authentication for protection of patient data privacy from potential breaches

pair of private and public keys, it protects the system from adverse assaults and malfunctions. Table 1.7 explains problems that occurred due to various attacks and recommended solutions.

1.7 CONCLUSION AND FUTURE SCOPE

WBANs have a wide range of uses, including health surveillance, assisted-living facilities, sports, military duties, search and rescue, security, and occupational health and safety. WBANs have grown in popularity among researchers and have a substantial impact on the healthcare industry. It is a secure system that stores patient information and transmits it to doctors without any loss. It prioritizes safety, reliability, speed, and, most important, security. The chapter examines the security and privacy needs of WBAN fields. It reveals that WBAN applications are susceptible to 11 different types of attacks, the most common of which are DoS, intrusion, and replay attacks. Data integrity, safety, privacy, extremely fine access control, simple encryption algorithms, and managing keys are the most frequently mentioned security requirements. More research is needed to tackle the overlapping challenges associated with the proposed new technology. The scope of the study does not provide a thorough discussion of the aforementioned difficulties; however, it will continue to do so in future scope. In future work, developing the WBAN technology requires a robust security system for authentication, especially in medical and military applications, as a secure authentication system is crucial for its widespread use.

REFERENCES

[1] Usman, M., Asghar, M.R., Ansari, I.S., Qaraqe, M. (2018), Security in Wireless Body Area Networks: From In-Body to Off-Body Communications. *IEEE Access*, 6.

[2] Ananthi, J.V., Jose, P.S.H. (2021), A Perspective Review of Security Challenges in Body Area Networks for Healthcare Applications. *International Journal Wireless Inf Networks*, 28, 451–466.

[3] Zeadally, S., Issac, J.T., Baig, Z. (2016), Security Attacks and Solutions in Electronic Health (E-health) Systems. *Journal of Medical Systems*, 40, 263.

[4] PwC Health Research Institute. (2016), *Medical Cost Trend: Behind the Numbers.* PwC: New York.

[5] Han, N.D., Han, L., Tuan, D.M., In, H.P., Jo, M. (2014), A Scheme for Data Confidentiality in Cloud-Assisted Wireless Body Area Networks. *Information Sciences*, 284.

[6] Farooq, S., Prashar, D., Jyoti, K. (2018), Hybrid Encryption Algorithm in Wireless Body Area Networks (WBAN). In *Intelligent Communication, Control and Devices; Advances in Intelligent Systems and Computing*; Singh, R., Choudhury, S., Gehlot, A., Eds. Springer: Singapore, Volume 624.

[7] Malik, M.S.A., Ahmed, M., Abdullah, T., Kousar, N., Shumaila, M.N., Awais, M. (2018), Wireless Body Area Network Security and Privacy Issue in E-Healthcare. *International Journal of Advanced Science and Computer Applications*, 9.

[8] Bedi, R.K. (2018), An Improved Energy Efficient TDMA Based MAC Protocol for WBAN. *International Journal of Computer Engineering*, 6, 34–39.

[9] Sagar, A.K., Banda, L., Sahana, S., Singh, K., Singh, B.K. (2021), Optimizing Quality of Service for Sensor-Enabled Internet of Healthcare Systems. *Neuroscience Informatics*, 1(3).

[10] Preethichandra, D.M.G., Piyathilaka, L., Izhar, U., Samarasinghe, R., De Silva, L.C. (2023), Wireless Body Area Networks and Their Applications—A Review. *IEEE Access*, 11, 9202–9220.

[11] Schukat, M., McCaldin, D., Wang, K., Schreier, G., Lovell, N.H., Marschollek, M., Redmond, S.J. (2016), Unintended Consequences of Wearable Sensor Use in Healthcare. Contribution of the IMIA Wearable Sensors in Healthcare WG. *Yearbook of Medical Informatics*, 1, 73–86.

[12] Hussain, S.J., Irfan, M., Jhanjhi, N.Z., Hussain, K., Humayun, M. (2020), Performance Enhancement in Wireless Body Area Networks with Secure Communication. *Wireless Personal Communications*, 1–22.

[13] Fotouhi, M., Bayat, M., Das, A.K., Far, H.A.N., Pournaghi, S.M., Doostari, M.A. (2020), A Lightweight and Secure Two-Factor Authentication Scheme for Wireless Body Area Networks in Healthcare IoT. *Computer Networks*, 107333.

[14] Haider, Z., Jamal, T., Asam, M., Butt, S., Ajaz, A. (2020), Mitigation of Wireless Body Area Networks Challenges Using Cooperation. *International Journal of Security and Its Applications*, 14(1), 15–30.

[15] Kasyoka, P., Kimwele, M., Angolo, S.M. (2020), Certificateless Pairing-Free Authentication Scheme for Wireless Body Area Network in the Healthcare Management System. *Journal of Medical Engineering & Technology*, 44(1), 12–19.

[16] Achour, M., Mohammed, M.A.N.A., Rachedi, A. (2020), On the Issues of Selective Jamming in IEEE 802.15. 4-Based Wireless Body Area Networks. *Peer-to-Peer Networking and Applications*, 1–16.

[17] Arya, K.V., Gore, R. (2020), Data Security for WBAN in E-Health IoT Applications. In *Intelligent Data Security Solutions for e-Health Applications*. Academic Press: London, pp. 205–218.

[18] Thamilarasu, G., Odesile, A., Hoang, A. (2020), An Intrusion Detection System for the Internet of Medical Things. *IEEE Access*, 8, 181560–181576.

[19] Dharshini, S., Subashini, M.M. (2020), DMASK-BAN Improving the Security of Body Area Networks. *Computer Fraud & Security*, 5, 13–19.

[20] Yaghoubi, M., Ahmed, K., Miao, Y. (2022), Wireless Body Area Network (WBAN): A Survey on Architecture, Technologies, Energy Consumption, and Security Challenges. *Journal of Sensor and Actuator Networks*, 11, 67.

[21] Qu, Y., Zheng, G., Ma, H., Wang, X., Ji, B., Wu, H.A (2019), Survey of Routing Protocols in WBAN for Healthcare Applications. *Sensors*, 19, 1638.

[22] Al-Janabi, S., Al-Shourbaji, I., Shojafar, M., Shamshirband, S. (2017), Survey of Main Challenges (Security and Privacy) in Wireless Body Area Networks for Healthcare Applications. *Egyptian Informatics Journal*, 18(2).

[23] Liu, X., Ge, Y., Zhu, Y.S., Wu, D. (2014), A Lightweight Integrity Authentication Scheme Based on Reversible Watermark for Wireless Body Area Networks. *KSII Transactions on Internet and Information Systems*, 8, 4643–4660.

2 Healthconnect
A Comprehensive Overview of WBAN Integration in E-Healthcare

Meena Gupta and Tanvi Mishra

2.1 INTRODUCTION

The notion of pervasive healthcare is an emerging technology that promises to amplify efficiency, precision, and accessibility in medical treatment [1]. Wireless Body Area Networks (WBANs) represent a vibrant realm of research and progress. They offer the potential for substantial enhancement in managing and monitoring healthcare provision. Latré et al. introduced WBANs, which are networks that comprise compact, smart biomedical sensors that are either implanted inside a patient's body or attached to it [2]. These sensor nodes manage the data and transmit it to a medical server. This data is then stored and evaluated by medical experts or physicians.

Recent advancements in wireless and information and communication technology (ICT) have significantly empowered the healthcare sector. These developments enable efficient control and delivery of diverse healthcare solutions and services. Enhanced ICT frameworks can potentially extend healthcare services beyond traditional hospital settings, reaching patients in their homes and workplaces. WBANs establish communication links with networks and various technologies, including ZigBee, Wireless Sensor Networks (WSNs), WiFi, Bluetooth, cellular systems, and Wireless Personal Area Network (WPAN) technology [3]. WBANs eliminate the need for daily hospital visits, offering patients the convenience of remaining in their homes while still receiving attentive medical care. Patient data is continuously monitored, allowing physicians to remotely analyze information at their convenience, ultimately improving patient care and operational efficiency in healthcare services.

Objectives of the chapter:

- To explore the significance of WBANs in E-healthcare
- To understand its components and architecture
- To obtain knowledge about technologies used in WBANs
- To know about its features and applications
- To understand security and privacy challenges and how to overcome them
- To learn about encryption and how it is implemented in WBANs

DOI: 10.1201/9781032635101-3

2.2 BACKGROUND

Every year, countless individuals succumb to chronic or fatal illnesses. The primary cause behind these fatal conditions often stems from the absence of timely diagnostic services. According to the findings of Movassaghi et al., a significant proportion of these diseases can be managed effectively through early detection [4]. The discovery highlighted that most ailments can be effectively managed by detecting them during their initial phases. This underscores the urgent requirement for proactive, accessible, and swift healthcare systems that enable ongoing health monitoring and prompt disease identification. The evolution of modern wireless technologies further accentuates the necessity for this research endeavor. According to Mohammed et al., the healthcare sector has been notably transformed by the influence of Internet of Things (IoT) technology and machine learning (ML). This transformation is evident in migrating routine medical procedures and healthcare services from hospitals to the comfort of patients' homes. This transition has simplified medical equipment utilization for both patients and healthcare providers [5]. The current state of the COVID-19 pandemic, coupled with unhealthy lifestyle habits, inadequate stress management, increasing healthcare expenses, and a progressively aging population, has collectively presented substantial hurdles for governments and healthcare institutions globally [6]. The escalating patient count has compelled the adoption of sophisticated technologies to empower physicians in remotely overseeing patients through sensors within WBANs.

While WBANs hold promise as a valuable asset for remote patient health monitoring, they are not devoid of constraints. WBANs encounter numerous obstacles in terms of data, network compatibility, security, and scalability. The imperative for WBANs is to securely, rapidly, and efficiently transmit patient information without altering the data while simultaneously ensuring minimal energy consumption. Security and privacy issues pose significant hurdles for the extensive acceptance of WBAN. The sensor nodes transmit highly critical and sensitive information. Any breach infringes upon patient privacy and could potentially jeopardize their well-being [7].

2.3 PROPOSED ARCHITECTURE

WBANs constitute an integral component within a multi-tier telemedicine framework. The standard architecture of WBAN encompasses three tiers of communication: intra-BAN communications, inter-BAN communications, and beyond-BAN communications [8].

a. Intra-BAN communications refer to the exchanges between wireless body sensors and the WBAN's master node. A WBAN monitors physiological signals through compact sensor nodes with wireless transmission capability placed either within or around a person's body. These nodes are employed to collect crucial health data during specific activities, be it medical, sports-related, or fitness-related. Each sensor node is capable of detecting, measuring, and processing at least one physiological signal. For instance, an electrocardiogram (ECG) sensor can track heart rate, an oxygen saturation sensor (SpO2) is utilized to measure oxygen levels, and blood pressure is monitored using a blood pressure sensor.

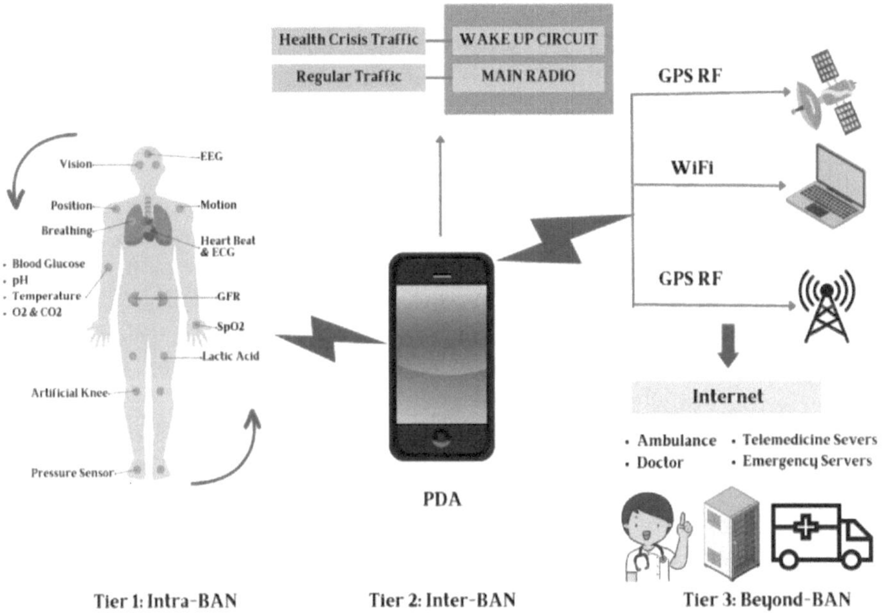

FIGURE 2.1 General architecture of WBAN communication.

b. Inter-BAN communications involve interactions between the master node and personal devices like laptops and home service robots. Tier 2 involves the implementation of a personal server (PS) application that operates on a portable device like a personal digital assistant (PDA), iPod, or similar mobile gadgets [9–11]. This application acts as a receiver for data from remote devices and subsequently transfers this data to a suitable computer once a communication interface becomes accessible.
c. Beyond-BAN tier connects these personal devices to the internet. These various communications are facilitated by technologies such as WiFi, Bluetooth, and IEEE 802.15.4. It contains multiple remote base stations responsible for storing patients' medical and nonmedical records, offering significant diagnostic recommendations.

2.3.1 Classification of Nodes

In a WBAN, a node is an independent device equipped with communication capabilities. Nodes can be categorized into three distinct groups, each of which assumes a role within the network based on its functionality and implementation.

2.3.1.1 Based on Functionality

a. ***Personal Device (PD):*** This device plays a pivotal role in interacting with other users. It collects data from sensors and actuators and then communicates this information to the user through an external gateway. The PD can display notifications through a screen or LED. In certain applications, it may be referred to as a body gateway, sink, or PDA.

b. ***Sensor:*** Sensors in WBANs are responsible for measuring specific parameters within or outside a patient's body. When a physical stimulus occurs, these nodes automatically gather and react to the data, processing vital information and sending responses wirelessly.
c. Some current examples of these sensors can be found in devices like wristwatches, smartphones, or headphones, allowing for wireless monitoring of individuals anytime, anywhere, and by anyone. Commercially available sensor types used in WBANs include blood pressure sensors, electromyography (EMG) sensors, spirometers, electroencephalogram (EEG) sensors, ECG sensors, temperature sensors, DNA sensors, CO_2 gas sensors, and more.
d. ***Actuator:*** Following the reception of data from sensors, the actuator engages with the user. Acting upon the sensor information, the actuator provides feedback within the system. An instance of this is the precise administration of medication dosages to a patient's body in applications like ubiquitous healthcare.

2.3.1.2 Based on IEEE Guidelines

The IEEE 802.15.6 has introduced an additional classification for nodes within a WBAN, particularly those implemented inside the human body [12]. This classification is outlined as follows:

a. ***Body Surface Node:*** In this category, nodes are positioned on the external surface of the body or placed at a distance from the patient, approximately 2 centimeters away.
b. ***External Node:*** Nodes classified as external are not in direct contact with the patient's body. Instead, they are positioned a few centimeters or around 5 meters away from the patient's body.
c. ***Implant Node:*** Implant nodes are those nodes that are inserted within the human body. They can be situated under the skin or within body tissues, either on the surface or internally.

2.3.1.3 Based on Type of Role

Nodes in WBANs are categorized based on their roles as follows:

a. ***Coordinator:*** The role of a coordinator node is similar to that of a gateway. It can resemble another WBAN or an access coordinator. All other nodes that need to communicate with each other utilize a PDA known as the WBAN coordinator.
b. ***End Nodes:*** Devices or end nodes in WBANs are designed to fulfil specific applications. However, they lack the capability to relay messages between nodes.
c. ***Relay:*** Also referred to as intermediate nodes, relay nodes have two types: parent and child nodes. These nodes serve as message transmitters.

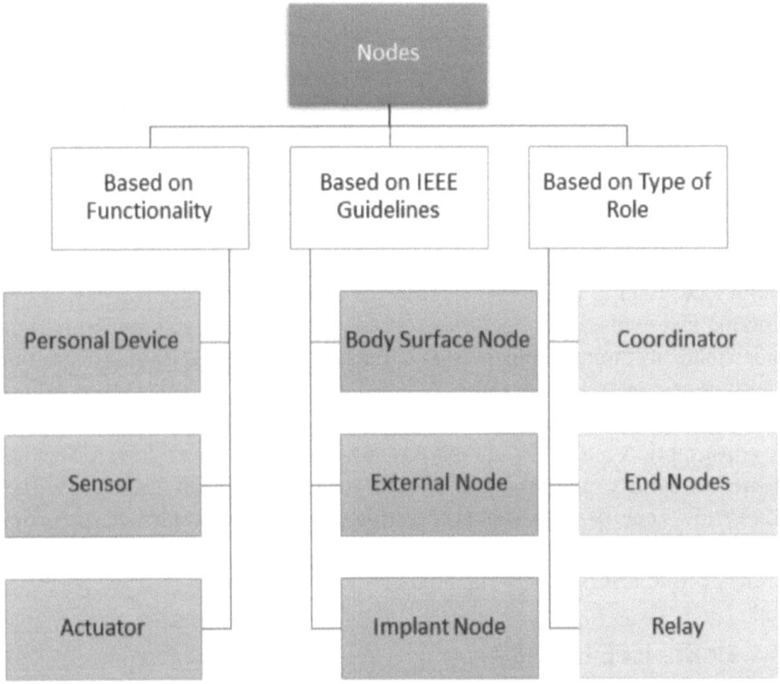

FIGURE 2.2 Classification of nodes.

Essentially, if one node reaches its transmission range limit, any data it needs to send must be passed through alternate nodes before reaching the PDA. Relay nodes are instrumental in data forwarding and can also be employed for data detection.

2.4 TECHNOLOGIES USED IN WBANS

Utilizing wireless technologies for transmitting data from sensors removes the necessity for physical cables connecting sensors to access points. The selection of the most suitable wireless technology holds significant importance within WBAN systems, as an inappropriate choice can lead to wastage of time and energy. In the following segment, we present the distinctive features of the primary proposed wireless technologies for WBAN [51].

2.4.1 Bluetooth

Bluetooth was designed to facilitate short-range wireless communication. Multiple Bluetooth devices establish a compact personal network referred to as a piconet [52]. A piconet includes a master device controlling up to seven slave devices. While slaves can't communicate directly with one another, they can with the master. Operating within the 2.4 GHz ISM band, Bluetooth devices cover distances from 1 to 100 meters. Data transmission rates peak at 3 Mbps, with approximately 100 ms needed for connection setup and data transfer. In WBAN setups, Bluetooth often serves as a medium between sensors and the PD in the first tier. It's widely supported and free to use but might not be optimal for power efficiency, especially for high-data-rate sensors.

2.4.2 Bluetooth Low Energy

Bluetooth Low Energy (BLE) is designed for connecting small, low-power devices with mobile terminals. BLE's network topology resembles a star, and a BLE device can participate in multiple piconets as a master or slave, facilitating communication between piconets [53,54]. BLE offers data rates of up to 1 Mbps [55], making it well-suited for WBAN applications. BLE's low duty cycle results in a 90% power consumption reduction compared with Bluetooth, extending sensor battery life. However, BLE's range is limited to 10 meters, constraining its use to the first WBAN tier. It excels in quick synchronization, essential for latency-sensitive devices in WBAN health monitoring. Operating at 2.4 GHz, BLE's connection setup and data transfer time is less than 3 ms [56].

2.4.3 ZigBee-IEEE 802.15.4

ZigBee, a wireless technology, prioritizes prolonged battery life, low data rates, and secure networking [57]. Easy to install and configure, ZigBee supports various network topologies and numerous nodes. Its devices can function for years before battery replacement, making it suitable for various medical WBAN applications. ZigBee emphasizes security, offering three security levels to safeguard data from unauthorized access or alteration [58]. With a 100-meter range, ZigBee operates at 2.4 GHz, 915 MHz, and 868 MHz frequency bands, with data rates of 250 Kbps, 40 Kbps, and 20 Kbps, respectively. ZigBee is particularly well-suited for WBAN medical applications requiring frequent measurements and low-power text-based data transmission, such as body temperature and pulse monitoring. Nonetheless, its low data rate makes it unsuitable for high-data-rate WBAN medical applications and susceptible to interference in the crowded 2.4 GHz band [59].

2.4.4 WiFi-IEEE 802.11

WiFi ensures reliable, secure, and high-speed communication, integrated into smartphones, tablets, and laptops. With standards like 802.11 a/b/g/n, WiFi operates within the 2.4 and 5 GHz bands, covering distances of up to 100 meters. It can achieve data rates of up to 600 Mbps, ideal for substantial data transfers, voice calls, and video streaming. In WBAN systems, WiFi's power consumption prohibits its use in the primary tier, but it serves communication between PDs and access points (APs) in the second tier for specific WBAN medical applications. Like Bluetooth and ZigBee, WiFi operates in the industrial, scientific, and medical (ISM) bands, leading to potential co-channel interference [59].

2.5 FEATURES OF WBANS

An ideal WBAN encompasses a range of features to ensure its effectiveness, reliability, and suitability for various applications. Some essential features of an ideal WBAN are as follows:

a. ***Low Power Consumption:***WBAN devices should consume minimal power to extend their operational lifespan. Efficient power management ensures prolonged use without frequent recharging, making them suitable for continuous monitoring over extended periods.
b. ***Small Form Factor:*** Compact and lightweight devices ensure user comfort and enable unobtrusive placement on or within the body. This promotes patient compliance and minimizes interference with daily activities.
c. ***Wireless Connectivity:*** Utilizing wireless communication protocols enables seamless data transmission between sensors and external devices. This wireless connectivity eliminates the need for physical connections and enhances mobility.
d. ***Real-time Data:*** Real-time data transmission is crucial for time-sensitive applications. It ensures that healthcare professionals can receive data promptly and make timely interventions in critical situations.
e. ***Reliable Data Transmission:*** Reliable data transmission with low packet loss and minimal latency ensures accurate and consistent data collection. This reliability is vital for accurate monitoring and diagnosis.
f. ***Interoperability:*** Interoperability ensures that devices from different manufacturers can communicate effectively. This compatibility expands the range of available devices and promotes flexibility in designing WBAN applications.
g. ***Security:*** Robust security measures, including encryption and authentication, protect sensitive patient data from unauthorized access. Strong security safeguards patient privacy and data integrity.
h. ***Scalability:*** A scalable network can accommodate the connection of multiple sensor nodes. It allows for the addition of devices without compromising the network's performance or efficiency.
i. ***Quality of Service:*** Quality of service (QoS) mechanisms prioritize specific types of data or applications, ensuring critical data is transmitted promptly. This is especially important in scenarios where urgent data requires immediate attention.
j. ***Low Interference:*** WBAN devices should operate on frequency bands with minimal interference from other wireless technologies. This minimizes signal disruption and ensures reliable data transmission.
k. ***Ease of Use:*** Devices should be designed for user-friendliness, allowing patients or users to easily wear, operate, and maintain them without extensive training. User-friendly devices encourage patient compliance and ease of adoption.
l. ***Cost-effectiveness:*** WBAN devices and infrastructure should be economically viable. Affordability widens access to these technologies, making them accessible to a broader range of patients and healthcare providers.

m. ***Remote Monitoring:*** WBAN's remote monitoring capabilities enable healthcare professionals to monitor patients from a distance. This reduces the need for frequent hospital visits, increases patient comfort, and lowers healthcare costs.
n. ***Data Analytics:*** Collected data should be suitable for analysis. This allows healthcare providers to derive valuable insights from patient data, aiding in diagnosis, treatment planning, and health management.
o. ***Compliance with Regulations:*** Adhering to medical and privacy regulations ensures patient safety and data security. Regulatory compliance is essential to maintain patient trust and uphold ethical standards.
p. ***Longevity and Durability:*** WBAN devices should be built to withstand everyday use, environmental conditions, and physical activities. Durable devices ensure reliable and consistent operation over time.

2.6 APPLICATIONS OF WBANS

WBANs facilitate a range of innovative and captivating applications that span various domains. These applications encompass diverse fields like smart healthcare, assisted elderly living, emergency response, and interactive gaming. As previously highlighted, research has categorized WBAN applications into medical and non-medical domains [13,14]. Ullah et al. differentiated between in-body and on-body applications, particularly for medical contexts. Applications of WBANs are demonstrating impressive efficiency. These applications extend beyond human healthcare monitoring to encompass diverse areas like sports, fitness, gaming, electronics, body position measurement, person location tracking, military, and more [15].

2.6.1 Medical Applications

2.6.1.1 Telemedicine and Remote Patient Monitoring

Telemedicine and remote patient monitoring have emerged as vital solutions in response to escalating healthcare expenses and the global aging population. These advancements are driving the establishment of robust telemedicine networks, facilitating the provision of diverse healthcare services. Telemedicine operates through the integration of health information systems and telecommunication technologies, enabling healthcare practitioners, physicians, and medical experts worldwide to extend their services to a larger patient base. WBANs play a significant role in various medical applications, enabling continuous health monitoring and facilitating necessary medical interventions for patients even outside traditional healthcare settings [16]. In-body sensor nodes monitor internal organ functions, such as the operation of pacemakers and implantable cardiac defibrillators, restoration of limb movement, and control of bladder function. On-body sensor nodes contribute to medical applications by monitoring parameters like blood pressure, heart rate, temperature, and respiration.

This wealth of data empowers remote medical professionals with real-time insights for accurate diagnosis and treatment recommendations. The integration of telecommunication technologies with healthcare systems is reshaping the healthcare landscape, offering efficient and accessible care delivery to patients regardless of geographical constraints.

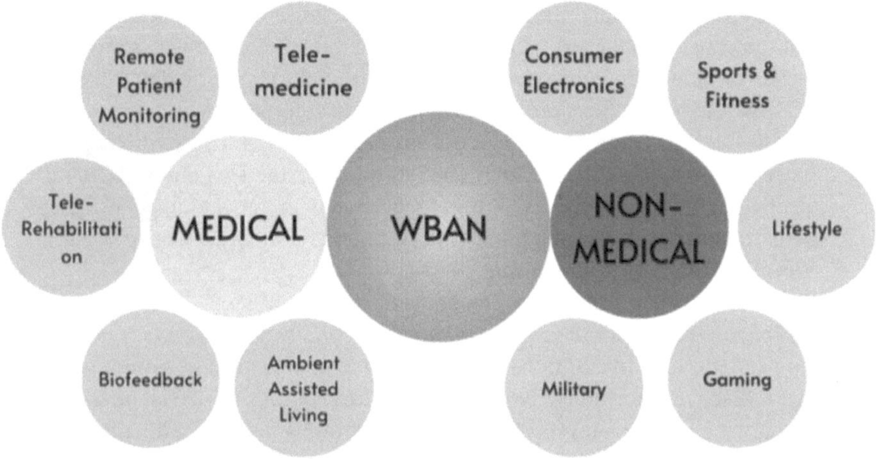

FIGURE 2.3 Applications of WBAN.

a. ***Cardiovascular Application***
This application focuses on monitoring cardiovascular diseases, which are leading causes of mortality worldwide. According to Ullah et al. [17], over 20 million individuals are affected by these diseases. WBANs can play a crucial role in reducing deaths by continuously monitoring patients' health. Any irregularities in heart rate or heart functions detected by sensor nodes can provide healthcare providers with valuable information, enabling proactive treatment.
b. ***Cancer Detection Application***
Cancer poses a significant threat to human lives, affecting millions of people, with cases increasing annually [18]. Small sensor nodes can detect nitric oxide, a substance often produced by cancer cells. These nodes can be strategically placed within the body's afflicted areas, allowing medical professionals to identify cancerous tumors without resorting to invasive biopsies.
c. ***Monitoring Blood Glucose Application***
Diabetes, a serious chronic disease, is on the rise globally. The World Health Organization indicates a steady increase in diabetes cases over the years, affecting more than 400 million adults. Poorly managed diabetes can lead to severe complications like blindness, stroke, kidney disease, heart disease, and high blood pressure [19]. WBANs have the potential to provide effective treatment for diabetes patients by continuously and accurately monitoring blood glucose levels. This monitoring can aid in timely interventions and

personalized treatment plans. Currently, the conventional method for blood glucose level measurement involves a finger prick and subsequent application of a blood drop to a test strip. However, this approach poses the risk of tissue damage over prolonged usage. To address this concern, wireless biomedical sensors can be implanted within the patient's body, offering a less intrusive means of monitoring glucose levels frequently throughout the day. Utilizing WBAN sensors, medical professionals can even automate insulin administration through actuator nodes. This automated process is triggered whenever glucose levels surpass a predetermined threshold [20]. This innovative strategy not only mitigates the discomfort associated with frequent finger pricking but also ensures precise and timely interventions for diabetes management.

d. ***Stress Monitoring Application***
Stress stands as a prominent precursor to various illnesses and disorders. Prolonged stress can trigger psychological conditions like severe depression and anxiety. In addition, it can contribute to elevated risks of coronary heart disease, morbidity, and mortality. Leveraging WBANs, real-time monitoring of stress levels in individuals becomes feasible. This real-time data assists healthcare professionals in accurately assessing stress and tailoring appropriate treatment approaches [21]. Nonetheless, the amalgamation of wearable sensors and IoT technologies has ushered in an era of continuous physiological signal monitoring. This advancement enables the early identification of stress-related issues and facilitates timely interventions to prevent the escalation of the condition [22]. By proactively addressing stress through real-time monitoring, these integrated technologies enhance the capacity for comprehensive health management and well-being.

e. ***Daily-Life Activity Monitoring***
This form of monitoring involves tracking the activities of patients as they go about their daily routines, particularly those with specific medical conditions. This approach provides insights into how these conditions affect their regular activities.

f. ***In-Hospital Monitoring***
In-hospital monitoring is centered around patients requiring intensive care and observation within a hospital setting. This form of monitoring is particularly pertinent for cases demanding prolonged hospital stays. During this period, various medical parameters are continually recorded to ensure comprehensive oversight.

g. ***Post-Surgery In-Home Recovery Monitoring***
Following a medical operation or surgery, patients often undergo a recovery phase at home. This phase is meticulously monitored to track the progress of recovery and ensure the patient's well-being. This monitoring approach bridges the gap between hospital care and the patient's independent recovery process.

These instances serve as illustrative examples, showcasing the extensive scope of telemedicine and how the incorporation of WBAN technology has facilitated its success in enabling close and remote patient monitoring.

2.6.1.2 Rehabilitation (Tele-Rehab)

Rehabilitation and therapeutic interventions aim to facilitate patients' return to normal functional capacity after their hospital discharge [23,24]. In essence, rehabilitation is a dynamic process that employs available resources to correct undesirable movement patterns, striving to achieve predefined expectations, such as an optimal position. Particularly for individuals recovering from conditions like strokes, the objective is to restore their independence and productivity. This necessitates ongoing monitoring and correction of patients' movements within a rehabilitation program to ensure accurate motion patterns. Consequently, the detection and tracking of human movements emerge as pivotal components, especially within a home-based rehabilitation framework. This underscores the significance of sensor diversity, real-time feedback for patients, multi-sensor data fusion, and integration with virtual reality technologies. These elements define rehabilitation as a distinctive research domain characterized by specific constraints and requirements [24].

2.6.1.3 Biofeedback

The realm of self-remote human body monitoring has now been opened through the utilization of WBANs to access data collected by embedded sensors. These sensors, whether implanted or positioned on the body, serve to monitor specific behaviors or medical conditions, thereby assisting patients in upholding their well-being. This is achieved by harnessing biofeedback mechanisms that encompass temperature analysis, blood pressure measurement, ECG, EMG, and more. In this context, biofeedback involves the quantification of physiological activities alongside other potentially informative parameters. These metrics are then relayed back to the user, enabling them to learn the art of controlling and modifying their physiological responses. The overarching objective is to enhance health and performance [25–27]. With roots tracing back to the early 1960s, biofeedback has proven its efficacy in regulating emotional states and involuntary bodily functions, such as migraines and blood pressure. Biofeedback devices encompass an array of monitoring mechanisms, spanning breathing, heart function, muscle activity, and brainwave patterns [25].

2.6.1.4 Ambient Assisted Living for Elderly

The aging demographic, coupled with escalating formal healthcare costs and the growing individual emphasis on maintaining independent living, collectively drives the innovation of assisted living technologies aimed at secure and autonomous aging. Applications in this domain elevate the quality of life, fostering a self-reliant lifestyle facilitated by home automation [28]. Significantly, assisted living facilities have emerged as an alternative housing option for individuals with disabilities and the elderly who fall between complete independence and the need for continuous medical care provided in nursing or retirement homes. Employing an ambient sensor network, parameters within the living environment are monitored and controlled, while body data is transmitted to a central hub through ongoing cognitive and physical surveillance. By assessing heart rate, blood pressure, and accelerometer data, the health status of these individuals can be approximated. This system can be linked to a healthcare center for continuous observation and immediate assistance during instances of notable parameter fluctuations or deviations from the established norms

[29]. This integrated approach holds the potential to significantly enhance the quality of life for the aging population while ensuring their safety and well-being.

2.6.2 Nonmedical Applications

2.6.2.1 Military

The potential applications of WBANs in military contexts are numerous and significant. In battlefield scenarios, WBANs can serve as a means of communication between soldiers, enabling the transmission of critical information such as actions such as attacks, retreats, or maneuvers to commanding officers at base. In addition, WBAN sensors can be employed to monitor the health condition, location, temperature, and hydration levels of soldiers in combat zones. These sensors, including cameras, GPS devices, monitoring sensors, and radio frequency components, can be seamlessly integrated into military uniforms. As a result, WBANs enhance accuracy, survivability, and connectivity in military operations.

2.6.2.2 Sports and Lifestyle

Numerous WBAN applications have been devised to monitor physical body positioning during activities such as gaming. For instance, a golfer can utilize this technology to determine the optimal body posture when striking a golf ball. By enhancing the accuracy of the player's stance, subsequent performances can improve. Moreover, these applications extend to tracking stress levels and heartbeats before, during, and after sports activities.

Further applications encompass monitoring an individual's workout regimen. This involves assessing the time taken for specific activities and tracking step count during walks. These applications can be implemented through sensors positioned on various body parts or even through wearable belts. Interestingly, these compact applications don't always require wireless functionality, as individuals can monitor their activities firsthand [30]. This illustrates the versatility of WBANs in catering to diverse needs, enhancing performance, and fostering self-awareness.

2.7 MITIGATING SECURITY AND PRIVACY CHALLENGES IN WBANS: STRATEGIES AND COUNTERMEASURES

The information generated, stored, and transmitted through BANs pertaining to patients' medical conditions is of paramount importance for accurate diagnosis and treatment. It cannot be overstated how critical it is to safeguard this data. The potential repercussions of data compromise or unauthorized access range from ineffective patient treatment to severe breaches compromising patients' well-being.

For widespread acceptance by patients as an integral component of the healthcare infrastructure, WBANs must also address privacy concerns. The inherent nature of data dissemination within WBANs introduces challenges in implementing robust security and privacy measures. These concerns remain pertinent even when a node is compromised or fails. Encryption and cryptography are emerging as effective tools to enforce access control and ensure patient data privacy.

2.7.1 Fundamentals of Security Concerns

The fundamental security concerns in WBANs are outlined below [31]:

a. ***Data Confidentiality:*** Data confidentiality necessitates shielding information from both unauthorized and authorized access. This extends to safeguarding data from other interconnected networks. Achieving this requires employing encryption algorithms and shared secret keys to maintain the privacy of patients' sensitive information during data transmission.
b. ***Scalability:*** Scalability is crucial as e-health systems handle a vast amount of patient-related data. A distributed access control system should seamlessly adapt to varying numbers of users. Low consumption and storage overhead are essential, along with easy setup and modification for optimal outcomes.
c. ***Data Integrity:*** Data integrity tackles the potential alteration of information during insecure network transmission. Ensuring data packets' accuracy during transmission involves error control mechanisms to prevent unauthorized modifications, even in the presence of adversaries.
d. ***Data Authenticity:*** Authentication is integral for both medical and non-medical applications. Nodes involved in communication must verify their identity for authenticity. This confirmation assures the coordinator node and member nodes that data is being transmitted to a legitimate destination and not intercepted by malicious entities for illicit purposes.
e. ***Data Availability:*** Data availability is imperative in healthcare systems to facilitate essential operations at any time, particularly during emergencies. Ensuring data availability is pre-planned to guarantee seamless functionality and patient safety. Network switching becomes pivotal to prevent disruptions in case of network degradation.
f. ***Data Security:*** Data security involves safeguarding the database from malicious forces, preserving its confidentiality. Unauthorized access or alterations are prevented by maintaining secrecy. Encryption techniques ensure secure data transmission, enhancing overall data security.
g. ***Encryption:*** Encryption algorithms such as Advanced Encryption Standard (AES), Data Encryption Standard (DES), and Advanced Data Encryption Standards (ADES) are employed to ensure data confidentiality during transmission.
h. **Data Privacy:** Data privacy dictates that data or information be accessible solely to authorized individuals. Unauthorized disclosure poses significant risks, especially given the sensitive nature of medical information. Privacy protection strategies, including non-cryptographic techniques and protocols like phantom routing, enhance the privacy of source location and medical data.

2.7.2 Security and Privacy-Related Threats and Solutions

Similar to any wireless network, WBANs face a spectrum of potential attacks. In this section, we delve into the threats and attacks relevant to WBANs.

a. ***Physical Layer:***
   - *Jamming:* This attack involves interfering with the radio frequencies used by network nodes, causing disruption [32]. Countermeasures against jamming encompass techniques like frequency hopping and code spreading.
   - *Tampering:* This attack occurs when a node is physically accessed by an intruder who gains access to sensitive data, such as cryptographic keys. Implementing tamper-proof packaging and design is an effective defense against this type of attack.

b. ***Data Link Layer:***
   - *Collision:* Collisions happen when multiple nodes attempt to transmit simultaneously on the same frequency, leading to data corruption. Error-correcting codes mitigate collisions [32].
   - *Exhaustion:* Attackers can provoke collisions repetitively, resulting in resource exhaustion [32]. Controlling transmission rates to limit excessive requests and energy drainage is a potential solution.
   - *Unfairness:* Attackers can manipulate transmission timing to gain an unfair advantage. Implementing minor frames can reduce the impact of such attacks.

c. ***Network Layer:***
   - *Selective Forwarding:* Malicious nodes selectively block or discard packets passing through them, causing data loss or misleading routing [33]. Employing multiple data paths and failure detection mechanisms enhances resilience.
   - *Sinkhole Attack:* Attackers attract network traffic to a compromised node, creating a figurative "sinkhole" at the center. This attack can exploit vulnerabilities introduced by selective forwarding [34].
   - *Sybil Attacks:* Attackers create multiple fake nodes, attempting to undermine fault-tolerant schemes. Authentication and encryption methods prevent outsiders from initiating Sybil attacks.
   - *Wormhole Attacks:* Attackers tunnel packets from one point to another within the network, bypassing regular communication paths. Detecting and preventing these tunnels are essential to thwart wormhole attacks [35].

d. ***Transport Layer:***
   - *Flooding:* Attackers initiate numerous connections, overwhelming network resources [36]. Requiring connection initiation authentication through puzzles can mitigate this threat.
   - *Desynchronization:* Attackers send messages with sequence numbers repeatedly, causing miscommunication. Requiring communication authentication between hosts counters this type of attack.

2.8 SIGNCRYPTION

By the year 2025, an impressive 100 billion IoT devices are projected to be in operation worldwide, holding a staggering economic value exceeding USD 11 trillion [38]. Among the standout technologies in this landscape, WBANs shine brightly in healthcare. These WSNs empower users to transmit real-time data across crucial domains such as remote health monitoring, sports, patient care, emergency response, and intrusion detection [39–42]. However, the current deficiency in robust data protection within this networking paradigm opens doors for malicious users to exploit sensitive medical information. This loophole results in grave consequences, including the compromise of sensitive data and user privacy, greatly affecting patients. To illustrate, consider a scenario where a patient experiences a heart attack—wearable devices or implanted wireless sensors could promptly detect it. Hence, in a public network setting, safeguarding both users and their data becomes imperative, enabling swift medical intervention by professionals [43,44].

Signcryption is a critical factor in secure communication; it is the first step toward secure communication and assists networks in decreasing unwanted users and avoiding deceptions [45]. Signcryption emerges as an optimal choice owing to its unique ability to seamlessly blend authentication and confidentiality into a single, streamlined process. This distinctive feature significantly streamlines the security protocol while enhancing its efficacy. In addition, signcryption's remarkable advantage lies in its minimal computational demands, rendering it an ideal fit for devices with constrained resources, such as sensors [46].

A hybrid key establishment technique for BANs based on symmetric cryptography with encryption was proposed by Amin et al. [47]. Their approach involved a consolidated process of cluster head selection and session key creation, aimed at achieving both reduced computational costs and minimized communication overhead. They asserted that their method effectively addressed various security

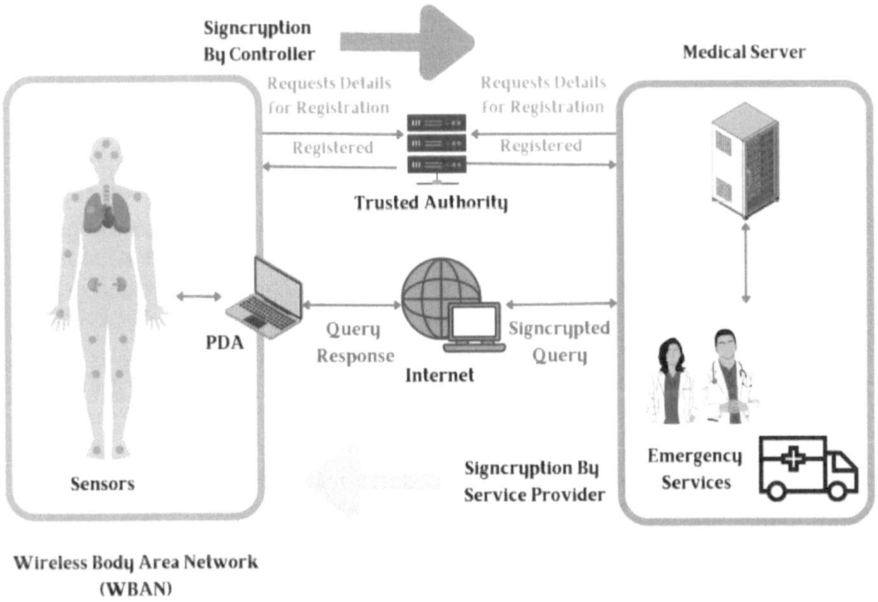

FIGURE 2.4 Signcryption model.

concerns, including confidentiality, protection against replay attacks, data integrity, and authentication. However, the scheme exhibited drawbacks such as high energy consumption and increased bandwidth usage due to the implementation of the elliptic curve cryptosystem (ECC). In addition, it faced challenges related to certificate renewal and revocation. Notably, the technique did not provide support for features like forward secrecy, public verifiability, mutual authentication, and nonrepudiation.

Wang and Liu [48] introduced a ring signcryption approach in the context of WBANs, employing an attribute-based cryptosystem. The robustness and efficiency of this method rely on the computational Diffie-Hellman (CDH) assumption using bilinear pairing. Their solution encompasses multiple security aspects, including authenticity, nonrepudiation, and confidentiality. Nevertheless, a significant limitation lies in the fact that the method does not effectively handle the key escrow issue. In this scheme, the Hospital Authority (HA) assumes the role of the Private Key Generation (PKG) center, responsible for generating private keys for both controllers and data users like doctors, researchers, and emergency personnel. This arrangement could potentially enable the HA to misuse the user's private key for signature forgery.

Saeed et al. [49] introduced a novel approach for the IoT using a combination of heterogeneous online/offline signcryption. In this method, the sensor nodes (acting as senders) leverage the capabilities of a certificateless infrastructure (CLI), while the server (acting as the receiver) utilizes services provided by a public key infrastructure (PKI). The researchers validated the security provisions of the scheme using a random oracle model, demonstrating its adherence to critical security requirements, including authentication, nonrepudiation, integrity, and confidentiality.

Adoption of a biometric approach to rectify the security and privacy issues within WBANs enhances authentication reliability, mitigates identity theft risks, and ensures robust protection of personal health data transmitted wirelessly. Biometric techniques leverage unique physiological or behavioral traits like fingerprints, retina scans, and palm prints, rendering them virtually unforgeable. This method aligns well with the essence of biometrics—a process predicated on the identification or verification of individuals through their distinct attributes. WBANs inherently harbor various security and privacy concerns encompassing data loss, authentication, and access control. In this regard, our proposal for leveraging biometric characteristics stands as a compelling solution that not only augments security and privacy but also offers pragmatic efficiency within WBANs. Biometrics, bolstered by its resistance against attacks, small key utilization, and operational efficiency, emerge as a potent tool for ensuring the integrity and privacy of sensitive healthcare data.

An emerging alternative to consider is the *HIGHT (high security and lightweight) block cipher*, which boasts enhanced security attributes and is better suited for energy-efficient operation. In addition, encryption methods such as Stream cipher, Digital signature, Message Authentication Code (MAC), Advanced Encryption Standard Counter Mode (AES-CTR), AES in Cipher Block Chaining mode with Message Authentication Code (AES-CBC-MAC), and AES in Counter with CBC-MAC (AES-CCM) have proven to be highly secure. However, their complexity presents challenges in implementation [50].

A particularly critical aspect that requires further refinement is user authentication and validation protocols, especially at the application layer. Introducing diverse user authentication mechanisms could serve as a formidable deterrent against unauthorized access to confidential patient medical data, which, if exposed, could lead to dire consequences.

2.9 REGULATIONS AND LAWS IN SECURITY AND PRIVACY

Security and privacy are paramount in E-healthcare, and various regulations influence healthcare providers globally. Regulations and acts significantly differ across countries. For instance, in the United States, the *Health Insurance Portability and Accountability Act of 1996 (HIPAA)* and the *Health Information Technology for Economic and Clinical Health Act (HITECH)* [37] play critical roles. HIPAA mandates the protection of health data collected by WBAN sensors to prevent unauthorized access and tampering. Violations can lead to severe penalties, including substantial fines and imprisonment.

2.10 CONCLUSION

There is a privacy and security challenge associated with all these uses and benefits. Concerns around unauthorized access, data breaches, and possible data exploitation have lately surfaced in relation to wireless transfer of health data. Owing to these restrictions, patients could be reluctant to use these technologies. As the obstacle emerges, so do strategies for conquering it. A multi-faceted strategy is needed to address privacy and security concerns. Patient data has been protected through the use of signcryption and other strong encryption techniques. Strict adherence to strict data protection legislation and the adoption of privacy-by-design principles to encourage patient confidence are two additional strategies for preserving the privacy and integrity of patient data. The importance of WBANs in E-healthcare, their applications, security and privacy issues, and solutions, has been covered in this chapter.

REFERENCES

[1] Hasan, K., Biswas, K., Ahmed, K., Nafi, N. S., & Islam, M. S. (2019). A comprehensive review of wireless body area network. *Journal of Network and Computer Applications*, 143, 178–198, ISSN 1084-8045. https://doi.org/10.1016/j.jnca.2019.06.016

[2] Latré, B., Braem, B., Moerman, I., Blondia, C., & Demeester, P. (2011). A survey on wireless body area networks. *Wireless Networks*, 17(1), 1–18. https://doi.org/10.1007/s11276-010-0252-4

[3] Malik, M. S. A., Ahmed, M., Abdullah, T., Kousar, N., Nigar, M. N., & Awais, M. (2018). Wireless body area network security and privacy issue in e-healthcare. *International Journal of Advanced Computer Science and Applications*, 9(4). https://doi.org/10.14569/IJACSA.2018.090433

[4] Movassaghi, S., Abolhasan, M., Lipman, J., Smith, D., & Jamalipour, A. (2014). Wireless body area networks: A survey. *IEEE Communications Surveys and Tutorials*, 16(3), 1658–1686. https://doi.org/10.1109/SURV.2013.121313.00064

[5] Mohammed, C. M., & Askar, S. (2021). Machine learning for IoT healthcare applications: A review. *International Journal of Science and Business*, 5(3), 42–51.

[6] Al-Barazanchi, I., Hashim, W., Ahmed Alkahtani, A., Rasheed Abdulshaheed, H., Muwafaq Gheni, H., Murthy, A., Daghighi, E., Shawkat, S. A., & Jaaz, Z. A. (2022). Remote monitoring of COVID-19 patients using multisensor body area network innovative system. *Computational Intelligence and Neuroscience*, 2022, 9879259. https://doi.org/10.1155/2022/9879259

[7] Hajar, M. S., Al-Kadri, M. O., & Kalutarage, H. K. (2021). A survey on wireless body area networks: Architecture, security challenges and research opportunities. *Computers and Security*, 104, 102211. https://doi.org/10.1016/j.cose.2021.102211

[8] Negra, R., Jemili, I., & Belghith, A. (2016). Wireless body area networks: Applications and technologies. *Procedia Computer Science*, 83, 1274–1281. https://doi.org/10.1016/j.procs.2016.04.266

[9] Chen, M., Gonzalez, S., Vasilakos, A., Cao, H., & Leung, V. C. M. (2011). Body area networks: A survey. *Mobile Networks and Applications*, 16(2), 171–193. https://doi.org/10.1007/s11036-010-0260-8

[10] Xing, J., & Zhu, Y. (2009, September). A survey on body area network. In *5th International Conference on Wireless Communications,Networking and Mobile Computing(WiCom "09)* (pp. 1–4). https://doi.org/10.1109/WICOM.2009.5302579

[11] Milenković, A., Otto, C., & Jovanov, E. (2006). Wireless sensor networks for personal health monitoring: Issues and an implementation. *Computer Communications*, 29(13), 2521–2533. https://doi.org/10.1016/j.comcom.2006.02.011

[12] Venkatasubramanian, K. K., Banerjee, A., & Gupta, S. K. S. (2010). PSKA: Usable and secure key agreement scheme for body area networks. *IEEE Transactions on Information Technology in Biomedicine*, 14(1), 60–68. https://doi.org/10.1109/TITB.2009.2037617

[13] Ullah, S., Higgins, H., Braem, B., Latre, B., Blondia, C., Moerman, I., Saleem, S., Rahman, Z., & Kwak, K. S. (2012). A comprehensive survey of wireless body area networks: On PHY, MAC, and Network layers solutions. *Journal of Medical Systems*, 36(3), 1065–1094. https://doi.org/10.1007/s10916-010-9571-3

[14] Tobon, D. P., Falk, T. H., & Maier, M. (2013). Context awareness in WBANs: A survey on medical and non-medical applications. *IEEE Wireless Communications*, 20(4), 30–37. https://doi.org/10.1109/MWC.2013.6590048

[15] Asif, A., & Sumra, I. A. (2017). Applications of wireless body area network (wban): A survey. *Engineering Science and Technology International Research Journal*, 64–71.

[16] Ragesh, G., & Baskaran, K. (2012). An overview of applications, standards and challenges in futuristic wireless body area networks. *International Journal of Computer Science Issues*, 9(1), 180–186.

[17] Ullah, S., Khan, P., Ullah, N., Saleem, S., Higgins, H., & Sup Kwak, K. S. (2009). A review of wireless body area networks for medical applications. *International Journal of Communications,Network and System Sciences*, 2(8), 797–803. https://doi.org/10.4236/ijcns.2009.28093

[18] Frangioni, J. V. (2008). New technologies for human cancer imaging. *Journal of Clinical Oncology*, 26(24), 4012–4021. https://doi.org/10.1200/JCO.2007.14.3065

[19] WH Organization. (2016). *Global Report on Diabetes*. https://www.who.int/publications/i/item/9789241565257

[20] Schwiebert, L., Gupta, S. K. S., & Weinmann, J. (2001). Research challenges in wireless networks of biomedical sensors. In *Proceedings of the 7thAnnual International Conference on Mobile Computing and Networking* (pp. 151–165). https://doi.org/10.1145/381677.381692

[21] Milenković, A., Otto, C., & Jovanov, E. (2006). Wireless sensor networks for personal health monitoring: Issues and an implementation. *Computer Communications*, 29(13–14), 2521–2533. https://doi.org/10.1016/j.comcom.2006.02.011

[22] Sun, W., Guo, Z., Yang, Z., Wu, Y., Lan, W., Liao, Y., Wu, X., & Liu, Y. (2022). A review of recent advances in vital signals monitoring of sports and health via flexible wearable sensors. *Sensors*, 22(20), 7784. https://doi.org/10.3390/s22207784

[23] Hadjidj, A., Souil, M., Bouabdallah, A., Challal, Y., & Owen, H. (2013). Wireless sensor networks for rehabilitation applications: Challenges and opportunities. *Journal of Network and Computer Applications*, 36(1), 1–15. https://doi.org/10.1016/j.jnca.2012.10.002

[24] Zhou, H., & Hu, H. (2008). Human motion tracking for rehabilitation – A survey. *Biomedical Signal Processing and Control*, 3(1), 1–18. https://doi.org/10.1016/j.bspc.2007.09.001

[25] Pereira, O., Caldeira, J. M. L. P., & Rodrigues, J. J. P. C. (2011). Body sensor network mobile solutions for biofeedback monitoring. *Mobile Networks and Applications*, 16(6), 713–732. https://doi.org/10.1007/s11036-010-0278-y

[26] Caldeira, J. M., Moutinho, J. A., Vaidya, B., Lorenz, P., & Rodrigues, J. J. (2010, February). Intra-body temperature monitoring using a biofeedback solution. In *eHealth, Telemedicine, and Social Medicine. ETELEMED2010. Second International Conference on, 2010* (pp. 119–124). IEEE Publications.

[27] Rodrigues, J. J. P. C., Pereira, O. R. E., & Neves, P. A. C. S. (2011). Biofeedback data visualization for body sensor networks. *Journal of Network and Computer Applications*, 34(1), 151–158. https://doi.org/10.1016/j.jnca.2010.08.005

[28] Rashidi, P., & Mihailidis, A. (2013). A survey on ambient-assisted living tools for older adults. *IEEE Journal of Biomedical and Health Informatics*, 17(3), 579–590. https://doi.org/10.1109/jbhi.2012.2234129

[29] Acampora, G., Cook, D. J., Rashidi, P., & Vasilakos, A. V. (2013). A survey on ambient intelligence in healthcare. *Proceedings of the IEEE. Institute of Electrical and Electronics Engineers*, 101(12), 2470–2494. https://doi.org/10.1109/JPROC.2013.2262913

[30] Qadri, S. F., Awan, S. A., Amjad, M., Anwar, M., & Shehzad, S. (2013). *Applications, Challenges, Security of Wireless Body Area Networks (WBANs) and Functionality of IEEE 802.15.4/ZIGBEE*. Science International, Lahore, Pakistan. https://www.ttsell.ir/ArticleFiles/ENARTICLE/50-46-400067.pdf

[31] Malik, M. S. A., Ahmed, M., Abdullah, T., Kousar, N., Nigar, M. N., & Awais, M. (2018). Wireless body area network security and privacy issue in e-healthcare. *International Journal of Advanced Computer Science and Applications*, 9(4). https://doi.org/10.14569/IJACSA.2018.090433

[32] Sen, J. (2012). Security in wireless sensor networks. In *Wireless Sensor Networks: Current Status and Future Trends*. Taylor & Francis.

[33] Venkatraman, K., VijayDaniel, J., & Murugaboopathi, G. (2013). Various attacks in wireless sensor network: Survey. *International Journal of Soft Computing and Engineering*, 3(1).

[34] Soni, V., Modi, P., & Chaudhri, V. (2013). Detecting Sinkhole attack in wireless sensor network. *International Journal of Application or Innovation in Engineering & Management*, 2(2).

[35] Singh, A. K., & Patro, D. B. (2019). A novel security protocol for wireless sensor networks based on elliptic curve Signcryption. *International Journal of Computer Networks & Communications (IJCNC)*, 11.

[36] Chaudhari, H. C., & Kadam, L. U. (2011). Wireless sensor networks: Security, attacks and challenges. *International Journal of Networking*, 1(1), 4–16.

[37] Li, M., Lou, W., & Ren, K. (2010). Data security and privacy in wireless body area networks. *IEEE Wireless Communications,IEEE Press*, 17(1), 51–58. https://doi.org/10.1109/MWC.2010.5416350

[38] Limbasiya, T., & Karati, A. (2018, January 10–12). Cryptanalysis and improvement of a mutual user authentication scheme for the Internet of Things. In *Proceedings of the 2018International Conference on Information Networking (ICOIN)* (pp. 168–173).Chiang Mai, Thailand. https://doi.org/10.1109/ICOIN.2018.8343105

[39] Limbasiya, T., & Karati, A. (2018, January 10–12). Cryptanalysis and improvement of a mutual user authentication scheme for the Internet of Things. In *Proceedings of the 2018International Conference on Information Networking (ICOIN)* (pp. 168–173). Chiang Mai, Thailand. https://doi.org/10.1109/ICOIN.2018.8343105

[40] Arif, A., Zubair, M., Ali, M., Khan, M. U., & Mehmood, M. Q. (2019). A compact, low-profile fractal antenna for wearable on-body WBAN applications. *IEEE Antennas and Wireless Propagation Letters*, 18(5), 981–985. https://doi.org/10.1109/LAWP.2019.2906829

[41] Sharma, A., & Kumar, R. (2019). A constrained framework for context–aware remote E–healthcare (CARE) services. *Transactions on Emerging Telecommunications Technologies*, e3649.

[42] Sharma, A., & Kumar, R. (2019). A constrained framework for context–aware remote E–healthcare (CARE) services. *Transactions on Emerging Telecommunications Technologies*, e3649.

[43] He, D., Ye, R., Chan, S., Guizani, M., & Xu, Y. (2018). Privacy in the Internet of Things for smart healthcare. *IEEE Communications Magazine*, 56(4), 38–44. https://doi.org/10.1109/MCOM.2018.1700809

[44] Shingala, M., Patel, C., & Doshi, N. (2018). An improve three factor remote user authentication scheme using smart card. *Wireless Personal Communications*, 99(1), 227–251. https://doi.org/10.1007/s11277-017-5055-9

[45] Hussain, S., Ullah, S. S., Uddin, M., Iqbal, J., & Chen, C. L. (2022). A comprehensive survey on encryption security mechanisms in wireless body area networks. *Sensors*, 22(3), 1072. https://doi.org/10.3390/s22031072

[46] Singh, A. K. (2014). A review of elliptic curve based signcryption schemes. *International Journal of Computer Applications*, 102(6).

[47] Amin, N. U., Iqbal, J., & Abbasi, A. R. (2014). Secure key establishment and cluster head selection for body area networks based on signcryption. *Journal of Applied Environmental and Biological Sciences*, 4, 210–216.

[48] Wang, C., & Liu, J. (2015). Attribute-based ring signcryption scheme and its application in wireless body area networks. In *International Conference on Algorithms and Architectures for Parallel Processing* (pp. 521–530). Springer, Cham.

[49] Saeed, M. E. S., Liu, Q., Tian, G., Gao, B., & Li, F. (2018). HOOSC: Heterogeneous online/offline signcryption for the internet of things. *Wireless Networks*, 24(8), 3141–3160. https://doi.org/10.1007/s11276-017-1524-z

[50] Hong, D., et al. (2006). *HIGHT: A New Block Cipher Suitable for LowResource Device,*CHES 06. *LNCS*, 4249. Springer, Berlin, Heidelberg.

[51] Taleb, H., Nasser, A., Andrieux, G., Charara, N., & Motta Cruz, E. (2021). Wireless technologies, medical applications and future challenges in WBAN: A survey. *Wireless Networks*, 27(8), 5271–5295. https://doi.org/10.1007/s11276-021-02780-2

[52] Negra, R., Jemili, I., & Belghith, A. (2016). Wireless body area networks: Applications and technologies. *Procedia Computer Science*, 83, 1274–1281. https://doi.org/10.1016/j.procs.2016.04.266

[53] Guo, Z., Harris, I. G., Tsaur, L.-F., & Chen, X. (2015). An on-demand scatternet formation and multi-hop routing protocol for BLE-based wireless sensor networks. In *IEEE Wireless Communications and Networking Conference (WCNC), 2015* (pp. 1590–1595). IEEE Publications.

[54] Siekkinen, M., Hiienkari, M., Nurminen, J. K., & Nieminen, J. (2012). How low energy is Bluetooth low energy? Comparative measurements with Zigbee/802.15. 4. In *IEEE Wireless Communications and Networking Conference Workshops (WCNCW), 2012* (pp. 232–237). IEEE Publications.

[55] Ghamari, M., Janko, B., Sherratt, R. S., Harwin, W., Piechockic, R., & Soltanpur, C. (2016). A survey on wireless body area networks for ehealthcare systems in residential environments. *Sensors*, 16(6), 831. https://doi.org/10.3390/s16060831

[56] Minakshi, J. A.-H. (2016). An overview of wireless body area network (wban) using ZigBee technology. *International Journal of Scientific Development and Research (IJSDR)*, 1, 2455–2631.

[57] Chung, Y. F., & Liu, C. H. (2013). Design of a wireless sensor network platform for telehomecare. *Sensors*, 13(12), 17156–17175. https://doi.org/10.3390/s131217156

[58] Cecilia, H., de Schatz, V., Medeiros, H. P., Schneider, F. K., & Abatti, P. J. (2012). Wireless medical sensor networks: Design requirements and enabling technologies. *Telemedicine and e-Health*, 18(5), 394–399.

[59] Arefin, M. T., Ali, M. H., & Haque, A. K. M. F. (2017). Wireless body area network: An overview and various applications. *Journal of Computer and Communications*, 5(7), 53–64. https://doi.org/10.4236/jcc.2017.57006

3 Evaluation of Wireless Body Area Networks

A Systematic Review on Security Concerns

Rahul Joshi, Suman Kumari, and Rhythma Badola

3.1 INTRODUCTION

Wireless Sensor Networks have emerged as an area for significant research and will become essential to human existence in the foreseeable future. It consists of sensor nodes strategically deployed in the surrounding environment. Each sensor node has been assigned a distinct function, independently executing and cooperating with other nodes. The primary purpose of these networks is to collect environmental data and transfer it to either the base unit or a distant server. The data submitted is after that subjected to a thorough examination. The wireless sensor, being the most diminutive component of a network, possesses distinct attributes that include the potential to facilitate extensive coverage, mobility, and dependability. This technology has extensive applications, including industrial process monitoring and control, device tracking, environmental tracking, traffic management, intelligent housing systems, defense and security measures, agricultural use, and several other sectors. One instance of Wireless Sensor Networks may be seen in medical treatment and wellness. The technology employs wearable sensors surgically implanted into the human body to observe and record physical movements and measure vital signs such as body temperature and respiration.

The sensors gather and transmit data to the base unit, which is then analyzed, stored, and processed. Subsequently, the data is securely transmitted to distant medical servers through the Internet or alternative communication channels. The importance of security lies in its critical role in safeguarding data that encompasses sensitive information derived from physiological variables. Ensuring the protection of personal data from unauthorized access is of utmost importance. These are distinct subsets of sensor networks that use wireless sensor nodes strategically placed on an individual's body to gather data about vital signs such as arterial pressure, body temperature, respiration rate, and glucose levels. This technological advancement enables the remote monitoring of a patient's health condition. It can be categorized as wearable or implanted. It aims to promote individual well-being by transmitting physiological data to medical servers through body-worn sensors. This facilitates healthcare professionals in comprehending the patient's health status [1].

DOI: 10.1201/9781032635101-4

These systems can potentially provide substantial support to medical professionals and individuals during emergencies. The services they provide include several functions, including monitoring and disseminating pharmaceuticals and medical information, optimizing patient data processing, managing household appliances, and facilitating data transfer. This technology has the potential to substantially decrease the expenses associated with patient treatment through the continuous monitoring of vital sign-related data over an extended duration. In contrast to Wireless Sensor Networks, body sensor networks employ a star topology architecture. Throughout this topology, there exists a designated location, commonly referred to as a sinkhole, which collects information from various nodes throughout the system. The sensor nodes within the network have limited utilization of energy sources. Physical sensor networks include deploying sensors, such as motion sensors, onto a person's body to monitor health indicators or detect movements. The body sensor network provides users with real-time feedback via continuous monitoring of a patient's health. This enables users to track the development of a patient's illness and implement appropriate preventive measures. The energy consumption of sensors within this network has significant relevance since the depletion of the energy source will reduce the network's longevity [2].

3.2 BACKGROUND

Recently, there has been a notable surge in academic research focused on examining Wireless Body Area Networks (WBANs). This research encompasses a wide range of topics, including the design considerations, barriers, and practical implementation challenges related to WBANs. The literature examined in references [3,4] covers a diverse array of subjects about the obstacles encountered in implementation and design, issues regarding privacy, models for diffusion, data routing, strategies for reducing energy consumption, ensuring satisfactory service, and practical issues to be considered. The primary objective of WBANs is to continuously monitor the health status of individuals by employing biosensors located within or near the human body. These biosensors are designed to identify potentially hazardous conditions and trigger an alert system accordingly. The study conducted by Adarsh and Kumar [5] showed a detailed analysis of the many sensors utilized in body network systems and assessed their performance. The research conducted in the WBAN domain predominantly focuses on the architectural configuration of WBANs, the incorporation of data within these networks, and the optimization of data transmission among the network units. The focus of the research generally revolves around the deployment of sensors, the consumption of energy, and the routing capabilities of node sensors within the network design framework. The primary focus of study in the data integration domain revolves around extracting features, data compression, and data classification.

Numerous scholars have directed their attention toward examining energy consumption within the context of WBANs [6,7]. Simultaneously, several researchers have investigated [8,9] to suggest appropriate routing algorithms for integrating sensors with medical equipment and other devices. To safeguard the confidentiality, integrity, and authenticity of various forms of data, security algorithms are employed to fortify the network infrastructure. These algorithms play a crucial role in upholding the privacy and trustworthiness of information. The study conducted by Zau and Wang [10] investigates various approaches to ensure the protection of personal data during the process of data collection, with a particular emphasis on upholding privacy. In data storage, it is imperative to ensure the confidentiality and security of the patient's information, safeguarding it from potential threats posed by unauthorized user nodes and collusion. Furthermore, the data concerning a patient must remain unaltered during the storage process. Alternatively, the data should undergo hashing using a series of distinct operations. One of the primary concerns is the need to authenticate the sender of the data in order to validate its integrity. It is crucial to prohibit the inclusion of any invalid data that does not originate from the WBAN. In addition, all data must remain readily accessible, even in the aftermath of a denial-of-service attack.

The privacy of security measures is crucial for ensuring the security of a secure communication session between biosensor nodes in a hostile atmosphere, as stated by Huanan and Jiannan [11]. This measure ensures that data remains accessible solely to persons who have been granted authorization. The studies examined did not collectively address the dual concerns of security and energy usage.

Naik and Samundiswary [12] comprehensively analyzed WBANs and Wireless Sensor Networks. Subsequently, the researchers thoroughly analyzed the security requirements specific to WBANs. Their study examines the importance, necessity, and difficulties related to privacy in collecting and transmitting critical patient health data, with a particular emphasis on the safety and confidentiality concerns involved. In addition, it underscored the importance of establishing a comprehensive and efficient security system that achieves a harmonious equilibrium between complexity and simplicity in its operational computations while prioritizing energy efficiency. This study recognizes the importance of energy conservation after adopting security measures. Nevertheless, it needs more comprehensive elucidation of energy-conservation tactics. Furthermore, a comprehensive analysis is conducted to systematically categorize the various forms of assaults targeting WBANs.

In their study, Ananthi and Jose [13] analyzed the security challenges associated with WBANs within medical care applications. Both aimed to analyze and assess the tactics and protocols utilized to enhance security and optimize the efficiency of WBANs. It delves into the challenges associated with WBAN security, explores various approaches employed to address these challenges, and provides an in-depth examination of the outcomes achieved by each strategy. The findings and implications of this study indicate that implementing clustering and adopting efficient router protocols can significantly enhance both the bandwidth and the network longevity within a secure and protected setting.

Ming Li et al. [14] argued that WBAN has emerged as an innovative technology within electronic healthcare. Small wearable or implantable sensors are utilized to capture patients' critical body parameters and motion data. This data is transmitted through short-range wireless communication systems. It has been reported that there has been an improvement in the quality of healthcare. Here, the primary objective is to ensure data security from WBANs. The target system studied by Lim et al. [15] was characterized by a scalable platform requiring minimal human participation in setup and monitoring procedures. The fundamental elements of the system consist of:

- Pairs of biosensors and transceivers.
- Hardware modules designed to establish the WBAN in an automated manner.
- A mechanism for delivering data to an Internet server.
- Automated processes for collecting, profiling, and reporting data.

The security risk assessment of the real-time remote health monitoring system is performed by evaluating the essential needs linked to acknowledged risks and threats. The authors have presented a case study on security threats assessment, focusing on a simple and adaptable wireless biosensor system that can grow effectively. The system is designed for real-time remote monitoring, and the authors have provided a comprehensive framework for its implementation. WBANs possess a dynamic network topology due to individuals' mobility, varying postures, and walking patterns. The diversity in the transmission of electromagnetic radiation through the tissues of humans is attributable to the discrepancies in body composition and thickness among individuals, leading to absorption and reflection phenomena occurring within the body. The diffraction phenomenon causes these waves to deflect as they encounter the human body rather than exhibiting transmission through it. Furthermore, the efficient distribution of packets is influenced by the mobility and posture of individuals.

The authors emphasized the security concerns that remain to be resolved to achieve widespread adoption of WBANs across many applications. The advent of this technology raises several concerns about scalability, the deployment and density of sensors, energy efficiency, security, privacy, and wireless technology. Consequently, the implementation of a robust security system becomes imperative for WBANs, with authentication being a crucial component thereto. Exploring a hybrid authentication technique is necessary to establish a robust security solution for the technology.

3.3 WBAN ARCHITECTURE

This technology is directly derived from existing Wireless Sensor Network technology. WBANs are based on the concept of placing small wireless sensors on the human body, leading to monitoring of various vital signs. The network helps provide real-time feedback to users and medical personnel. In a WBAN, individual medical sensors monitor different vital signs such as but not limited to temperature, blood pressure, and electrocardiogram (ECG) readings. Jamal and Mendes stated, "The system consists of a series of sensor nodes that monitor body movements and heart rate. Also, the network manager and server work on personal digital assistants or workstations" [16]. Figure 3.1 signifies the components of WBAN.

The WBAN design typically operates on a three-tiered system, accommodating medical and nonmedical uses. Tier 1 comprises BAN Nodes (BNs), located both in body and on body. These BNs include the ECG, which is used for monitoring the heart's electrical activity. Jamal and Mendes in their other paper stated, "The Oxygen saturation sensor (SpO2) is employed to measure the quantity of oxygen, and Electromyography (EMG), is utilized for monitoring muscle activity"

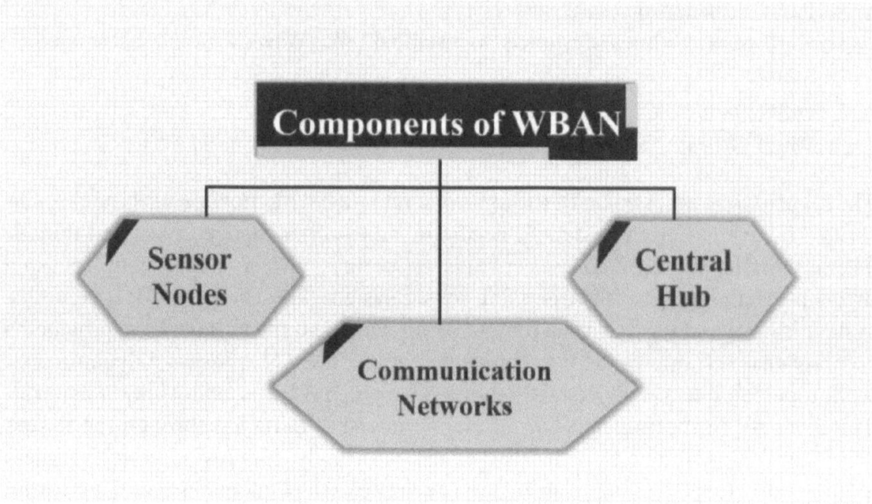

FIGURE 3.1 Components of WBAN.

[17]. Tier 2 comprises a BAN Network Coordinator (BNC) responsible for collecting essential patient data from the BNs and facilitating communication with the base station. Tier 3 contains a series of geographically dispersed base stations that store both medical and nonmedical data of patients while also offering pertinent suggestions. Traffic may be classified into three categories: on-demand, emergency, and routine. The BNC initiates on-demand traffic to get specific information. Emergency traffic is activated by the Base Nodes (BNs) when they surpass a certain threshold. "Routine traffic" refers to data flow that occurs under typical conditions without any time-sensitive or on-demand activities. The BNC is responsible for collecting and processing standard data. The BNC includes electronic design, primary radio, and security circuit, all connected via the data interface. The electronic design cycle is used to manage demand and transportation needs in emergencies. The purpose of the security circuit is to prevent bad contact with the WBAN [18].

3.4 FEATURES OF WBAN

The medical sensors are categorized by IEEE 802.15.6 into many types. Each sensor operates independently by processing and relaying data via radio waves to another node. The primary components used in a WBAN system are as follows [19]:

- ***Coordinator Nodes:*** The coordinator node, sometimes called the personal digital assistant (PDA), is a central hub for data collection from various sensors. These sensors may include personal computers, smartphones, or laptops, and the coordinator node facilitates communication among all other nodes.

- ***End Nodes:*** The end nodes in a WBAN are constrained in their ability to execute their embedded applications. They cannot transmit messages from other nodes. This device is sometimes called a body control unit (BCU) [20].
- ***Relay:*** In the context of communication networks, a relay, often referred to as an intermediary node, forwards messages to another node before they arrive at the PDA. Furthermore, the security architecture implemented for the WBAN system must comply with the predetermined physical limitations of the node sensors outlined in the reference [21].
- ***Bandwidth and Space Utilization:*** A sensor is a compact device with a significantly constrained bandwidth. Therefore, due to limited storage capacity, insufficient space is available to store an entire code base, such as an encryption mechanism.

3.5 APPLICATIONS OF WBAN

WBAN has a diverse range of applications, mainly in the healthcare and medical sectors, though their use also extends to different domains. Several applications of WBANs include:

- ***Healthcare:*** Individuals diagnosed with chronic conditions such as diabetes, hypertension, and heart disease might undergo monitoring via the use of WBANs. The use of these tools enables healthcare providers to effectively manage patients' conditions by monitoring vital signs, medication adherence, and activity levels. WBANs have the potential to provide remote patient monitoring, hence obviating the need for in-person consultations and hospitalization. WBANs represent a burgeoning technological advancement that has significantly transformed the healthcare industry. This groundbreaking innovation enables the remote monitoring of a patient's vital signs and biometric data, surpassing the limitations of traditional healthcare practices. The system comprises small, energy-efficient remote sensor hubs that can be affixed to the patient's body or integrated into clothing or accessories. These sensor hubs gather physiological data like ECG, electroencephalogram (EEG), temperature, and respiratory rate. The collected data is transmitted wirelessly to a centralized facility or remote server for further processing and analysis.
- ***Sports:*** Sports are widely recognized as a prevalent physical exercise and amusement, including structured events and physical effort. These activities span a broad spectrum. WBANs can monitor athlete performance and reduce injury risks in sports. To optimize training programs and detect potential symptoms of weariness or injury at an early stage, coaches and trainers might choose to use technological devices for monitoring vital signs, biomechanical data, and hydration levels. WBANs can support the creation of individualized rehabilitation approaches for athletes while also enabling the evaluation of their progressions over some time.

- ***Fitness:*** The term "fitness" refers to an individual's condition of physical wellness and capacity to do physical tasks efficiently. WBANs have the potential to effectively monitor several aspects of individuals' health and wellness, including exercise routines, sleep habits, and food preferences, particularly among fitness enthusiasts. The provision of real-time feedback by fitness devices has the potential to serve as a motivating factor for users in their pursuit of achieving their fitness goals. Using WBANs may also assess developmental progress and identify areas for prospective improvement.
- ***Rehabilitation:*** It refers to the application of therapeutic interventions with the objective of reinstating patients' typical functional abilities. Effective rehabilitation treatments and therapeutic procedures can restore independent functioning in persons who have suffered a stroke. Continuous monitoring is implemented on the patients to ensure the preservation of a precise movement pattern. The main use of WBANs in this field involves the diversification of sensors, fusion of data, provision of real-time feedback, and facilitation of home-based rehabilitative health using devices that continuously monitor physiological activity. This will facilitate a heightened comprehension of physiological mechanisms.
- ***Biofeedback:*** It refers to the methodology that facilitates the remote surveillance of the human body through the utilization of WBANs. Sensor technology facilitates data acquisition, permitting gathering significant physiological signs from the human body. Patients possess the capacity to effectively regulate and uphold their overall health by employing biofeedback methodologies, including but not limited to temperature analysis, blood pressure monitoring, ECG, and analogous approaches. Biofeedback is the utilization of technological tools to continuously monitor physiological activity with the aim of maintaining and improving overall well-being. This will facilitate the dissemination of knowledge regarding distinct physiological systems.
- ***Supportive Care Facilities:*** Assisted living services play a significant role in promoting the overall well-being and quality of life for persons. Assisted living technologies enable the surveillance and supervision of elderly and disabled adults in the comfort of their own homes. It is anticipated that the implementation of this strategy will lead to a decrease in healthcare costs. These devices and technologies provide precise assessment of people's health status.

3.6 SECURITY ATTACKS IN WBAN

Sensor nodes rely on low-power radios for communication, so WBAN applications are vulnerable to many attacks and threats. End users can use the open Internet to create links for applications. Including this available connection feature results in the expansion of the potential vulnerabilities that malicious users might exploit. The occurrence of these violations has the potential to impact the operational efficiency and accessibility of the service, hence posing a risk to individuals' lives and, in extreme cases, resulting in fatalities [22]. Figure 3.2 illustrates the security attacks on WBANs.

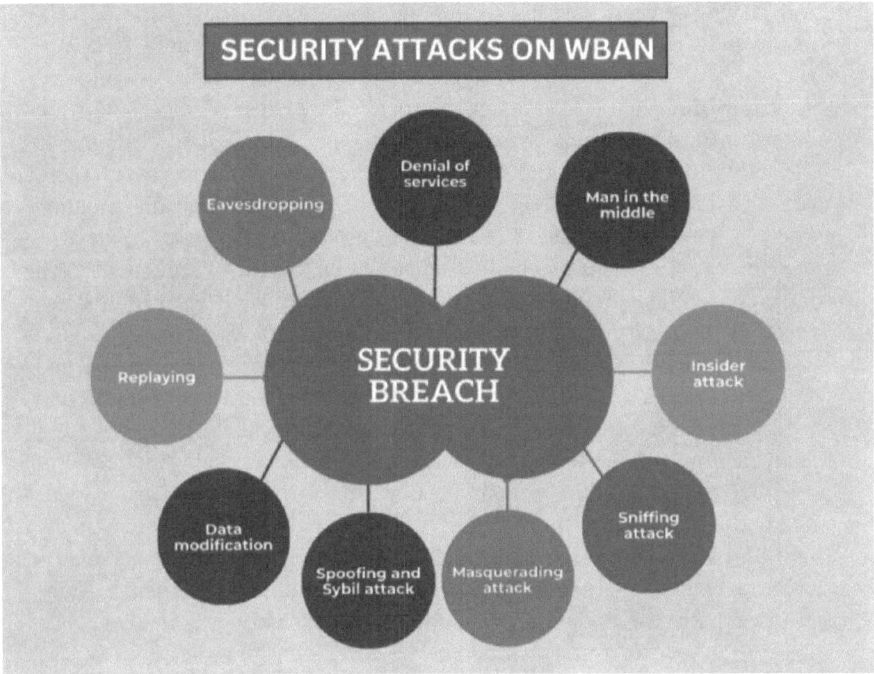

FIGURE 3.2 Some wireless security breaches in WBANs.

Some of the major wireless security breaches include the following:

- ***Eavesdropping:*** By intercepting communications between the sensor node and a base station, someone with malicious intent can get a patient's medical data and device details. An adversary may use this data to present oneself as an authorized individual and carry out a phishing attack [23] or disrupt the critical exchange process during the pair-up phase [24]. WBANs face a significant risk of eavesdropping, which can compromise the privacy and security of patient information. The gateway that transmits data to healthcare providers or monitoring systems is particularly vulnerable to interception. Despite the implementation of robust security measures for transmitting data to a central server, there is a potential vulnerability to unauthorized interception during the transit of data between the gateway and the server. These breaches may also result in legal and ethical transgressions that undermine the confidentiality between patients and doctors and violate data protection legislation. To address the possible risk of unwanted interception in WBANs, it is essential to build effective security mechanisms. The recommended protocols should include encryption, authentication, access control, and user awareness. Through the implementation of these steps, the preservation of data privacy and the maintenance of network integrity may be achieved.

- ***Denial of Services:*** The use of low-frequency wireless media in WBAN applications renders the radio signal susceptible to disruption by an adversary who may provide noisy signals, decreasing the signal's continuity. A denial-of-service (DoS) attack can potentially cause service failure for other sensor nodes already operating inside the network, hence impacting the ability to access the channel [25]. To address this problem, several security measures may be used, such as the utilization of intrusion detection systems, the implementation of rate-limiting approaches, and the deployment of traffic-filtering mechanisms. In addition, the use of redundancy and failover techniques may ensure continuous operation, even in the occurrence of a security breach. Ensuring the security of WBANs, especially within the healthcare industry, necessitates the adoption of a proactive and watchful stance. It is paramount to ensure the safeguarding of patients and their personal information against possible DoS attacks.
- ***Man in the Middle:*** Individuals with malicious intent can be intermediaries or servers inside a communication session, including sensor nodes and end-user applications in WBANs. The perpetrator can surreptitiously intercept and alter the communication in a live setting without detection by either the sender or the recipient. During an assault, it is possible to compromise patient confidentiality, manipulate data, and put patients at risk. Delayed or incorrect information may be communicated to practitioners in the healthcare field. Implementing strong encryption, reliable authentication methods, and secure communication protocols is of utmost importance to reduce the likelihood of Man-in-the-Middle (MitM) attacks in WBANs. By implementing these security measures, we can ensure the privacy and protection of health information during its transfer, thereby safeguarding patients' welfare and maintaining the integrity of the data.
- ***Insider Attack:*** A malicious entity can carry out an insider attack on WBAN applications. This kind of attack entails using a node that has been physically compromised, hence granting unauthorized access to the system. Using this hacked node, the assailant can partake in various operations, including but not limited to the intentional dropping, alteration, and misdirection of data packets. These malicious actions are intended to disrupt the regular performance of the network [26].
- ***Sniffing Attack***: External attackers may use network-sniffing techniques to extract critical information from network data packets, such as source and destination addresses, port numbers, and protocol types, provided the packages are not encrypted [27]. To mitigate the risks posed by sniffing attacks, it is imperative to use robust encryption techniques, implement secure communication protocols, and maintain vigilant monitoring of network activity inside WBANs.
- ***Masquerading Attack:*** In the context of a WBAN application, an adversary can assume the identity or replicate an authorized device to execute a masquerade attack. The primary objectives of such an assault include:
  - The unauthorized acquisition of data.
  - The injection of falsified information.
  - The attainment of unrestricted access to healthcare applications.

The attacker can manipulate or remove personal health record (PHR) data from the medical database when granted complete access to the application [28].

- ***Spoofing and Sybil Attack:*** It refers to an evil node assuming the identity of a genuine entity inside the system to disrupt the network while evading discovery. This specific attack also facilitates the implementation of a broader Sybil attack, in which the integrity of several nodes within the network is compromised.
- ***Data Modification:*** The intercepted information is partially or entirely altered or substituted by the perpetrators. The limited data is sent to the initial recipient to fulfill illicit objectives.
- ***Replaying:*** Replay attacks include the interception of legitimate information by a hacker, who later transmits this information back to the original recipient, achieving a similar objective in a different context. The perpetrator captures the data packets during transmission and retransmits them to the network, either in their original state or with malevolent modifications. This phenomenon might give rise to the recurrent dissemination of antiquated or altered health-related data, which may give rise to erroneous assessments of unsuitable interventions and jeopardize the well-being of patients. It is crucial to implement robust security measures to prevent replay attacks in WBANs. These measures should include data timestamping, message freshness checks, and secure authentication mechanisms. By incorporating these measures, healthcare providers and monitoring systems can ensure that the data they receive is current and authentic. In addition, these security measures help prevent data tampering, thereby maintaining the integrity and reliability of patient health information.

3.7 SECURITY REQUIREMENTS IN WBAN

The sensor nodes receive the patient's vital signs from the Body Control Unit (BCU). Hence, the connectivity between network participants is susceptible to frequent disruptions, exerting a substantial influence on the patient's physical well-being. The reason why the system is vulnerable to hackers is the lack of security mechanisms. The enemy can listen and manipulate information transmitted through sensors and PDAs. Hence, it is essential to include suitable safeguarding mechanisms into the architecture of WBANs to guarantee the preservation of data honesty, anonymity, and secrecy. Here are some important security requirements for WBANs:

- ***Data Originality:*** Both verification and licensing services are necessary for both medical and nonmedical uses. To ensure the reliability of data provided by sensors in WBAN medical care applications, it is required to use authentication measures for both the sensors and the base station.
- ***Data Confidentiality:*** The presence of an open wireless WBAN channel allows for the potential eavesdropping on radio communication between nodes by inactive attackers, hence leading to the unauthorized disclosure of information. Consequently, it is essential to encrypt data prior to its transmission [29].

- ***Data Integrity:*** An attacker can modify the data that has been intercepted and afterward transmit it to the designated recipient to accomplish unauthorized objectives. This might lead to the system's failure and inflict harm on patients. Consequently, it is essential to verify the integrity of the data [30].
- ***Data Availability:*** The concept of availability pertains to the assurance that services and information are accessible and ready for use at the required time. The presence of medical sensor nodes guarantees the continuous availability of medical information for medical purposes. If a person with proper authorization seizes control of a sensor node, data access will be recovered. Hence, it is essential to ensure the perpetual preservation of accessibility in healthcare apps.
- ***Data Freshness:***The freshness approach safeguards against the reuse of outdated data by prohibiting an attacker node from recording, replaying, and publishing.
- ***Data Authentication:*** To provide secure communication, each node must be able to recognize and verify the identity of the nodes from which it receives data via the authentication mechanism. It is possible to achieve this objective using techniques such as symmetric authentication [31].
- ***Secure Management:*** To ensure the preservation of authenticity in the processes of decoding and authentication across nodes, it becomes imperative to use a secure and reliable mechanism for the distribution and removal of cryptographic keys.
- ***Reliability:*** A security system needs to possess a high level of reliability. To ensure the integrity and reliability of a WBAN system, it must be able to restore accurate information effectively. The restoration of corrupted information may be achieved by using error coding methods [32].
- ***Safe Positioning:*** Due to the dynamic nature of the WBAN and the continuous movement of patients, it is essential to regularly update all apps responsible for recording the precise physical position of the patient. Nevertheless, it is necessary to acknowledge that these movements and updating the patient's location might create a vulnerability, allowing an attacker to introduce counterfeit data and signals into the geographical database.
- ***Accountability:*** When considering healthcare and medical facilities responsible for recording and maintaining patient information, it is imperative to acknowledge that the staff's duty to preserve patient information is of utmost significance. The medical center and its personnel are responsible when a patient's personal information is misused [33].
- ***Flexibility:*** In the case of a patient's necessity or desire, it is plausible for the emergency information of such a patient to be sent to an alternate individual or medical facility. In the given context, the additional party and the healthcare facility must possess the requisite capability and legal permission to gather the patient's information.
- ***Data Authenticity:*** Authenticity serves as a means of ensuring that the transmission of information originates from the purported sender. The use of both public and private keys is advantageous in the pursuit of guaranteeing the validity of data.

- ***Data Authorization:*** Implementing authorization to limit users' access to a system's resources and services to a certain extent is a pragmatic approach. Integrating an Access Control List (ACL) with an access policy allows for precise control over user access to various resources and services.

3.8 SAFETY RISK ASSESSMENT

As of now, three threats have been identified. The incorporation of additional risks will be avoided until the implementation of the system and the provision of services are complete.

- ***Qualitative Security Threats Assessment:*** A risk assessment is made in the context of current practice and test results. The test considers the user's daily activities when using the device for remote healthcare. Monitoring services include many processes such as data collection, processing, transmission, storage, and sharing. Users' daily activities and data exchange constitute two important security factors. Communication lines related to IEEE 802.15.6, IEEE 802.11g, and Internet Protocol are considered security risks. The transmission of crucial signals from a user's wireless device may be intercepted by unauthorized individuals who use the IEEE 802.15.6 bandwidth to synchronize with the network's sensor transmission node. It raises concerns about privacy and usability. The unauthorized alteration of critical signals inside the sensor transmitting network results in misinterpreting the user's crucial physiological indicators. This misunderstanding between the machine learning software of the remote portal system and healthcare professionals has led to ethical problems.
- ***Suggested Security Features:*** The protection of security and privacy during the transmission of information necessitates the monitoring of individuals' bodily indicators. The features included by this definition are registration, veracity, restriction of access, denial, and encoding. The system must have principles of scalability and compliance to improve its security measures. This can be done using elliptic curve cryptography and using shared proof and group consensus techniques. It is important to incorporate security aspects into cryptographic protocols and data structures. Evaluating the security content of the system requires checking the utility and functionality.
- ***Enabling Technology Resolving Security:*** This system provides a wireless platform that is cost-effective, adaptable, and capable of handling large amounts of data from biosensors. To maintain robust security across all applications within the system, it is advisable to incorporate cryptographic functions through a plug-and-play (PnP) device. This device should be designed to facilitate quick and easy setup, even by individuals who need more professional expertise. By implementing such a solution, it becomes feasible to enable pervasive monitoring without causing any disruptions to patients' daily routines. According to Chipcon Products, Texas Instruments, "To provide a high degree of security, it is necessary to include the Advanced Encryption Standards (AES) 128 cryptographic algorithm inside the hardware accelerator of the sensor node module used by the user" [34]. In the realm of research, a novel block cipher has emerged to cater to

low-resource devices. Among them is the HIGHT block cipher, also known as high security and lightweight. This cipher boasts a block size of 64 bits and a key length of 128 bits [35]. Liu and Ning stated,

> The cryptographic technique used in TinyOS includes the elliptic curve cryptographic algorithm (ECIES) for encryption and the key distribution protocol (ECDH) for secure key exchange. Additionally, the digital signature algorithm (ECDSA) is utilized to verify messages' authenticity and integrity. It is essential to do an examination and feasibility study about the practical use and field testing of security measures in the remote health monitoring system about the cryptographic algorithms employed.

3.9 EXISTING SECURITY MECHANISMS

Security frameworks refer to a set of established methods designed to recognize, mitigate, and endure breaches of security (see Figure 3.3). This subsection examines the issues about current security systems, as outlined next:

- ***Cryptography:*** Cryptographic features provide patients with enhanced privacy and security measures, effectively safeguarding against many forms of hostile violations. In addition, the choice of a cryptography system is contingent upon the computational and communication capabilities of the sensor nodes. There is a contention that asymmetric cryptographic systems are often seen as too expensive for medical sensors, while interchangeable cryptographic methods are perceived as needing more versatility.
- ***Key Management:*** Key management methods are essential requirements for the establishment of a secure application. The methods mentioned previously are used to establish and disseminate diverse cryptographic keys to nodes present inside the network.
- ***Secure Routing:*** Lorincz and Malan argued,

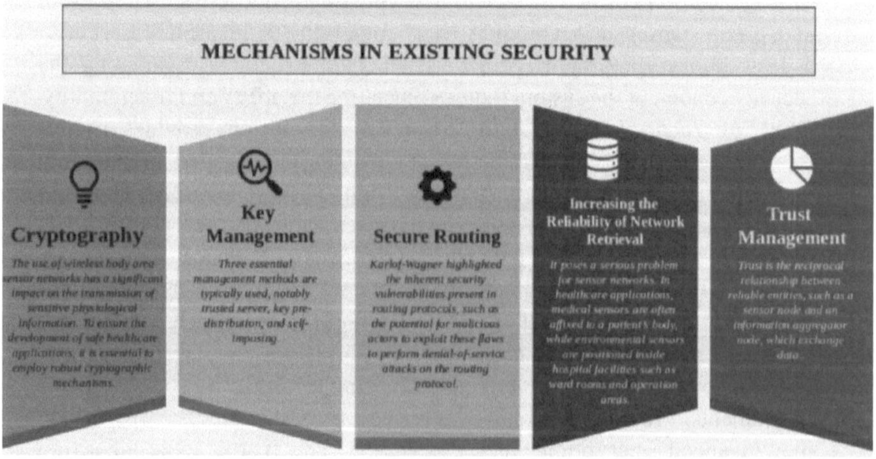

FIGURE 3.3 Existing safety mechanisms in WBANs.

In-home monitoring or emergency situations, a sensor device may need to send its data to other devices outside the local radio network [36]. Therefore, the provision of routing and messaging services facilitates end-to-end communication. So far, several policies have been proposed for sensor networks.

- It is also worth noting that this process was not specifically designed with an emphasis on security measures.
- ***Increasing the Reliability of Network Retrieval:***It is important to note that these locations may be vulnerable to unauthorized access by potential attackers. Therefore, an adversary can gain control of a sensor node, extract its cryptanalytic information, and afterward modify the programming of the sensor. Subsequently, the individual will integrate thc hacked node into the network, potentially jeopardizing the program's overall success. The existing cryptographic functions, including those related to node authentication and identification, can identify and mitigate node-compromised violations to some extent. However, according to Hong, Sung and Lim, "It is essential to note that these compromised node attacks cannot be instantaneously detected. This is a significant concern, particularly in the context of healthcare applications" [37].
- ***Trust Management:*** According to Boukerche and Ren [38], trust management refers to the degree to which a network node may be regarded as reliable, safe, or trustworthy in any interaction with such network. In their study, they evaluated trust within the context of mobile healthcare systems.

3.10 KEY AGREEMENT INVOLVED IN WBAN

The primary focus in ensuring the security of WBANs is vital agreement methods. To effectively construct a secure network, the negotiation of keys must occur among all the nodes involved in the network. Key agreement is the cryptographic protocol sensor nodes use to establish mutual authentication by exchanging secret keys. The essential key agreement process consists of many fundamental elements outlined in Figure 3.4.

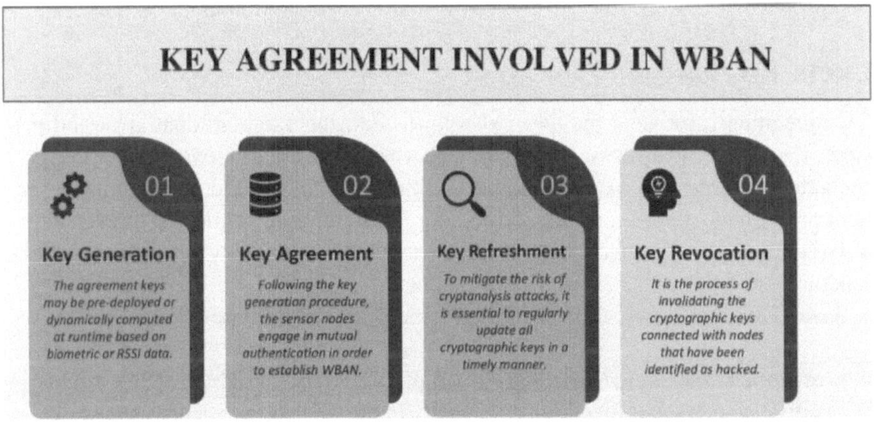

FIGURE 3.4 Key agreements in WBANs.

Traditional cryptography techniques may be classified into two main categories: symmetric key methods and asymmetric key methods. The integration of a symmetric key approach is facilitated by its low memory demand and efficient execution speed. The symmetric key technique is often used, while securely transferring a shared key between both parties remains difficult. Researchers have suggested many techniques for transferring the key between parties throughout the years. The use of an asymmetric key encryption and decryption technique is characterized by slower processing speeds and a higher demand on energy and memory resources. Consequently, this renders it unsuitable for implementation within the resource-constrained environment of a WBAN. Authentication of the nodes within a WBAN is a prerequisite for the establishment of the network. The safe distribution of secret keys to all network nodes is crucial. According to Ali and Khan [39], "The key agreement systems are categorized as: physiological value-based, non-physiological value-based, and hybrid key agreement schemes," whereas M. Kompara and M. Hölbl comment that [40] "[t]he key agreement systems are categorized into four distinct classes: classic, physiological value-based, hybrid, and signal-based secret key agreement methods." The techniques used by aforementioned researchers are identical, with the only one exception. The non-physiological value-based method, as described by Ali and Khan, may be categorized into two distinct types: conventional and signal-based secret key agreement, as outlined by Kompara and Hölbl. These two schemes vary fundamentally from each other.

In classic key agreement methods, pre-distributed keys are installed prior to the establishment of the network. One benefit of this approach is that it reduces the execution time and processing efforts involved in key calculations. However, it necessitates the allocation of extra storage space for key storage. Biological factors are used in the computation of the shared secret key inside a physiological-based key scheme. Since the same biometric characteristics are used to compute a secret key at all deployed nodes, it is anticipated that the key values at each node would be identical. The hybrid key scheme is a combination of conventional and physiological values-based systems. The integration of two methodologies yields superior outcomes. The fourth strategy, known as signal-based secret key creation, may be compared with the key agreement method based on physiological data. This approach leverages the properties of the transmission channel and the biometric qualities of the human body to create key values.

3.10.1 Key Agreement Protocols

Key agreement protocols play a vital role in securing the confidentiality and privacy of sensitive healthcare information that is transmitted between wearable and implanted sensors and external devices within WBANs. Several essential agreement protocols have been developed to address the unique challenges presented by WBANs, including devices with limited resources, constantly changing network structures, and rigorous security requirements.

Several essential protocols for key agreements are as follows:

- ***Elliptic Curve Cryptography:***The Elliptic Curve Cryptography (ECC) algorithm is well recognized for its capacity to provide strong security measures while using reduced vital sizes. This characteristic renders it particularly well-suited for WBAN devices that possess restricted resources. Given its efficacy in mathematical computations and its ability to alleviate computational strain, this technology presents a viable alternative for devices with limited processing capabilities and energy resources. ECC in a WBAN facilitates the safe transmission of cryptographic keys between internal sensors and external devices. This process establishes a robust framework for secure communication, mitigating the potential threats of eavesdropping and unauthorized access to sensitive health data.

- ***Diffie-Hellman Key Exchange:*** The Diffie-Hellman Key Exchange is a significant cryptographic protocol that has been modified for WBANs. The Diffie-Hellman protocol, in its conventional manifestation, requires substantial computer resources. Nevertheless, there have been modifications made to this protocol in order to optimize its functionality for WBAN environments. The protocols above provide secure key exchange between devices in WBANs, hence allowing the construction of secure communication channels. This is particularly beneficial in scenarios where there are constraints on computational resources. Establishing dynamic encryption keys is a crucial aspect of WBANs characterized by frequent network topology alterations. In response to this need, researchers have devised modified versions of the Diffie-Hellman protocol that provide safe mechanisms for exchanging cryptographic keys while being capable of successfully accommodating dynamic scenarios.
- ***Quantum Key Distribution:*** It is an advanced methodology for establishing secure key agreements within WBANs. Quantum Key Distribution (QKD) utilizes the fundamental principles of quantum mechanics to provide a very robust level of security. The proposed solution is based on the basic principles of quantum entanglement and the uncertainty principle, making it theoretically impervious to any attacks from quantum computers. Although QKD provides exceptional levels of security, it requires more resource allocation and operational complexities, making it more suitable for specific high-security WBAN applications.
- ***Pairing-Based Cryptography:*** The methodology used in this technique employs the mathematical principle of pairings to create secure cryptographic keys across multiple devices. Pairing-based encryption offers a favorable balance between security and efficiency, making it well-suited for WBANs with constrained computing and memory capabilities. Through this methodology, WBAN devices can secure critical agreements while maintaining a performance level consistent with their limited resources.

3.11 CONCLUSION

The advancement of medical science is intricately interconnected with other disciplines of research and technology. It is characterized by the significant impact that sporadic breakthroughs in many scientific domains have on human well-being and the delivery of healthcare services. If we talk about India, we will see that in the 21st century, India has immense potential for progress in every field. Progress has doubled in all sectors, including significant breakthroughs in the healthcare industry. After the advent of WBAN technology, many things have made the path of Indian hospitals easier. In a country like India, healthcare professionals' most significant problem is collecting patient's physiological data. This technology has made this easy, too. The emergence of body sensor networks signifies a considerable advancement. These networks represent a novel and stimulating field within remote health monitoring. In addition, this technology facilitates the evaluation of patients' conditions during their hospitalization and post-discharge periods at their residences and workplaces. Moreover, it enables healthcare professionals or intensive care units to conduct health monitoring, thereby allowing for the implementation of treatment protocols when deemed necessary, particularly in the case of elderly individuals.

Though the primary concern, like any other country, is data security, cybercrimes related to healthcare have increased, and several cases of data theft have been recorded, like the recent case of data theft in All India Institute of Medical Sciences, New Delhi. However, with the growing technology and innovations, the healthcare industry will make significant advancements, and data security-related challenges will be minimized.

3.12 SCOPE FOR FURTHER STUDY

The classification proposed in this study provides a potential framework for developing a robust intrusion detection system. Consequently, integrating additional sensors enables capturing and analyzing more precise and targeted data. These principles will have practical relevance within military operations in dynamic environments. The use of this technology has the potential to provide cost-effective healthcare services to underserved rural regions. Furthermore, it will facilitate the establishment of an extensive database that enables clinicians to diagnose individuals from diverse geographical locations and cultural backgrounds. The chapter serves as a forum for advancing Internet of Things in the healthcare industry. In the context of developing nations such as ours, the implementation of creative and economically efficient initiatives of this kind has the potential to enhance the trajectory of technological advancements in the future.

REFERENCES

[1] Bangash, J. I., Abdullah, A. H., Anisi, M. H., & Khan, A. W. (2014). A survey of routing protocols in wireless body sensor networks. *Sensors*, *14*(1), 1322–1357.

[2] Mainanwal, V., Gupta, M., & Upadhyay, S. K. (2015, March). A survey on wireless body area network: Security technology and its design methodology issue. In *2015 International Conference on Innovations in Information, Embedded and Communication Systems (ICIIECS)* (pp. 1–5). IEEE.

[3] Tavera, C. A., Ortiz, J. H., Khalaf, O. I., Saavedra, D. F., & Aldhyani, T. H. (2021). Wearable wireless body area networks for medical applications. *Computational and Mathematical Methods in Medicine, 2021.*

[4] Javaid, N., Khan, N. A., Shakir, M., Khan, M. A., Bouk, S. H., & Khan, Z. A. (2013). Ubiquitous healthcare in wireless body area networks-a survey. *arXiv preprint arXiv:1303.2062.*

[5] Adarsh, A., & Kumar, B. (2020). Wireless medical sensor networks for smart e-healthcare. In *Intelligent Data Security Solutions for e-Health Applications* (pp. 275–292). Academic Press.

[6] Qu, Y., Zheng, G., Ma, H., Wang, X., Ji, B., & Wu, H. (2019). A survey of routing protocols in WBAN for healthcare applications. *Sensors*, *19*(7), 1638.

[7] Abidi, B., Jilbab, A., & Mohamed, E. H. (2019). Wireless body area network for health monitoring. *Journal of Medical Engineering & Technology*, *43*(2), 124–132.

[8] Rady, A., El-Rabaie, S., Shokair, M., & Abdel-Salam, N. (2021). Comprehensive survey of routing protocols for Mobile Wireless Sensor Networks. *International Journal of Communication Systems, 34*.

[9] Xu, G., & Wang, M. (2014). An energy-efficient routing mechanism based on genetic ant colony algorithm for wireless body area networks. *Journal of Networks*, *9*(12), 3366.

[10] Zou, S., Xu, Y., Wang, H., Li, Z., Chen, S., & Hu, B. (2017). A survey on secure wireless body area networks. *Security and Communication Networks, 2017.*

[11] Huanan, Z., Suping, X., & Jiannan, W. (2021). Security and application of wireless sensor network. *Procedia Computer Science*, *183*, 486–492.

[12] Naik, M. R. K., & Samundiswary, P. (2016, December). Wireless body area network security issues—Survey. In *2016 International Conference on Control, Instrumentation, Communication and Computational Technologies (ICCICCT)* (pp. 190–194). IEEE.

[13] Ananthi, J. V., & Jose, P. S. H. (2021). A perspective review of security challenges in body area networks for healthcare applications. *International Journal of Wireless Information Networks*, 1–16.

[14] Li, M., Lou, W., & Ren, K. (2010). Data security and privacy in wireless body area networks. *IEEE Wireless Communications*, *17*(1), 51–58.

[15] Lim, S., Oh, T. H., Choi, Y. B., & Lakshman, T. (2010, June). Security issues on wireless body area network for remote healthcare monitoring. In *2010 IEEE International Conference on Sensor Networks, Ubiquitous, and Trustworthy Computing* (pp. 327–332). IEEE.

[16] Jamal, T., & Mendes, P. (2013). *Cooperative Relaying for Dynamic Networks.* http://eprints.rclis.org/33674/1/Cooperative%20Networking%20Final.pdf

[17] Jamal, T., & Mendes, P. (2013). *802.11 Medium Access Control in MiXiM.* https://www.researchgate.net/profile/Paulo-Mendes-18/publication/262727209_80211_Medium_Access_Control_In_MiXiM/links/00b495389f0f27897b000000/80211-Medium-Access-Control-In-MiXiM.pdf

[18] Jamal, T., Alam, M., & Umair, M. M. (2017, March). Detection and prevention against RTS attacks in wireless LANs. In *2017 International Conference on Communication, Computing and Digital Systems (C-CODE)* (pp. 152–156). IEEE.

[19] Movassaghi, S., Abolhasan, M., Lipman, J., Smith, D., & Jamalipour, A. (2014). Wireless body area networks: A survey. *IEEE Communications Surveys & Tutorials*, *16*(3), 1658–1686.

[20] Fouad, H. (2014, September). Continuous health-monitoring for early detection of patient by web telemedicine system. In *International Conference on Circuits, Systems and Signal Processing, Saint Petersburg State Politechnical University, Russia* (pp. 23–25). IEEE Xplore.

[21] Walters, J. P., Liang, Z., Shi, W., & Chaudhary, V. (2007). Wireless sensor network security: A survey. *Security in Distributed, Grid, Mobile, and Pervasive Computing*, *1*(367), 6.

[22] Kotz, D. (2011, January). A threat taxonomy for mHealth privacy. In *2011 Third International Conference on Communication Systems and Networks (COMSNETS 2011)* (pp. 1–6). IEEE.

[23] Kompara, M., & Hölbl, M. (2018). Survey on security in intra-body area network communication. *Ad Hoc Networks*, *70*, 23–43.

[24] Hosseini-Khayat, S. (2011, March). A lightweight security protocol for ultra-low power ASIC implementation for wireless implantable medical devices. In *2011 5th International Symposium on Medical Information and Communication Technology* (pp. 6–9). IEEE.

[25] Omoogun, M., Seeam, P., Ramsurrun, V., Bellekens, X., & Seeam, A. (2017, June). When eHealth meets the internet of things: Pervasive security and privacy challenges. In *2017 International Conference on Cyber Security and Protection of Digital Services (Cyber Security)* (pp. 1–7). IEEE.

[26] Sajid, A., & Abbas, H. (2016). Data privacy in cloud-assisted healthcare systems: State of the art and future challenges. *Journal of Medical Systems*, *40*(6), 155.

[27] Langone, M., Setola, R., & Lopez, J. (2017, July). Cybersecurity of wearable devices: An experimental analysis and a vulnerability assessment method. In *2017 IEEE 41st Annual Computer Software and Applications Conference (COMPSAC)* (Vol. 2, pp. 304–309). IEEE.

[28] Zeadally, S., Isaac, J. T., & Baig, Z. (2016). Security attacks and solutions in electronic health (e-health) systems. *Journal of Medical Systems*, *40*, 1–12.

[29] Kargar, M. J., Ghasemi, S., & Rahimi, O. (2013). Wireless body area network: From electronic health security perspective. *International Journal of Reliable and Quality E-Healthcare (IJRQEH)*, *2*(4), 38–47.

[30] Fatema, N., & Brad, R. (2014). Security requirements, counterattacks and projects in healthcare applications using WSNs-a review. *arXiv preprint arXiv:1406.1795.*

[31] Singh, A. K., & Patro, B. D. K. (2020). Signcryption-based security framework for low computing power devices. *Recent Advances in Computer Science and Communications (Formerly: Recent Patents on Computer Science)*, *13*(5), 845–857.

[32] Ferdous, M. S., Chowdhury, F., & Moniruzzaman, M. (2007, January). A taxonomy of attack methods on peer-to-peer network. In *Proceedings of the 1st Indian Conference on Computational Intelligence and Information Security (ICCIIS, 07)* (pp. 132–138). Allied Publishers.

[33] Sharma, D. (2013). Wireless health care monitoring system with data security and privacy. *International Journal of Research in Electronics and Computer Engineering*, *2*(2), 1–2.

[34] CC2420, C. (2006). 2.4 GHz IEEE 802.15. 4/ZigBee-ready RF transceiver. *Chipcon Products from Texas Instruments.* https://www.ti.com/lit/ds/symlink/cc2420.pdf

[35] Hong, D., Sung, J., Hong, S., Lim, J., Lee, S., Koo, B. S., & Chee, S. (2006). HIGHT: A new block cipher suitable for low-resource device. In *Cryptographic Hardware and Embedded Systems-CHES 2006: 8th International Workshop, Yokohama, Japan, October 10–13, 2006. Proceedings 8* (pp. 46–59). Springer Berlin Heidelberg.

[36] Lorincz, K., Malan, D. J., Fulford-Jones, T. R., Nawoj, A., Clavel, A., Shnayder, V., & Moulton, S. (2004). Sensor networks for emergency response: Challenges and opportunities. *IEEE Pervasive Computing*, *3*(4), 16–23.

[37] Kavitha, T., & Sridharan, D. (2010). Security vulnerabilities in wireless sensor networks: A survey. *Journal of Information Assurance and Security*, *5*(1), 31–44.

[38] Boukerche, A., & Ren, Y. (2009). A secure mobile healthcare system using trust-based multicast scheme. *IEEE Journal on Selected Areas in Communications*, *27*(4), 387–399.

[39] Ali, A., & Khan, F. A. (2015). Key agreement schemes in wireless body area networks: Taxonomy and state-of-the-art. *Journal of Medical Systems*, *39*, 1–14.

[40] Kompara, M., & Hölbl, M. (2018). Survey on security in intra-body area network communication. *Ad Hoc Networks*, *70*, 23–43.

4 A Review of Forensics Security Hazards and Challenges in WBAN and Healthcare Systems

S. Karthika and K. Juliana Gnanaselvi

4.1 INTRODUCTION

A collection of internal or implanted devices that function as actuators, sensing devices, and transceivers inside and outside the human body is referred to as a Wireless Body Area Network, or WBAN. WBANs are becoming a fascinating topic of study in numerous fields due to the significant developments in Internet of Things (IoT) research. Many applications, including those related to health and wellness, medicine, sports, military, animal husbandry, interactive gaming, analytics for security, live chat, and private data communication, rely on WBANs and the World Wide Web of Things. As more sensors become available for IoT applications, WBAN adoption is becoming cost-effective for many industries.

Furthermore, the WBANs' ability to collect as much data from field workers as they do would not be conceivable without wireless connectivity. People are frequently observed and instructed in specific tasks using this knowledge. Network challenges will surely develop as WBAN usage grows; it should go without saying. If WBAN is to be utilized to its maximum capacity in the future, these problems must be solved. The WBAN produces and transmits biological data that can be used to identify particular persons uniquely; hence, network security is crucial. Power management for the WBAN, the creation of various sensors with low energy consumption, and network characteristic optimization for optimum power management are additional significant issues [1]. Obtaining energy from the environment and from the movement of people is an important area of study for WBAN, and researchers have proposed various methods. Choosing the appropriate antenna for a particular application is essential since transmitting intra- and inter-WBAN data requires significant power. The antenna design heavily influences how much power is needed, how much channel loss is there, and how much throughput is there. Since sensor flexibility is another crucial issue, creating flexible sensors is a tough research topic.

Future WBAN antennas, energy harvesting plans, and tactics, wearable WBAN sensors, privacy problems about WBAN, authentication process approaches, and

DOI: 10.1201/9781032635101-5

WBAN functionalities across many sectors are all examined in this analysis. It also looks recently suggested WBAN designs and communication protocols that are widely used. The goal is to provide in-depth information on every topic of interest that has already been covered.

A particular type of sensor network called a WBAN uses biological nodes with wireless sensors to gather physiological indications including the body's temperature, blood pressure, heart rate, and glucose level. This allows for the possibility of tracking patients remotely. Wearable and integrated WBANs are accessible. The primary objective of WBANs is to protect public health by implanting detectors that transmit metabolic data to medical servers, providing doctors with information about their aliments of patients. In addition, by collecting data on a patient's vital signs over time, WBANs can significantly lower medical care costs. In contrast to wireless networks, networks with body sensors are designed using a star topology.

Body Area Wireless Network: In order for portable and implantable devices to work, multiple sensors must be employed. The sensors are intended to track numerous physiological and environmental parameters, such as temperature, blood pressure, heart rate, saturation of oxygen, and movement. The sensors collect data and wirelessly transmit that data to a centrally located monitoring system [2].

Data is gathered from various sensors by the central monitoring system, which then analyzes it to look for trends, warning indications, and anomalies. Analyses can be run in batches or in real time. The study's findings are then put to use in order to offer patients, athletes, and fitness fanatics specialized advice, alerts, and treatments.

4.1.1 Wireless Body Sensor Networks

Wireless sensor networks have sensor nodes that work together, are cheap and energy-efficient, and have wireless connectivity. An electrical system that observes its environment and notifies another electrical system of occurrences or changes is known as a sensor. Some of the main limitations of sensor networks include memory size, processing speed, and energy use. Because of its improved data-gathering efficiency, the wireless sensor network has been used in the medical industry and for environmental control, disaster reporting, military and security surveillance systems, and traffic control. Applications for different types of wireless sensor networks need different levels of quality (including latency, dependability, bandwidth, and security). The QoS (quality of service) must be ensured in every area. Because wireless sensor networks struggle with problems including constrained bandwidth power, inconsistent communications, and node vulnerabilities, quality of service requirements are essential.

Wireless sensor networks are used to collect data on the health of hospital patients and monitor the development of rare diseases. Wireless sensors can also be used to monitor an agent's effectiveness. These networks are used for various functions, such as managing medications in a hospital, finding patients and physicians in a clinic, and remotely gathering physiological data about a person's body. Medical personnel can be made aware of any potential medical emergency, allowing the healthcare team to take rapid action [3]. This is possible because of the use of the data produced by this technology outside of hospitals and the significant advancements made in wireless sensor network technology in the medical industry. Accessibility, privacy, secrecy, consistency, service quality, and mobility should all be considered when selecting how to provide data via telemedicine networks.

4.1.2 Using Sensors to Manage Health

When experiencing strange health problems, most patients visit a doctor, and in many dangerous situations, prompt treatment is necessary. Even if the patient's care may be very costly in certain circumstances, the patient needs to be treated as soon as feasible. Utilizing a network of sensors, however, can help further save costs and risks throughout the treatment process by keeping people physically fit and spotting the early stages of sickness.

Sensor networks will be widely used in the healthcare sector over the next several decades. Modern hospitals are already using many simple sensor types to track patient physiological data, deliver medication schedules, and keep an eye on personnel and patients. In this domain, elderly nursing is one application of sensor networks. This is achieved using sensor cameras, which work as a sophisticated network to recognize muscle contraction. This network monitors older people and is capable of detecting falls, inattention, vital signs, nutrition, and activities. In the future, experts believe these networks will be very beneficial for determining the body's actual health status because of advancements in technology and the utilization of broadband technology. Real-time health diagnostics made possible by this technology have the potential to save lives, reduce medical costs, and stop the neglect of degenerative diseases. Currently, health exams can be performed virtually in some circumstances, albeit these methods are not flawless [4]. As previously indicated in prior studies, heat from the body generates the small, portable, ultra-wide sensor band monitoring sensors used today. These little devices can transmit a lot of data, which significantly enhances medical services and healthcare, reduces treatment costs, and aids in illness prevention. It can also collect data on insulin levels, cholesterol levels, the density of bones, and cardiovascular health. This device not only monitors a patient's health and prevents illness, but it can also diagnose a patient with an arrhythmia and predict the onset of a heart attack. This gadget must be able to store and transmit massive amounts of data, be accessible to anyone, and be reasonably priced.

4.1.3 IoT Healthcare

An innovative technology known as IoT involves intelligent things with physical components like sensors that can comprehend the internal state of an event or the environment and take appropriate action in response to the information gathered. Intelligent objects' sensors produce data that can be further analyzed to aid in decision-making. The embedded software in these intelligent things manages all of its numerous components and events. The IoT paradigm, which combines the benefits of multiple communication technology domains such as edge, cloud, fog, WBANs, and automated computing, opens up new avenues and opportunities for many enterprises. By 2025, 75 billion connected gadgets are anticipated to be made possible by IoT. IoT has an impact on a number of industries, including business, manufacturing, retail, healthcare, and defense/security [5]. The detection of disasters and fires in forests is one application of IoT. Additional uses include trash management, intelligent parking, intelligent lighting, intelligent crop management, and remote senior home care.

4.2 WBAN ARCHITECTURE AND COMMUNICATION PROTOCOLS

Many research groups have released multiple WBAN Architectures for various purposes. Performance is enhanced through application-specific design and communication protocols. The ad hoc, machine-to-machine (M2M), and body-to-body (B2B) variations are less common than the three-tier WBAN architecture.

4.2.1 WBAN Architecture

In the WBAN architecture, each sensor is a node, and nodes can be either attached to or detached from the human body. Each node in the WBAN functions as a sensor to collect and sample human physiological vitals, transmit data, and communicate with other nodes and mobile devices that operate as gateways. The network features a three-tiered communication architecture. Up to 12 nodes can fit in this network configuration (see Figure 4.1).

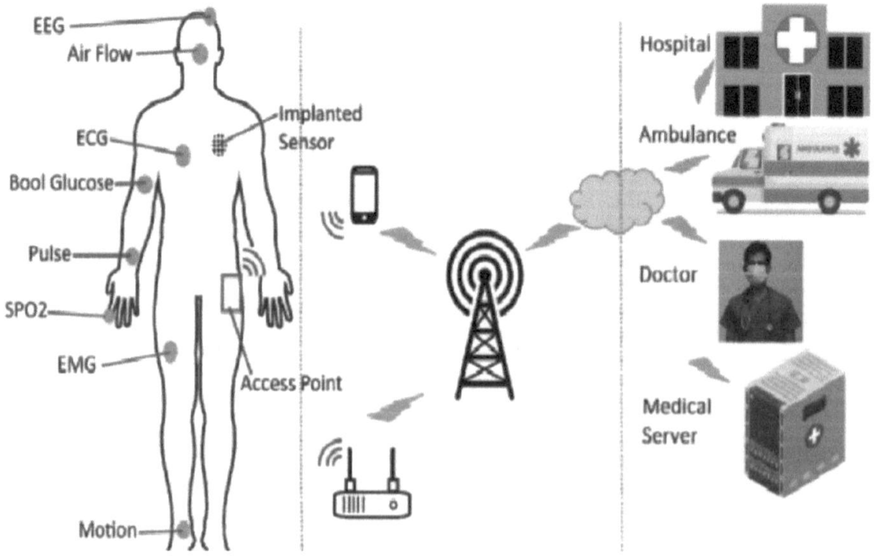

FIGURE 4.1 Architecture diagram of WBAN.

4.2.1.1 Tier 1: Intra-WBAN

The Tier 1 WBAN component consists of several body sensor nodes that are located inside the human body, on clothing, and on the skin. The body nodes process all seen data in addition to continuously monitoring their surroundings and essential physiological indicators including heart rate, blood pressure, temperature, and electrocardiogram. Through their contractual link, the coordinator receives the processed records from these sensor nodes. Every perceived piece of information from every node is collected, analyzed, and sent to the tier of the hierarchy beneath the coordinator [6]. When used properly, the WBAN can prove to be a highly efficacious technique for individuals who may require periodic exams.

4.2.1.2 Tier 2: Inter-WBAN

Tier 2 has a gateway-functioning wireless networking configuration. Promising WBAN technologies include wireless local area networks (WLAN), Bluetooth, ZigBee, Wi-Fi, Global System for Mobile Communications (GSM), 3G, and 4G wireless networking protocols. The mobile network or a portable device may also act as a gateway. The majority technologies, save for cellular networks, are frequently useful for narrow-band communication.

4.2.1.3 Tier 3: Extra-WBAN

The Tier 3 decision control unit (DCU) finishes all necessary activities and chooses the optimal course of action for a patient's health based on the results. Because this layer also maintains the patient information, any healthcare professional, including the doctor, has constant access to the patient's medical history. They can see the online prescriptions that the doctor has written for the patient's treatment. In addition, there are emergency services available, including ambulances.

4.3 NETWORK FORENSICS TECHNIQUES

One area of network security called "network forensics" usually focuses on detecting and preventing community assaults. It talks about how having certain investigative skills is essential. Examining the neighborhood and its surroundings will benefit organizations. In addition, it is crucial for police force investigations. The location of the intrusive party and the point of intrusion can be determined using network forensic techniques. In addition, by employing these strategies, attacks that are imminent can be avoided. Many obstacles must be solved by contemporary network forensic techniques in order to advance the forensic methodologies. The primary obstacles include fast storage, the need for ample storage space, data integrity, data secrecy, IP address accessibility, and the location of data extraction. Potential solutions as well as the complexities of these problems are investigated. In general, these forensic network issues need to be resolved before any tools or methods for network forensics are created [1,2].

4.3.1 Categorization of Contemporary Forensic Techniques

This section presents a proposed taxonomy for network forensic techniques, together with information on how these approaches are evaluated, applied, and critically analyzed in relation to other fields (see Figure 4.2).

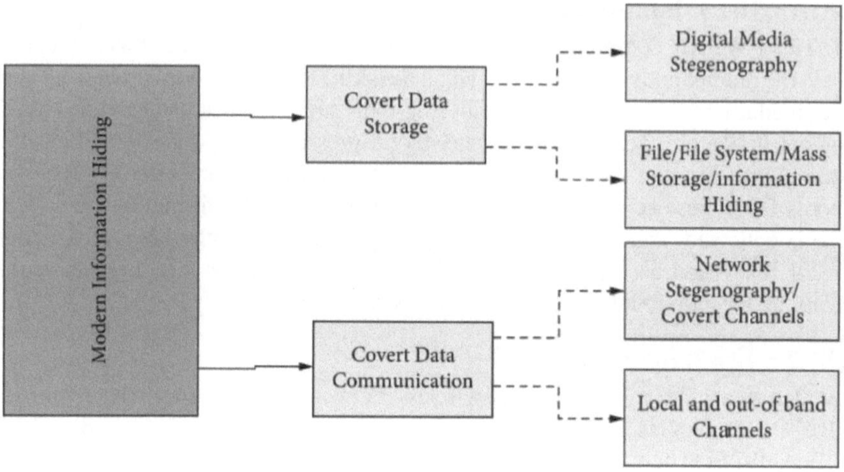

FIGURE 4.2 Categorization of forensics techniques.

4.3.2 The Network Forensic Technique Based on Traceback

When it is necessary to ascertain the network sources of a packet, the term "traceback" is commonly employed. One helpful technique for determining and assessing network threats is IP tracing. By determining the device from which packets are generated, this technique finds the assault's origin. For tracing packet sources in scenarios like distributed denial-of-service (DDoS) attacks and spoofing, the traceback approach comes in handy. Since distribution networks are the main targets of DDoS and botnet activities, the traceback approach is helpful in these attacks.

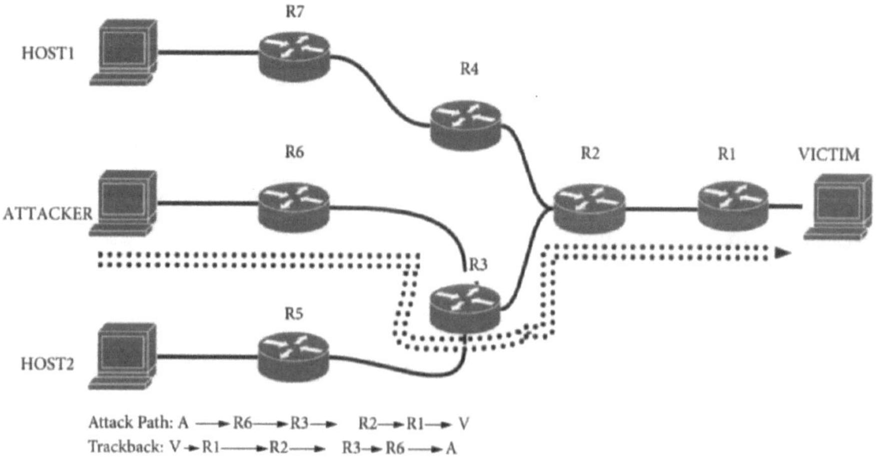

FIGURE 4.3 Traceback technique.

The network requires a number of traceback techniques, including the ones mentioned later, in order to successfully thwart various attacks: In order to gather, combine, identify, analyze, correlate, evaluate, and record evidence from unquestionably electronic, unquestionably processing, and transmitting digital resources, network forensics employs procedures that have been scientifically validated (see Figure 4.3). This helps to provide light on the intended use or estimated success of unauthorized actions intended to disrupt, corrupt, or compromise system components and to provide information useful for responding to incidents [7,8]. Period of observation: Security for forensics and real-time network monitoring are covered. This also includes real time.

4.3.3 Combining Network-Based and Network Forensic Methods

Digital evidence housed inside converged networks can be found with great assistance, thanks to convergence network-based network forensic tools. Voice over Internet Protocol (VoIP) communication is one particular application of a converged network (see Figure 4.4). VoIP is prone to several vulnerabilities, security risks, and mishaps since it depends on a data transmission medium. VoIP frames are created by splitting up communication signals, and they are subsequently added to data packets as voice codes. These data packets follow the same protocol that voice packets use to go across an IP network. When audio packets are sent from a transmitter to a destination in an unmodified and uninterrupted manner, they are referred to as 181 normal voice packets. Typically, SIP and H.3231 are used to send voice packets over

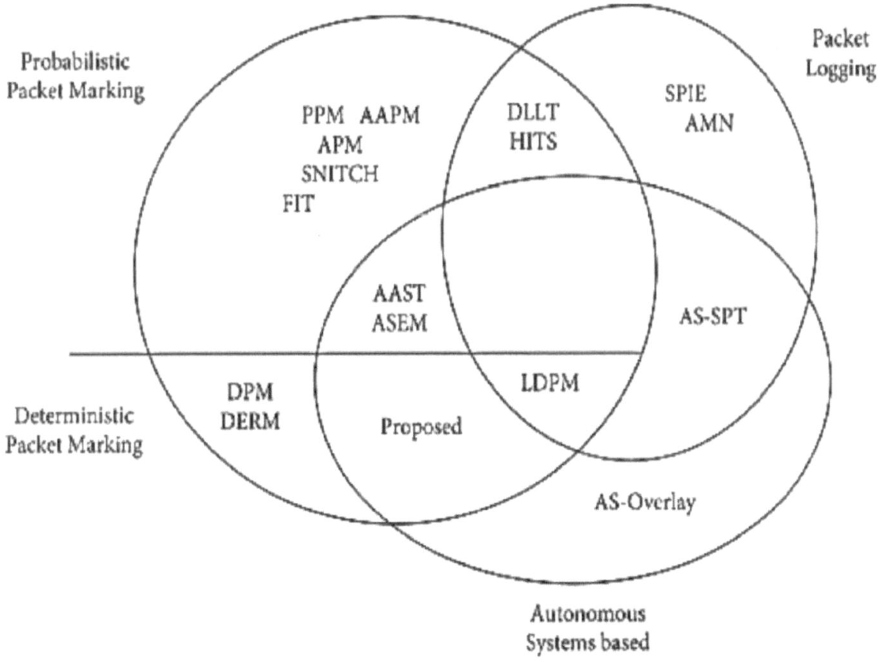

FIGURE 4.4 Converge network-based technique.

IP. The IP address and port information required to initiate the connection protocols are contained in the voice packets. The communication protocols are session control number 184 protocols. The related ports and IP addresses in the voice packets are not encrypted since the address translation devices need to translate the speech packets [3]. The voice packets are not encrypted, and therefore, intrusion attempts are possible. The hacker is able to change 188 legitimate voice packets into invalid ones by manipulating speech packets as they are being transmitted. Numerous techniques, such as VoIP device exploitation, call integrity degradation, privacy leakage, eavesdropping, man-in-the-middle attacks, and buffer overflows, can result in malicious voice packets.

4.3.4 An Attack Graph-Based Method for Network Forensics

The attack graphs are used in attack graph-based network forensic techniques to identify all possible attack paths that an attacker might have taken (see Figure 4.5). Analyzing networks, hosts, and other security tools is necessary for this process. Each vertical axis in an attack graph, which consists of vertices, represents a potential assault node. The edges show the differences in the statuses of several attack

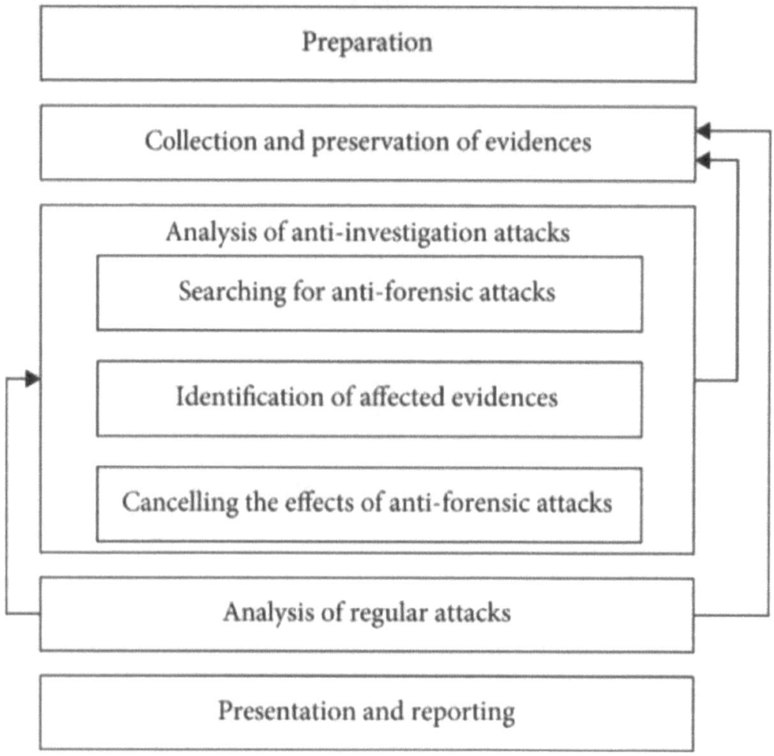

FIGURE 4.5 Attack graph-based network.

nodes. Assault graphs are highly helpful in network forensics because they show which nodes are vulnerable to attack and which are the worst paths with the highest likelihood of being attacked [9]. The network administrator can schedule the security measures in advance by identifying these nodes. The attack graphs can also be used for cost–benefit security hardening, effect analysis, evidence gathering, and the identification of multistage network attacks [10].

4.3.5 Method for Distributive-Based Network Forensics

To address scalability issues, the forensic network servers and data agent systems can be distributed using distributive-based network forensic approaches. All of the data server agents that are scattered throughout the network are collected for analysis by the forensic network servers. The distributive network forensic technique is employed to gather information, locate the attack's origin, and carry out an inquiry. To protect the server from hackers, the distributive network forensic technique raises overhead. The forensic servers are vulnerable to hacking since they are dispersed throughout the network. Leverage the distributed framework to efficiently manage multiple network devices and systematically monitor their network logs. The network logs are stored in multiple locations in a network environment. The usual practice is to manually examine host and packet logs. Manual analyses have a lot of problems because of their slow response times, poor logging techniques, problems with data synchronization, and low reaction times [10]. The majority of the issues listed earlier are resolved by the For Net network forensic design, which enhances evidence gathering. The pieces of evidence are obtained from the network hardware, such as switches and routers, by installing a program on each network. This application is called SynApps.

4.3.6 Using Network Devices for Data Storage

For the purpose of the inquiry, a sizable quantity of data must be gathered and kept on the network. It's challenging for forensic investigators to retrieve pertinent evidence from these networks due to the massive volume of data stored on them. Since the storage capacity of interconnectivity devices is limited, the collected data packets need to be saved in a significant amount of storage space. By creating a framework for machine-based data collection, the issue is also overcome. Inquiry times are shortened, and large storage rates are not required with this architecture.

4.3.7 Location of the Extracted Data

The distributive structure of networks and their virtualized features make network forensics a difficult task. The device and the best place for data recovery become harder to locate for forensic specialists. For networks with thousands of linked devices and millions of data packets passing through each device per second, managing all linkages and connected devices is nearly impossible. Since these networks have the potential to breach user privacy or data integrity at any time, network forensics finds it very difficult to collect data from any point in these networks. Furthermore, a wide range of network-connected devices, such as routers, firewalls, packet sniffers, protocol analyzers, and tools for network forensic analysis, are designed to gather and examine data for network forensics [11]. Determining the best location for these devices to collect and process data from the data network is one of the main problems for network forensics.

4.4 NETWORKS FOR FORENSICS IN MOBILE CLOUD COMPUTING

Mobile Cloud Computing (MCC) network cloud services are available to smart device users who are linked to long-term evolution networks via WLAN, 3G/4G/5G, and Wi-Fi. To send user requests to parallel computing clouds and provide results to users of connected devices, these networks must be incredibly quick and secure. On the contrary, attackers use network attacks to target these networks in an attempt to connect and get access. Firewalls and intrusion detection systems (IDS) are used as part of the current security strategy to identify and detect attack patterns. But cunning attackers use these defenses to spread malicious network activity. Protection mechanisms need to be smart enough to identify threats that could compromise the system. The network position shows the network address that links two MCC entities. Typically, MCC includes intercloud networks, data centers, and cloud access [12]. Every component of the communication paths inside the MCC network adds to the process's usefulness. Network Exposure Function (NFF) is required to monitor risky actions via network packets if a user of a smart device is connected to mobile clouds or data centers that are connected to or linked to other cloud data centers. Since Network Function Interfaces (NFIs) have limited or no access to analyze various network susceptibilities, MCC users should always have access to the forensic investigation using secure cloud resources and channels. The linking network between two or more clouds is called an intercloud network. This network makes use of a fiber optic system with a high line rate and capacity.

4.5 WBAN IN FORENSICS SECURITY AND PRIVACY REQUIREMENTS

WBAN systems must implement specified security measures in order to maintain the security, privacy, data integrity, and confidentiality of a patient's medical records [13]. A supporting WBAN infrastructure must implement specific security measures in order to guarantee all of these characteristics. Security and patient data privacy are each WBAN system's two most important elements. Security refers to protecting data from unauthorized users while it is being transported, collected, processed, and stored. On the other side, the idea of privacy implies the right to control the gathering and use of one's personal information (see Figure 4.6).

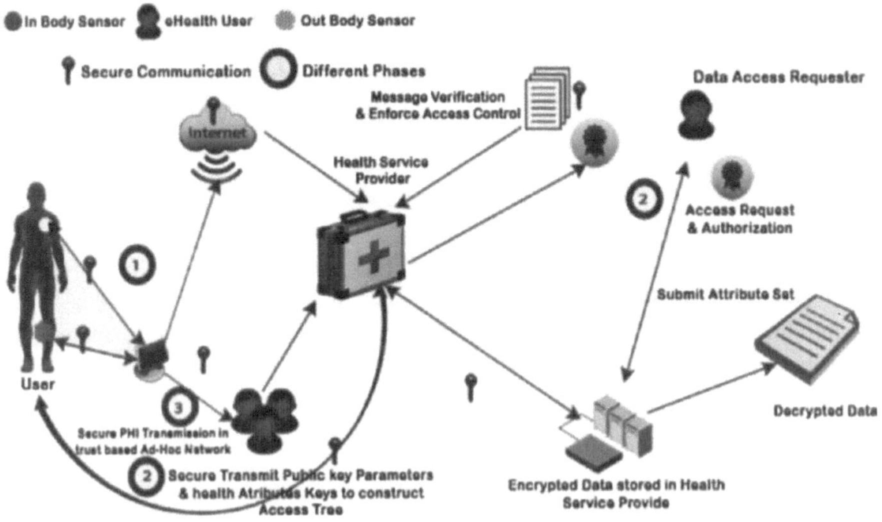

FIGURE 4.6 Security and privacy architecture.

4.5.1 Data Confidentiality

Protecting personal information from public view is known as data confidentiality, and it is one of the main concerns of a WBAN. Since WBAN nodes are used in medical contexts to transmit confidential and sensitive health information, it is critical to protect patient data from unauthorized access that could pose a threat to their life. This important information could be "overheard" during transmission, which could be detrimental to the patient, the medical provider, or the system as a whole. Encryption helps protect this sensitive data by allowing a shared key to be sent across a secure communication channel between coordinators and secured WBAN nodes [14].

4.5.2 Data Accuracy

The procedures used to protect a message's accuracy, consistency, and content are referred to as data integrity. Both individual messages and message streams can use it. However, since an opponent can easily alter patient information submitted to an insecure WBAN before it reaches the network coordinator, data confidentiality does not protect information from outside alterations [15]. More specifically, modifications can be carried out by patching together a few fragments, modifying certain information within a packet, and forwarding the packet to the PS.

4.5.3 Freshness of Data

By using data freshness techniques, the WBAN coordinator can be effectively protected from an attacker's attempt to capture and replay the data, compromising its secrecy and integrity. It ensures the accuracy and non-reuse of outdated data frames.

Currently, we use two types of data freshness: strong freshness, which guarantees delay in addition to both, and weak freshness, which is limited to frame ordering [16]. Strong freshness is required for synchronization when a beacon is being relayed to the WBAN coordinator; WBAN nodes with low duty cycles employ weak freshness.

4.5.4 Network Accessibility

It implies that a doctor has quick and easy access to a patient's records. This kind of system contains vital, extremely sensitive, and possibly lifesaving information, so it is critical that the network is always open for patients to use in an emergency. For this reason, in the event of an availability loss, it is imperative to move the operations to another WBAN.

4.5.5 Verification of Data

For both nonmedical and medical purposes, data authentication could be required. Consequently, nodes inside a WBAN need to be able to verify that the data is indeed coming from a reliable trust center and not malicious sources. Consequently, the network and coordinator nodes employ a shared secret key for all data in order to determine the Message Authentication Code (MAC). Accurate MAC code calculation reassures the network coordinator that a trustworthy node is sending the message.

4.5.6 Safeguarding Operations

For key distribution to be transferred to a WBAN, the coordinator must supervise the encryption and decryption process securely. The coordinator is responsible for overseeing WBAN node additions and deletions under node affiliation and disassociation.

4.5.7 Dependability

The system needs to be trustworthy and reliable. Another serious issue with WBANs is the failure to retrieve the correct data, which could endanger the patient's life. Techniques for error-correcting code can be applied to solve this problem.

4.5.8 Secure Localization

The bulk of WBAN applications require accurate patient location estimation. In the event that tracking tools are not accessible, an attacker may be able to convey misleading information, for example, by answering with a phony signal that reveals the patient's location.

4.5.9 Responsibility

Healthcare providers are required by law to secure patient health information in the medical industry. If a supplier fails to maintain this data or, worse, abuses their authority over it, they ought to be held accountable in order to stop similar abuses in the future.

4.5.10 Flexibility

The patient must be able to choose which APs can access their medical records under a WBAN. For instance, in an emergency, consent can be granted on demand to a different doctor who isn't on the list of those who are authorized to interpret patient data.

4.6 WBAN ROUTING PROTOCOLS

Some of the most current routing protocols developed in the last 10 years include those that are temperature sensitive, postural mobility aware, quality of service (QoS) sensitive, cluster oriented, and cross-layered (see Figure 4.7).

4.6.1 Temperature-Sensitive Routing Protocols

Radio broadcasts generate magnetic and electrical forces that raise temperatures, especially in implanted nodes, because wireless communications employ radio antennae and node power. Long-term retention may hinder the development of the supporting tissues and other essential organs. The radioactive intensity volume that the tissue absorbs per unit weight is known as the specific absorption rate, or SAR. Furthermore, all temperature-sensitive routing protocols aim to restrict the increase in temperature of embedded body nodes.

4.6.2 QoS-Sensitive Routing Protocols

In order to optimize the network's overall throughput, QoS-based routing is expected to effectively direct network traffic. An increasingly common tactic is to always choose the shortest path among the feasible options, as a longer path makes greater use of network resources.

Achieving QoS criteria including packet delivery ratio, performance, packet failure ratio, and latency is the primary challenge in WBANs. The patient's medical information is categorized into four groups: normal (heartbeat and temperature), emergency (electrocardiography [ECG] and electroencephalogram [EEG]), delay-responsive (video streaming), and reliability-sensitive (pH and breathing monitoring).

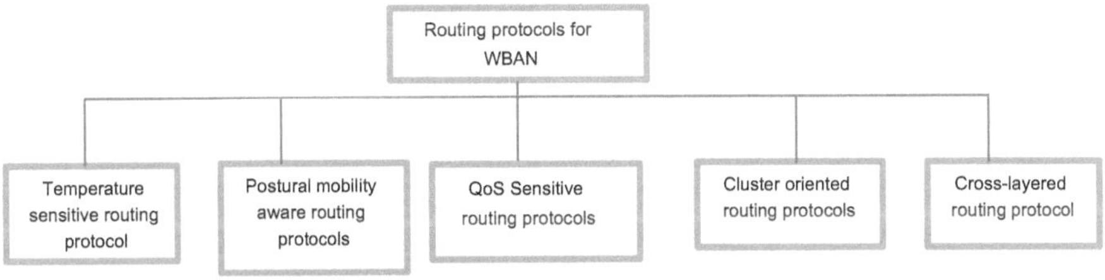

FIGURE 4.7 Categories of WBAN routing protocols.

4.6.3 Cluster-Oriented Routing Protocols

For each cluster, a cluster head is chosen to oversee membership data documentation. Based on predefined criteria, a cluster head is chosen by cluster-oriented routing protocols from a group of body nodes that have been divided into multiple clusters. All of the data sensed by every node connected to their cluster must be gathered by each designated cluster head, who then forwards the created information to the coordinator. There is no direct link between any of the body nodes; all communication is routed via the cluster head.

4.6.4 Postural-Mobility-Aware Routing Protocols

The partitioning of the WBAN design may occur due to brief RF transmission ranges and/or changes in body posture. Routing of data problems and failed links between nodes might result from network segmentation. Numerous academics have developed a range of methods based on the dynamic communication cost factor to address these issues [17]. The ratio of average packets sent to the sink to the total energy utilized by the network is the energy consumption used to compute the communication cost factor. These protocols aim to minimize the expense of transmitting data packets from the patient's body to the coordinator.

4.6.5 Cross-Layered Routing Protocols

These protocols aim to improve the overall network efficiency by concurrently addressing problems and constraints at the media access control and network layers. To solve the problems of collision, delay, dependability, and energy, several media access control protocols have been created especially for WBAN. media access control protocols are built using a range of access strategies. Two main categories into which the various WBAN media access control algorithms can be categorized are time division multiple access (TDMA) and carrier sense multiple access (CSMA). A collision is likely to happen when two sensors try to access the same channel at the same time because of the contention-based architecture of the CSMA protocol [18]. In an effort to have another chance to send their data, the sensors wait for a predetermined period of time before attempting again. Sensors deliver their data at certain time periods as part of the schedule-based method used by TDMA to avoid collisions.

4.7 PROCESS OF PRACTICAL IMPACT ASSESSMENT

When assessing the effects of WBAN security concerns, the Practical Impact Assessment (PIA) is a helpful tool. To demonstrate how earlier technologies fell behind WBAN, each test will be carried out in real time by inserting security risks into the system. The Descriptive Report of the Convention on Cybercrime served as the foundation for a selection of wireless security threats. The United States of America, the United Kingdom, Australia, and other 51 countries have recognized the Convention on Cybercrime's methodology (see Table 4.1).

TABLE 4.1
Attack Names of PIAm

m	PIAm	Name of the Attack
1	PIA1	Eavesdropping
2	PIA2	Denial-of-service (DoS)
3	PIA3	Authentication bypass
4	PIA4	Role bypass

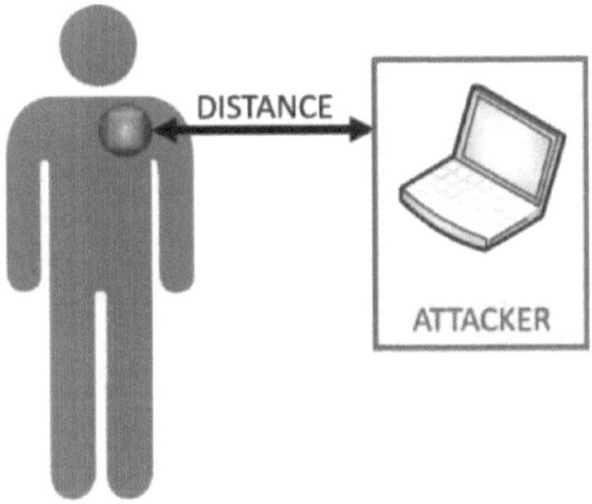

FIGURE 4.8 Attacker's distance from the WBAN target system.

The PIA process will measure two different metrics. First, the attacker's distance from the WBAN target system is measured; this is the amount of time it requires to carry out a successful PIAm execution (see Figure 4.8). To ascertain the minimum and greatest distance required to execute a successful PIAm, the PIA was conducted from a number of perspectives. The only set of parameters that successful PIAm logs include the specific privacy risks identified, the mitigation measures implemented, the stakeholders involved, and the compliance status of the privacy impact assessment.

The entire procedure will be repeated if the PIAm was unsuccessful. Using the logs that the WBAN target system kept while the PIAm was being executed, the success of the PIAm was confirmed. The logs will also be used to identify the affected impact. The final phase of the PIA process will determine the impact on WBAN security based on the CIA. CIA means confidentiality, integrity, and availability [19,20].

Results for all successful PIAm are shown in Table 4.2. Table 4.2 successfully demonstrates that it is possible to impose additional wireless technology security hazards to WBAN. For every successful PIAm executed, a maximum distance of 5 meters was recorded between the attacker and the WBAN target system. The WBAN

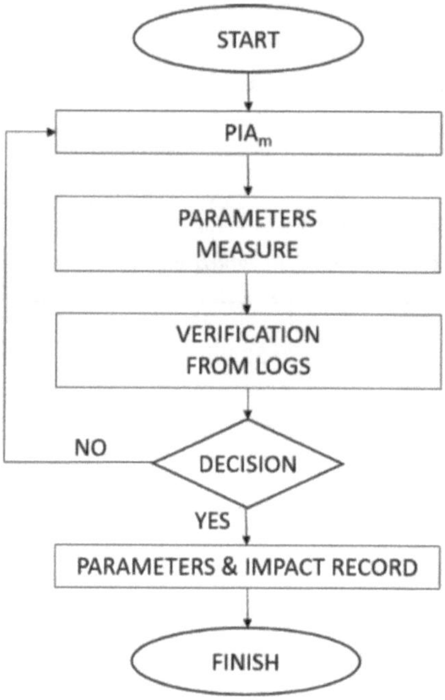

FIGURE 4.9 Flow chart for PIA.

TABLE 4.2
PIA Result

		Measurement of Parameters		Impact Parameters		
m	**PIAm**	**Maximum Recorded Distance (m)**	**Time to Execute (s)**	**C**	**I**	**A**
1	PIA1	5.0	85.50	1	0	0
2	PIA2	5.0	123.23	0	0	1
3	PIA3	5.0	366.35	1	1	1
4	PIA4	5.0	529.66	1	1	1

target system and the PIA setup were separated by a clear line of sight, with no barriers or obstructions in the way. The execution times of PIAs 1, 2, 3, and 4 were, according to Table 4.2, 85.5 seconds, 123.23 seconds, 366.35 seconds, and 529.66 seconds, respectively. There are several reasons for the time discrepancy, but the complexity of the different security risks, or PIAm, is the main one. See Figure 4.9.

In PIA1, the WBAN target system's confidentiality was compromised. PIA2 had a negative impact on the WBAN target system's availability. When the WBAN target system's confidentiality, integrity, and availability were compromised, PIA3 and PIA4 each had different consequences.

4.8 READINESS IN FORENSICS FOR WBAN

A system is said to be forensically ready if it can investigate post-incident occurrences and is forensically prepared (see Figure 4.10). The hospital's internal network is strengthened with the addition of a forensic server and drone to the wireless network. The drone employed in wireless networks can record conversations taking place in other wireless clients' networks and communicate such recordings to the forensic server while remaining hidden from other wireless clients.

The network's internal configurations indicated by this study are another set that will help protect the link between the hospital's internal network and the WBAN system. A successful safety risk can be deployed up to 5 meters away from the WBAN target system, according to the trial results, according to the PIA. Present recommendations state that setting up a WBAN system should take place no sooner than 6 meters after the installation of any fresh wireless equipment. To ensure the security of the WBAN system, a protected area covering 113.14286 m^2 will be set up at a minimum distance of 6 meters [21].

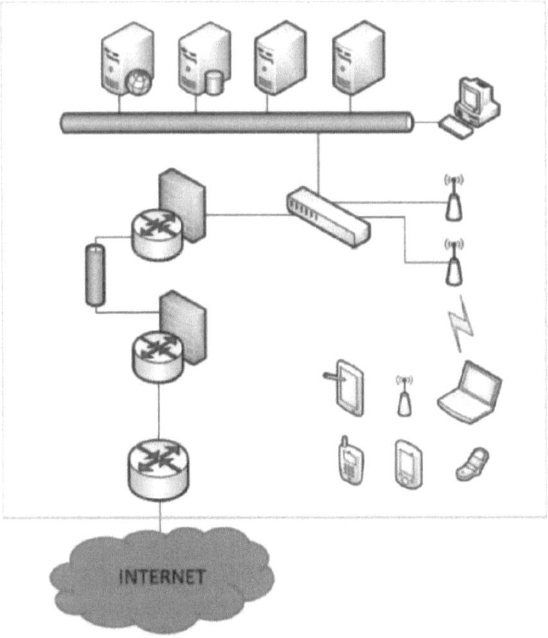

FIGURE 4.10 Healthcare internal network forensics preparedness.

4.9 SECURITY ISSUES

Security is one of the most important parts of any system. Security is described in many different ways because various individuals have different viewpoints on it. Security and system safety are two ideas that are generally associated. When security measures are put in place and kept up to date, they guarantee that there is no protection against fraudulent activity or impacts on the website. This is the condition of security.

Wireless communication is mostly used in sensor network applications in healthcare. As a result, these systems may be exposed to various security concerns. A person using wireless sensing devices may experience severe social isolation as a result of these threats and assaults. In some instances, such as when tracking the whereabouts of a patient or other individual, a security compromise could have detrimental effects. Individuals with ill purpose may exploit private information against them [22].

Security issues have always been a main topic of continuing research in sensor network applications for the medical industry. Recent research has turned an enormous amount of emphasis on security concerns in wireless sensor networks. Many of them are. Others have talked about security concerns related to healthcare applications. In the part that follows, we try to draw attention to some of the threats and attacks and talk about them along with some recommended defenses.

4.10 CONCLUSION

WBAN collects vital signs and transmits them to any mobile device connected to a back-end server or database, which can store patient information and offer critical diagnostic guidance. The security and privacy risks associated with the various IoT-based smart hospital communication strategies are highlighted. We carried out a quality inspection to make sure the techniques were appropriate for the subject of the study. Given the complexity of WBAN, a small number of research directions are thought to be viable. High security predictions have been proven by a number of study methodologies, some of which are very intricate and challenging. Still, some research methodologies produced less effective security fixes. The security keys for WBAN in this work have been created by reviewing several prior studies. Despite the fact that many various information security strategies are covered in this literature, only a select few are thought to be particularly successful for information security. Several methods that are now in use, as explored in the literature, are used to update the security for patient health data. In order to test the WBAN vulnerability in a real-world environment, this work was purposefully done without testing for interference. Human capabilities strongly influence the execution time metric. It is common knowledge that the initial level of competent performance will drastically diminish in the absence of sufficient human motor skill exercise. Although the PIA was performed multiple times to generate accurate and reliable data, the staff members who do it have a significant influence on the final product's quality. To completely understand the PIA, more research on these topics is strongly advised. The objective of this project is to construct a secure framework for WBAN deployment in hospital internal networks using a practical methodology.

REFERENCES

[1] Cusack, B., & Kyaw, A. K. (2012). Forensics readiness for wireless medical devices. *Published in the Proceedings of the 10th Australian Digital Forensics Conference.* http://ro.ecu.edu.au/adf/108

[2] Ngobeni, S., Venter, H., & Burke, I. (2010). Forensics readiness for wireless networks. *IFIP Advances in Information and Communication Technology*, *337*, 107–117. ISBN 978-3-642-15506-2

[3] Council of Europe. *Explanatory Report of Convention on Cybercrime.* http://conventions.coe.int/Treaty/en/Reports/

[4] Ngobeni, S., Venter, H., & Burke, I. (2010). Forensics readiness for wireless networks. *IFIP Advances in Information and Communication Technology*, *337*, 107–117. ISBN 978-3-642-15506-2

[5] Tufail, F., & Islam, H. (2009). Wearable wireless body area networks. *Information Management and Engineering, ICIME '09* (pp. 656–660). https://ieeexplore.ieee.org/abstract/document/5077116

[6] Keong, H. C., & Yuce, M. R. (2009). Analysis of a multi-access scheme and asynchronous transmit-only UWB for wireless body area networks. *Proc. EMBC'09* (pp. 6906–6909). IEEE.

[7] Ananthi, J. V., & Jose, P. S. H. (2021). A perspective review of security challenges in body area networks for healthcare applications. *International Journal of Wireless Information Networks*, 1–16.

[8] Jovanov, E., Milenkovic, A., Otto, C., & De Groen, P. C. (2005). A wireless body area network of intelligent motion sensors for computer-assisted physical rehabilitation. *Journal of NeuroEngineering and Rehabilitation*, *2*(1), 1–10.

[9] Halperin, D., Heydt-Benjamin, T. S., Fu, K., Kohno, T., & Maisel, W. H. (2008). Security and privacy for implantable medical devices. *IEEE Pervasive Computing*, *7*(1), 30–39.

[10] Chen, M., Gonzalez, S., Vasilakos, A., Cao, H., & Leung, V. C. (2011). Body area networks: A survey. *Mobile Networks and Applications*, *16*, 171–193.

[11] Al Ameen, M., Liu, J., & Kwak, K. (2012). Security and privacy issues in wireless sensor networks for healthcare applications. *Journal of Medical Systems*, *36*, 93–101.

[12] Meingast, M., Roosta, T., & Sastry, S. (2006, August). Security and privacy issues with health care information technology. In *2006 International Conference of the IEEE Engineering in Medicine and Biology Society* (pp. 5453–5458). IEEE.

[13] Ahmad, N., Shahzad, B., Arif, M., Izdrui, D., Ungurean, I., & Geman, O. (2022). An energy-efficient framework for WBAN in health care domain. *Journal of Sensors, 2022.*

[14] Tavera, C. A., Ortiz, J. H., Khalaf, O. I., Saavedra, D. F., & Aldhyani, T. H. (2021). Wearable wireless body area networks for medical applications. *Computational and Mathematical Methods in Medicine, 2021.*

[15] Rahman, A. F. A., Ahmad, R. (2013). Hybrid Method to Measure Vulnerability in Wireless Body Area Network. In: *Proceedings of the 6th International Conference on Sensor Asiasense*, August 23–25; Malacca, Malaysia.

[16] Pandi, G. S., Shah, S., & Wandra, K. H. (2020). Exploration of vulnerabilities, threats and forensic issues and its impact on the distributed environment of cloud and its mitigation. *Procedia Computer Science*, *167*, 163–173.

[17] Wu, H., Ding, Y., Winer, C., & Yao, L. (2010, November). Network security for virtual machine in cloud computing. In *5th International Conference on Computer Sciences and Convergence Information Technology* (pp. 18–21). IEEE.

[18] Cameron, L. (2018). Future of digital forensics faces six security challenges in fighting borderless cybercrime and dark web tools. https://publications.computer.org/security-and-privacy/2018/03/01/digital-forensics-security-challenges-cybercrime/

[19] Hajar, M. S., Al-Kadri, M. O., & Kalutarage, H. K. (2021). A survey on wireless body area networks: Architecture, security challenges and research opportunities. *Computers & Security, 104*, 102211.

[20] Zhong, L., He, S., Lin, J., Wu, J., Li, X., Pang, Y., & Li, Z. (2022). Technological requirements and challenges in wireless body area networks for health monitoring: A comprehensive survey. *Sensors, 22*(9), 3539.

[21] Siva Bharathi, K. R., & Venkateswari, R. (2019). Security challenges and solutions for wireless body area networks. In *Computing, Communication and Signal Processing: Proceedings of ICCASP 2018* (pp. 275–283). Springer.

[22] Vignesh, M. R., & Sivakumar, S. (2021). Healthcare sensors issues, challenges & security threats in wireless body area network: A comprehensive survey. *International Journal of Trend in Scientific Research and Development, 5*, 989–997.

Part 2

Security and Privacy of Wearable Devices

5 Challenges and Current Trends in Wearable Devices for the Assessment of Vital Parameters

A Review

R.M. Bommi, G. Uganya, P. Muthu Krishnammal, and A. Mary Joy Kinol

5.1 INTRODUCTION

The pervasive integration of millions of novel machine learning techniques propels technology into a transformative growth phase. Wearable devices, rooted in machine learning advancements, have become omnipresent in contemporary society. This surge is driven by an increased awareness of health issues, coupled with substantial developments in health science. Wearables, exemplified by armbands and smartwatches, leverage machine learning to gather personalized health data, offering a tailored user experience. The proliferation of wearable technology is fueled by consumers' growing inclination to monitor their well-being actively. Originally conceived as medical tools, wearables have transcended their initial purpose and found applications in diverse fields. Their ability to collect rich data has led to tailored solutions across healthcare, wellness, impairment, mobility, workplace, banking, entertainment, and more.

The transformative potential of wearable technology lies in its capacity to provide proactive strategies for addressing a spectrum of issues. Figure 5.1 illustrates the comprehensive structure of this chapter, highlighting the multifaceted impact of wearable technology across various domains.

Wearable technology, or "wearables," encompasses smart gadgets worn as shoes, integrated into clothing, implanted in the body, or tattooed on the skin. These hands-free devices are powered by microcontrollers, capable of transmitting and receiving data over the internet. Among the innovations in this field, the triboelectric nanogenerator (TENG) stands out as a power-generating system and specialized sensing technology. TENG, developed in 2012, utilizes mechanical motion for energy generation and

DOI: 10.1201/9781032635101-7

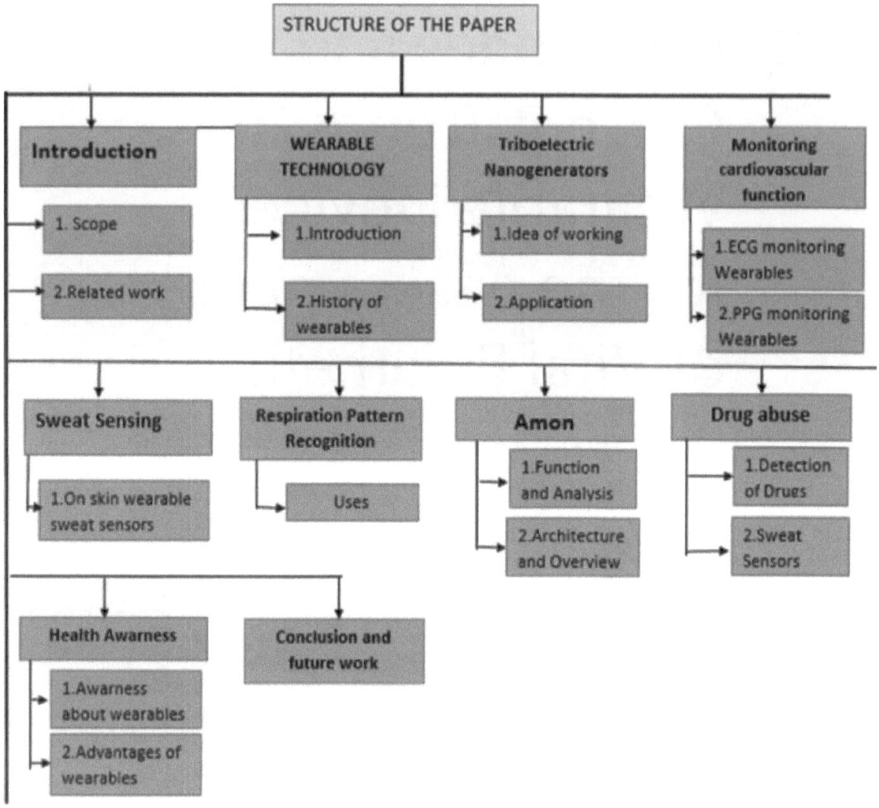

FIGURE 5.1 Structure of the chapter.

has gained attention for its self-sufficiency, high output, lightweight design applicability, cost-effectiveness, stability, and durability. Researchers are drawn to TENG's potential in energy harvesting and its applications in various wearable technologies [21].

The AMON [13] project, funded by the EU's FP5 IST program, introduces a noninvasive wearable device for high-risk cardiac patients that monitors real-time blood–oxygen saturation, blood pressure, and cardiac electrical activity [1]. Nanomaterial processing enables a wearable sensor to detect illicit narcotics in sweat, offering a noninvasive alternative for drug detection [15]. In addition, a portable wireless mask system for respiratory monitoring integrates various parameters, providing a comprehensive understanding of the respiratory system in natural situations.

The wearable mask device employs Principal Component Analysis (PCA), Micro-Electro-Mechanical Systems (MEMS), and a gyroscope-assisted baseline monitoring algorithm to accurately compute respiratory parameters [27]. This noninvasive device, resistant to motion artifacts and cross-contamination, opens new avenues for illness diagnosis and management [27]. Chronic diseases, responsible for 60% of fatalities, necessitate inconspicuous monitoring devices for timely intervention and prevention of severe deterioration [8].

In virtual reality (VR), integrating modular electronics for health monitoring offers opportunities for remote and wearable treatment. Flexible electronics combined with VR enable real-time physiological data monitoring and remote interaction between patients and physicians. Despite global advancements in wearable technology, limited awareness and high costs remain challenges. Affordable production is crucial to ensure accessibility for individuals, addressing the urgent need for wearable gadgets for health monitoring.

5.2 WEARABLE TECHNOLOGY

Compared with smartphones and tablets, wearables, such as those from Jawbone, Garmin, Fitbit, and Polar, excel in a broader range of tracking capabilities, including biofeedback and acute physiological functions like biometrics [19,10]. Originally intended for medical use, wearables now play a crucial role in disease detection, monitoring, and treatment, offering continuous health surveillance and the ability to predict diseases for preventive measures [24,40]. These high-definition smart gadgets enable proactive health management, measuring parameters like heart rate and activity levels to improve health habits [30].

The Cicret Bracelet, illustrated in Figure 5.2, is a sleek wristband featuring a built-in projector light. With the pre-installed Android operating system, this smart bracelet is a viable substitute for smartphones or tablets. It projects the screen directly onto the user's skin, enabling viewing and interaction through touch gestures on the skin's surface. Functions include making phone calls, sending SMS, watching movies, and video calling without needing an additional device [38]. The impressive projection size provides a substantial screen for various activities with the Cicret Bracelet.

5.2.1 The Future of Wearables

In addition, warning indications of an impending potential hazard are given, in addition to warning sounds or vibrations. The most significant challenge associated with these tools is convincing and educating personnel about their necessity. See Table 5.1.

FIGURE 5.2 Cicret Bracelet: A wristband that turns your arm into a projection.

TABLE 5.1
History of Wearables

Centuries	Wearables	How Was It Used?
Before 20th century	Spectacles	R. Bacon's spectacles were the first smart glasses since they were meant to be worn comfortably while increasing eyesight.
Early 20th century	Walkie-talkie	For usage in the field, the first mobile wireless systems were redesigned. They were enormous and heavy, and military horses first transported them. Donald Hings' "packet" gadget, later known as a "walkie-talkie," was a breakthrough in portable radio in 1937.
Mid-20th century	Stereo cassette player	It was the first premium wearable with an attractive design and a leather casing.
Late 20th century	MBracelete	It was a computer that could be worn on the wrist and used with an Automatic Teller Machine (ATM) to conduct financial transactions.
Early 21st century	Solar power jacket to charge the phone	The jacket was outfitted with sewn-in solar batteries connected to a battery concealed in one of the patch pockets on the front of the coat. For example, a cell phone or tablet might be linked to the battery.

Research in smart healthcare has proven effective in emergencies like heart attacks and asthma, utilizing monitoring for blood pressure and breathing. Challenges in rural healthcare require further investigation. Cardiovascular reviews, with a more extensive body of evidence, highlight complexities hindering disease treatment. User-friendly apps, particularly for responsive workers, are crucial for safety and operational efficiency [31]. A significant discovery suggests integrating wearables on construction sites with an analytical system collaborating with multiple data repositories. This study explores solutions to challenges in existing literature.

5.3 TRIBOELECTRIC NANOGENERATORS

Nanogenerators (NGs) are electronic devices that convert mechanical and thermal energy from small physical oscillations into electric power. Utilizing Maxwell's displacement current as the driving force, NGs play a crucial role in applications like wearable gadgets, self-powered sensors, contemporary medicine, artificial intelligence (AI), and robotics. There are three main types of NGs: piezoelectric, triboelectric, and pyroelectric, each harnessing different approaches to generate electricity from minute deformations caused by mechanical or thermal factors [17]. This chapter focuses on the application of NGs using the triboelectric method in the realm of human healthcare.

5.3.1 Basic Idea of Working

When materials with different electro-negativities come into contact, they generate opposite charges. Mechanical pressure, separating these charges, creates a potential that induces electron flow across an external circuit, converting mechanical force into electrical power. Triboelectric sensors (TSs), or TENGs, are adaptable for various applications and can be constructed differently. Figure 5.3 illustrates the four fundamental TENG operating modes: freestanding triboelectric-layer mode, linear-sliding mode, contact-separation mode, and single-electrode mode. Each mode has distinct characteristics and specific use cases, chosen based on requirements in practical applications [32].

5.3.2 Applications of Triboelectric Nanogenerators

The concern of power requirements has become particularly significant as wearable devices have gained prominence. Due to their simple and ubiquitous design, TENGs have become an indispensable tool for converting low-frequency mechanical power into electrical power. Wearable TENGs are projected to collect energy from various activities and power real-time wearable electronics. We may also access the richness of data buried beneath the output signals utilizing this approach of energy translation [28].

5.3.3 Role of TENG in Pacemakers

Using TENG self-powered pacemakers offers an alternative to traditional battery-dependent models. In 2014, Zheng et al. introduced the world's first self-powered pacemaker (iTENG), employing micro-nanostructured aluminum and pyramid-patterned polydimethylsiloxane (PDMS). Subsequent research by Ouyang et al. in 2019 presented an implanted symbiotic pacemaker (SPM) using iTENG, featuring a core-shell architecture and in vivo testing in a Yorkshire pig. Challenges regarding stability, power, and size must be addressed for potential clinical applications [9].

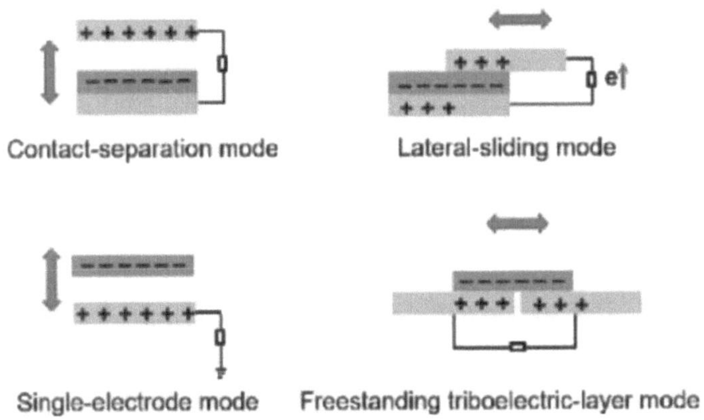

FIGURE 5.3 Operating modes of TENG.

5.3.4 TENG's Function in Neural Prosthesis

A crucial tool in neurological rehabilitation, a brain prosthesis aids in recovering neurological injuries by substituting for impaired sensory-motor or cognitive functions. In 2013, Zhang et al. developed a sandwich-shaped TENG using a freestanding aluminum film between two micro/nanoscale layers of PDMS. This TENG powered a 3D microelectrode array brain prosthesis without rectification or energy storage. Platinum electrodes were fabricated using a dual-layer lift-off method on a flexible and biocompatible perylene C substrate. While promising, there is room for improvement in compatibility, integration, and adaptability [33]. Figure 5.4 illustrates some important applications of TENGs.

5.3.5 Cardiovascular Care

Ambulatory monitoring devices allow a new paradigm in healthcare by gathering and analyzing long-term data for reliable diagnosis. These gadgets are becoming more prevalent for continuous monitoring of cardiac conditions. Recent technical advancements have enabled the creation of cost-effective and dependable systems, allowing vulnerable persons to be protected. They identify critical physiological processes early, leading to timely warnings to seek medical assistance [22].

Due to their user-friendly nature and cost-effectiveness, wearable devices are well-suited for long-term monitoring of physiological indicators in patients with chronic illnesses. Research has led to diverse technologies capable of monitoring crucial physiological data [4]. In residential care for chronic conditions, wearables can detect symptoms and signs of deterioration, including urgent monitoring of

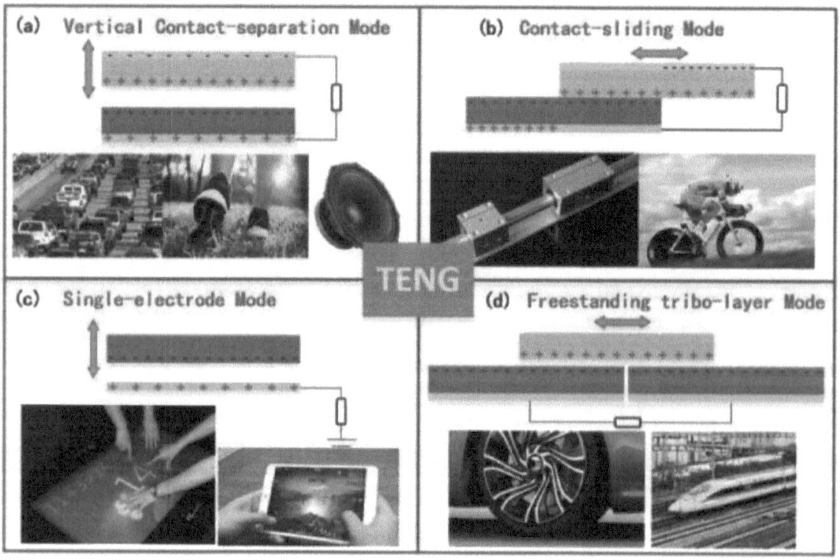

FIGURE 5.4 Applications of TENG.

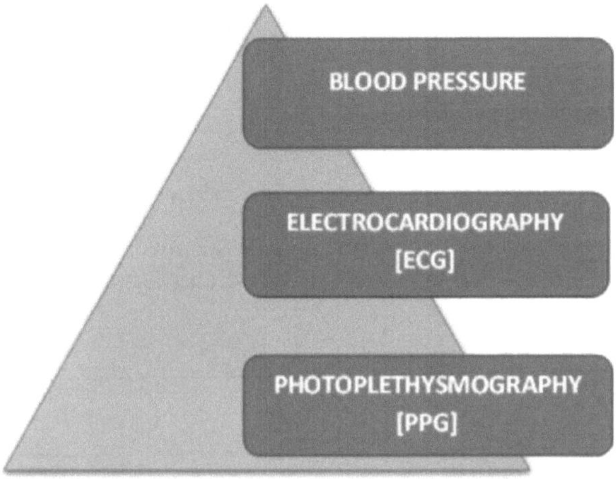

FIGURE 5.5 ECG monitoring variables.

cardiovascular and respiratory functions (Figure 5.5). Digital electrocardiogram (ECG) analysis is beneficial for patients with arrhythmia, heart failure, and other cardiac disorders.

5.3.6 Wearables for Monitoring Cardiovascular Functions

Cardiovascular diseases, marked by high mortality rates and potential chronicity, pose a threat to both financial and psychological well-being. Wearable technologies and algorithms directly monitor crucial clinical parameters like heart rate and heart rate variability. In addition, machine learning and indirect monitoring capabilities enable the computation of blood pressure, blood oxygen saturation (SpO_2), and blood glucose levels. These indicators play a pivotal role in clinical decision-making, providing early warnings and detecting cardiac issues for improved diagnosis and therapy [34].

5.3.7 ECG Monitoring Wearables

ECG is the gold standard for cardiac monitoring, measuring ventricular electrical activity projected onto the body's surface through the potential difference between two locations. Innovations in user-friendly ECG monitoring include integrating electrodes into riding helmets and earphone-shaped devices, ensuring reliable connection quality during head movement. Wristbands, commonly integrated into wearable ECG devices like watches, enable routine monitoring with just two fingertips. However, challenges in long-term continuous monitoring exist due to limitations in traditional bracelet systems, requiring users to touch the band for a cardiac cycle [29]. A more optimal approach involves single-sided wristbands with specially designed electrodes, enhancing precision. Various prototypes for wearable ECG monitoring exist, using materials like dry metal or textile electrodes instead of traditional hydrogel electrodes (see Figure 5.6). Textile electrodes, valued for their skin-friendly properties, are explored as alternatives. The application of multi-lead ECGs has received attention, with recent advancements in fabric-printed circuits and analogue buses addressing challenges associated with traditional 12-lead electrode systems.

5.3.8 Photoplethysmography Monitoring Wearables

Photoplethysmography (PPG) is a clinical technique involving the injection of photons into human body cells, with analysis based on transmitted or absorbed light.

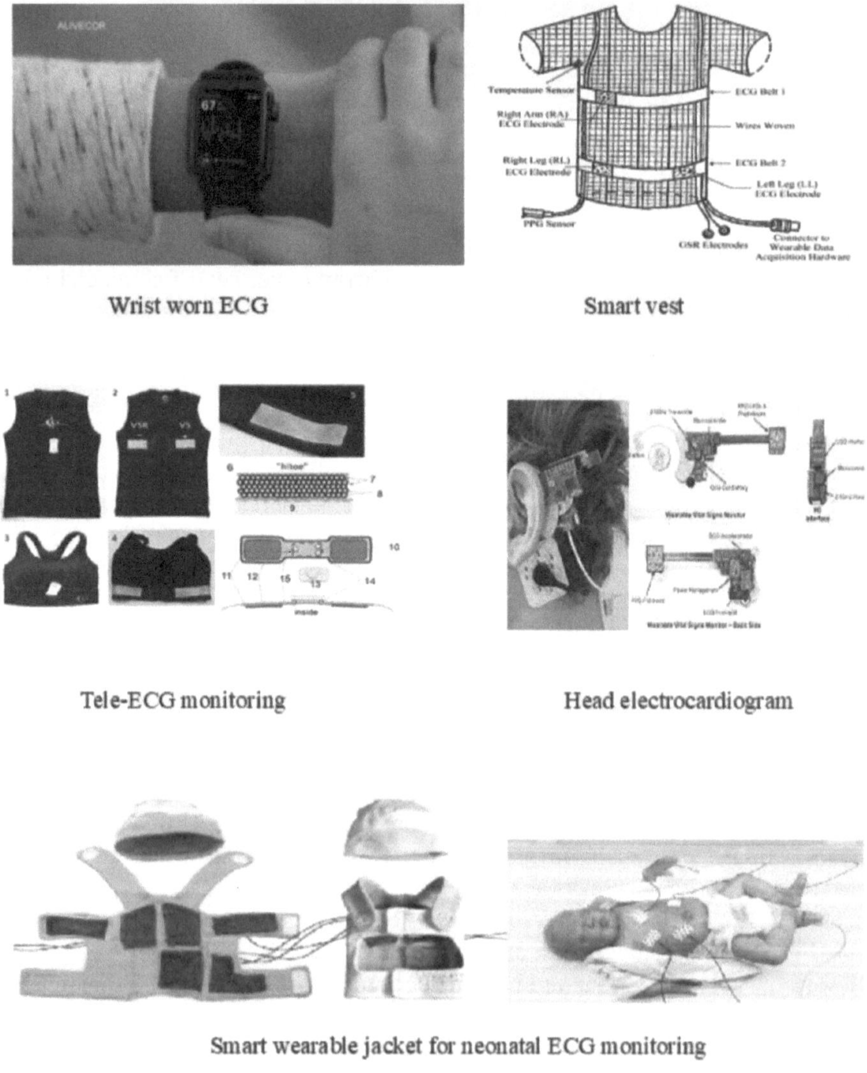

FIGURE 5.6 Prototypes for ECG monitoring.

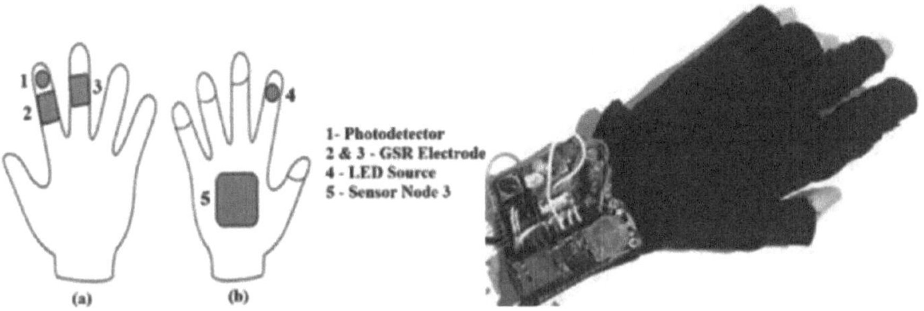

FIGURE 5.7 Gloves-type PPG monitor.

Reflective-type PPG, commonly used in wearables, allows monitoring from any body part with sufficient subdermal vascular perfusion, as in Figure 5.7. Wrist-worn PPG, integrated into fitness bracelets, has gained attention, but motion artifacts hinder its reliability during intense exercise [11]. Recent research has successfully addressed this issue, enabling accurate heart rate extraction and waveform information restoration from flawed PPG signals, suitable for medical use. Beyond ECG and PPG, biomechanical data of the cardiac cycle is deciphered using approaches such as ballistocardiography (BCG) and phonocardiography (PCG).

BCG, traditionally obtained from static sources, is now facilitated in wearables using delicate accelerometers linked to the torso. PCG has also been demonstrated in wearables with effective heart sound segmentation through signal processing. However, both BCG and PCG are susceptible to environmental artifacts, limiting their everyday use. Despite their potential for chronic cardiovascular disease monitoring, extracting higher-level physical traits from raw data is crucial before practical medical applications can be realized.

5.4 SWEAT SENSING

Wearable biochemical sensors, such as sweat sensors, offer noninvasive and continuous monitoring of ions and molecules in biological fluids like sweat. These sensors, utilizing flexible polymeric substrates and various fabrication techniques like screen printing and optical lithography, provide insights into physiological dynamics. Sodium, a key element in sweat, plays a crucial role in osmotic pressure and pH correction [12]. Wearable sweat sensors are particularly beneficial for athletes engaged in high-intensity sports, allowing for health monitoring without hindering daily activities [35].

5.4.1 Skin-Wearable Sweat Sensors

On-skin sweat sensors, including sweatbands, patches, and tattoos, enable the collection and analysis of sweat samples in close proximity, reducing dissipation. These sensors detect biomarker variations quickly before sample degradation due to their proximity to the collection site. While sweatbands offer a convenient platform for wearable components, patches and tattoos provide flexibility for monitoring sweat variables on any body part. However, they require higher tensile stability, resilience, and skin adhesion, especially during high-intensity physical activity.

5.4.2 Wearable Sweat Sensors

Near-skin sweat sensors use fluidic channels to transport sweat to electrodes, preventing pooling and infection. Textiles and garments, exemplified by wearable perspiration sensors, integrate sensors into a textile matrix through various designs. Incorporation methods include turning yarns into conducting substrates with detecting membranes and incorporating electrodes into retail garments. Low-cost, highly programmable materials like polymers are effective for prolonged sweat physiology monitoring by binding to fabrics or skin. However, the durability of wearable sensors embedded in fabrics over multiple washing and drying cycles remains uncertain.

5.5 RESPIRATION PATTERN RECOGNITION

The wearable mask device offers consistent and precise respiratory monitoring using nanoscale Venturi tubes and MEMS sensor-based fluid subsystems. It employs gyro sensors to adjust for pressure sensor signal variance and eliminate motion-induced artifacts, ensuring accurate monitoring. The device utilizes a gyroscope-assisted baseline detection approach to address pressure sensor signal drift and motion-induced errors. Data processing, based on PCA, reliably detects individual users' specific breathing rates. Constructed with computer numerical control machining and Solidworks design, the mask incorporates a microprocessor, variable pressure MEMS sensor, thermistor, barometer, memory, Bluetooth module, and switch. The device focuses on wearability, precision, and self-containment to achieve its primary goals in respiratory monitoring.

5.6 AMON: A WEARABLE MEDICAL MONITORING AND ALERT SYSTEM WITH MULTIPLE PARAMETERS

5.6.1 AMON Functionality

The AMON system offers continuous monitoring and logging of various parameters, including pulse, blood oxygen saturation, temperature, and activity level. It analyzes the user's health online and takes additional measurements like ECG and blood pressure if deviations from the norm are detected. The system minimizes false alarms through intricate analysis, and users can manually trigger alerts. In case of anomalies, the device notifies a nearby health center via integrated cell phone link, providing collected data for accurate diagnosis. Medical professionals can communicate with patients through text messages, and the system can be regulated and instructed by healthcare professionals

5.6.2 AMON Architecture

As shown in Figure 5.8, the essential elements of the system consist of a user interface architecture, a network component, a digital data processing unit, a power supply and control subsystem, and a sensor/analog unit.

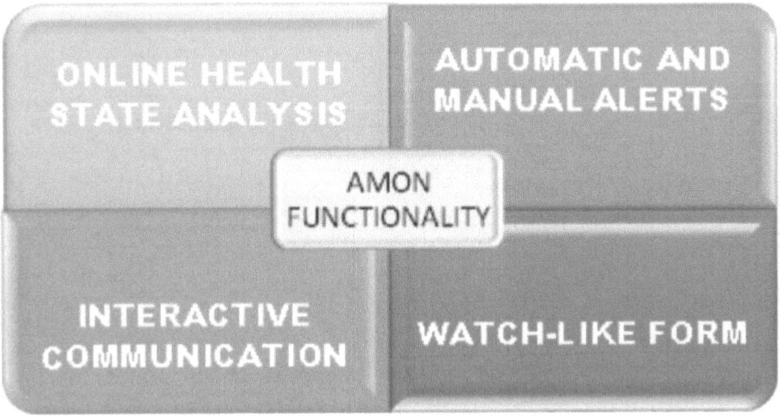

FIGURE 5.8 AMON functionality.

5.6.2.1 Analog Subsystem/Sensors

The AMON detector developed by SPO Inc. for the research project is a crucial component, monitoring oxygen saturation levels and heart rate. It calculates the ratio of infrared to red light absorption in blood. The system includes a thermopile for skin temperature monitoring and a commercial blood pressure meter with a wrist strap. The working prototype features silver electrodes, one placed under the electrocardiograph housing and the other in contact with the patient's hand. Custom-designed amplification and filter circuits, along with an 11-bit analog-to-digital converter (ADC), process the analogue signals.

5.6.2.2 User Interface Design

The user interface design prioritized simplicity for effective use, especially in emergencies. A grayscale LCD panel with a resolution of 128 × 64 and four large buttons was employed (Figure 5.9). For communication with the medical center, a Siemens TC 35 Global System for Mobile Communications (GSM) module was utilized, known for its compact size (36 × 55 × 7 mm) and minimal power consumption (35 mA at 3.3 volts) (Figure 5.10).

5.6.2.3 Digital Processing

The digital data processing subsystem not only handles the user interface but also manages health data interpretation, signal transmission to and from the medical center, and the coordination of sensors and the ADC. A specialized AMX integrated multipurpose operating system is responsible for task coordination. Given that only occasional bursts of computing power are required for odd ECG and blood pressure results, the system's core, an ARM7 microprocessor with 136 KB of on-chip RAM, operates efficiently at low power (1.8 volts) and speeds ranging from 1 megahertz to 40 megahertz (3.3 volts).

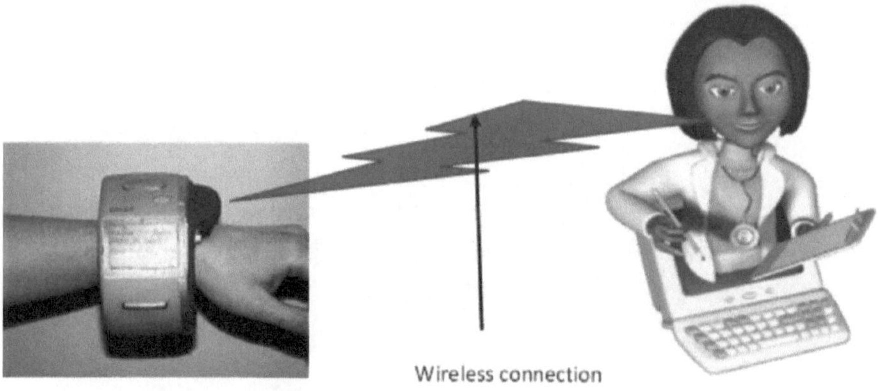

FIGURE 5.9 Remote monitoring.

Patient's requirement of medical monitoring

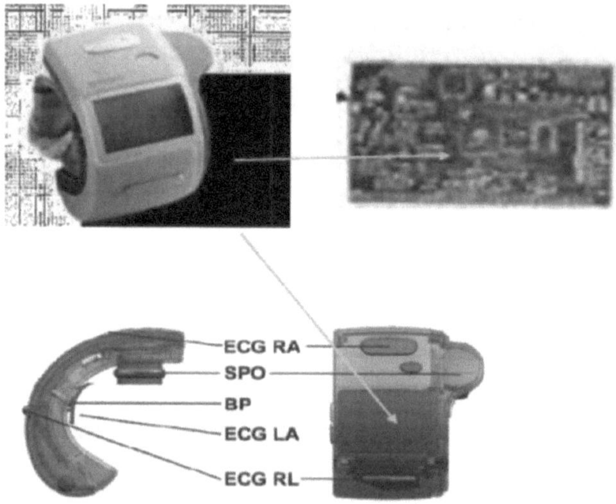

FIGURE 5.10 AMON process, outlook, parts, and circuit.

5.6.2.4 Power Management

Power efficiency was a critical consideration in the AMON design. Housed in a compact wrist-mounted enclosure, the system is designed to operate for a minimum of 24 hours without recharging. It utilizes a Toshiba LGR 18500P 1000mAh battery and three DC–DC converters for voltage supply (5V, 3V, and variable 1.8–3V). Extreme power control, featuring low switching frequencies and effective CPU voltage and frequency management, is implemented for optimal energy conservation.

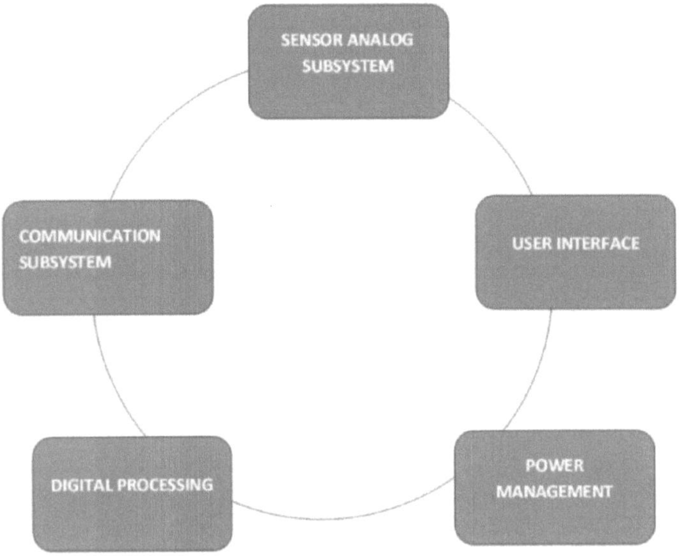

FIGURE 5.11 The main system components.

5.6.2.5 System Overview

The AMON system integrates various health monitors into a compact wristband, representing significant advances in wrist-wearable technology (see Figure 5.11). It allows continuous measurement of pulse, SpO_2, skin temperature, and activity levels. With an on-board GSM transceiver, the system facilitates mobile communication. Medical professionals at the telemedicine center utilize the wrist-worn device for data collection and assessment. The device, with its unconventional sensor placement and integration, has reduced in size for wrist compatibility. It operates through a JAVA server platform and Windows workstation, enabling secure access to patient information. The system performs essential digital signal processing, calculates medical values, and conducts initial evaluations, utilizing data from blood pressure and single lead electrocardiogram measurements, with additional insights from ECG data [36].

5.6.2.6 AMON Analysis

During the initial assessment, the patient's pulse and oxygen saturation are compared with predefined criteria. Three possible outcomes include:

1. All measurements are fine, indicating normal conditions.
2. If a parameter is outside its normal range, additional measurements are conducted, and if the issue persists, a warning is issued to the user. Examples include SpO_2, blood pressure, and ECG.
3. If multiple parameters indicate abnormalities, an automatic alarm is sent to the telemedicine center for analysis. A manual verification by a doctor follows, leading to further actions based on the severity. Noncritical data is

transmitted to the center three times daily, ensuring continuous monitoring. In case of communication issues, the device stores up to four days of data and attempts re-establishing connectivity. The telemedicine center is equipped to handle multiple remote devices, and users are advised to stay in areas with good GSM coverage for optimal system performance [37].

5.7 DRUG ABUSE

Detecting illegal drug use is critical in addiction medicine, with relapse being a crucial predictor of outcomes. Current methods involve biological specimen testing and self-reporting. Traditional techniques, like urine tests and chromatography–mass spectrometry, have limitations in sensitivity. A flexible, body-worn sweat patch, as depicted in Figure 5.12, offers rapid and highly sensitive drug detection by tracing the optical signal of drugs after exposure to light. Wearable biosensors play a vital role in disease diagnosis and treatment. Advances in big data and machine learning enable wearables to predict early sickness detection. These sensors, available in various forms and sizes, can noninvasively detect drugs of abuse. A pilot observational study involving emergency room patients demonstrated the feasibility of sweat-based drug detection devices. Challenges include real-time measurement accuracy, time-consuming sample collection, and environmental factors affecting analysis results. Despite these challenges, sweat analysis finds applications in drug detection, diabetes monitoring, sports performance analysis, and illness identification, contributing to innovative drug abuse treatment strategies.

In an observational study, wristbands were worn by 15 individuals who had consumed cocaine recreationally, allowing researchers to monitor physiological "signatures" associated with drug use for 30 days post-hospital discharge. The rising trend in drug-related deaths, responsible for more preventable illnesses and fatalities, emphasizes the severity of drug addiction. Wearable biosensors show promise in complementing drug abuse treatment trials, providing more accurate analysis of relapse prevention programs. Ongoing research explores larger datasets, natural usage scenarios, and diverse drug exposure histories. Wearable biosensors, increasingly

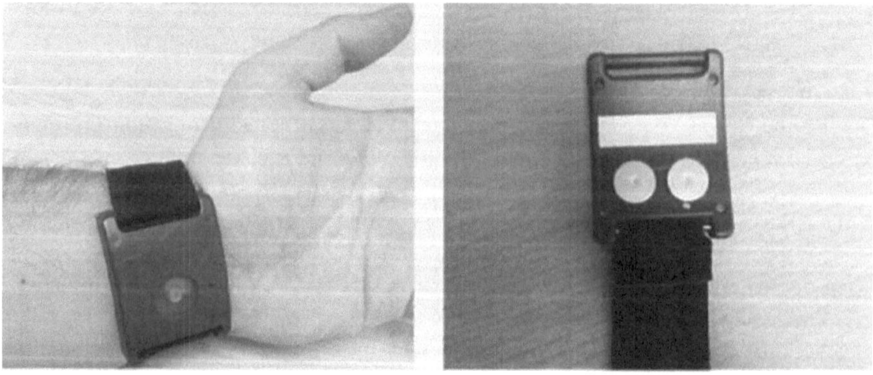

FIGURE 5.12 Front and back view of the affective Q sensor.

popular, have the potential to revolutionize continuous biomarker assessment for drug use and other disorders, attracting interest from major companies like Google, Microsoft, and Novartis.

5.8 HEALTH AWARENESS

The systems discussed in this chapter require additional clinical testing before being authorized for medical use. Major companies, including Google, Microsoft, and Novartis, along with smaller firms, are showing interest in biosensors, highlighting substantial market potential for innovative ideas in personal mobile health. The pandemic has exposed vulnerabilities in the traditional medical structure, emphasizing the role of smart and connected wearables, which can collect unprecedented data through Internet of Things (IoT) advancements. A life-course approach to health, considering social, environmental, and biological factors, is crucial for disease prevention. Physical fitness, encompassing cardiorespiratory fitness, strength, endurance, body composition, and motor abilities, is vital for overall health [14]. Recognizing early warning signs of mental illness is essential, and psycho-education, community support, and awareness campaigns play key roles. Overcoming cultural and linguistic barriers is crucial for promoting mental health awareness, especially in remote areas. Individuals with comorbidities face double stigma, and support is necessary for achieving mental health awareness at a sufficient level.

5.9 CHRONIC DISEASE MANAGEMENT

5.9.1 Emergent Themes in Health Information Systems for Chronic Disease Management

Contagious illnesses such as sepsis, malaria, dengue fever, and emerging viruses elicit fever and systemic involvement through an innate inflammatory process [18]. The 2014 West African Ebola epidemic emphasized the importance of robust and prepared healthcare services, crucial in addressing the 2019 COVID-19 outbreak [18]. In health IT and analytics research, hospital-based information systems, online health communities, health social media analytics, and mobile and robotic assistive systems are significant areas of exploration, with a focus on novel approaches and implementations [2].

5.9.2 Wearable Respiration Sensor

Respiration, a fundamental physiological process, serves as a continuous biological information source for monitoring individual health. Respiratory diagnostics, particularly the detection of ammonia (NH_3) concentration, holds promise in health monitoring. Elevated NH_3 levels, associated with disorders like renal disease, can be indicative of health conditions. Traditional detectors for biomedical signals faced limitations in design, portability, and power requirements. To address these issues, a triboelectric self-powered respiration sensor (TSRS) was developed. This sensor, specifically a Ce-doping variant at 0.004M, demonstrated superior selectivity and sensitivity, particularly at trace NH_3 levels (0.1–1 ppm). The TSRS, activated by chest expansion during breathing, measures not only respiratory patterns but also NH_3 biomarkers in exhaled gases. The study introduces a conceptual framework based on permittivity changes during petrol molecule adsorption, validated through finite element modeling. This research not only proposes a wearable breathing rate surveillance device for personal health but also explores theoretical modeling of triboelectric self-powered gas sensing in terms of permittivity variation [39].

5.9.3 Emotion and Stress Detection

In recent years, stress has become a buzzword that affects a large number of individuals all over the world, regardless of gender, age group, or career. Stress, like emotions, is ephemeral, and physiology plays a significant influence in both sensations and worry. Many of the negative emotions are just symptoms of a high level of stress. Various automated/semi-automated clinical assessment methods related to human physiology have become increasingly vital and widespread in recent years. Physiological characteristics have a variety of characteristics that contribute to system precision, dependability, and resilience [20].

5.9.4 Emotional Smart Wristband

The Emotional Smart Bracelet is designed to identify the emotional responses of a group and relay the social emotion values to the iGenda. In settings like retirement homes, maintaining a peaceful atmosphere is crucial despite an expanding resident population. The iGenda, functioning as a visual interface and information center, utilizes data from the Emotional Smart Bracelet to induce specific moods, fostering psychic abilities to facilitate open communication and address problems subtly. Monitoring physical changes, such as pupil dilation and muscle tension, provides accurate insights into an individual's mental state. The Emotional Smart Wristband (ESW) is a wearable transmitting data to the iGenda, allowing proactive planning based on the user's emotional state, vital for intelligent virtual situations [5,7,8].

5.10 CURRENT CHALLENGES

The widespread adoption of smart wearable medical monitoring devices faces several challenges [16]:

- Prolonged wearability is hindered by potential skin irritation caused by developing technologies.
- Contact resistance between electrodes and the skin can change during use, for instance, drying gels in electroencephalogram (EEG) electrodes.
- The size and weight of current data holsters, resembling Walkman dimensions, make them bulky and inconvenient.
- Machine intelligence may increase device dimensions and density, affecting user comfort.

- Connectivity and sensor routing are crucial, especially in applications requiring numerous analog input channels (e.g., ECG, EEG, sleep studies).
- Concerns about the device's accuracy and potential legal implications for medical outcomes, emphasizing the need for reliability and accountability.

5.11 CONCLUSION

The use of wearable "smart gadgets" has surged in popularity globally, with companies like Apple, Fitbit, and Samsung offering accessible health monitoring technologies, from heart rate to arrhythmia detection [23]. Dietary plans are increasingly recognized for their therapeutic benefits in promoting health and longevity [14]. As the aging population grows, there is a shift in focus from life-prolonging measures to improving the quality of life for the elderly [26]. Nurses, positioned in healthcare, can engage with patients using health tracking devices, fostering health discussions [7].

Smart health aims to simplify the lives of individuals dealing with health issues, reducing hospital visits and associated costs [25]. The rise of activity tracking devices aligns with global efforts to prioritize well-being for cost-effective healthcare [3]. The increasing data collection from wearables, environmental sensors, and mobile apps facilitates valuable insights for patient-physician collaboration in chronic illness management [6]. This evolving physician–patient interface holds potential for enhanced telemedicine and increased patient engagement in healthcare.

REFERENCES

[1] Anliker, Urs, Jamie A. Ward, Paul Lukowicz, Gerhard Tröster, François Dolveck, Michel Baer, Fatou Keita, et al. (2004), "AMON: A Wearable Multiparameter Medical Monitoring and Alert System." *IEEE Transactions on Information Technology in Biomedicine* 8 (4): 415–427. https://doi.org/10.1109/TITB.2004.837888.

[2] Bardhan, Indranil, Hsinchun Chen, and Elena Karahanna (2020), "Connecting Systems, Data, and People: A Multidisciplinary Research Roadmap for Chronic Disease Management." *MIS Quarterly: Management Information Systems* 44 (1): 185–200. https://doi.org/10.25300/MISQ/2020/14644.

[3] Bryson, David (2009), "Smart Clothes and Wearable Technology for the Health and Well- Being Market." In *Smart Clothes and Wearable Technology*. Woodhead Publishing Limited. https://doi.org/10.1533/9781845695668.4.335.

[4] Carreiro, Stephanie, David Smelson, Megan Ranney, Keith J. Horvath, R. W. Picard, Edwin D. Boudreaux, Rashelle Hayes, and Edward W. Boyer (2015), "Real-Time Mobile Detection of Drug Use with Wearable Biosensors: A Pilot Study." *Journal of Medical Toxicology* 11 (1): 73–79. https://doi.org/10.1007/s13181-014-0439-7.

[5] Costa, Angelo, Jaime A. Rincon, Carlos Carrascosa, Vicente Julian, and Paulo Novais (2019), "Emotions Detection on an Ambient Intelligent System Using Wearable Devices." *Future Generation Computer Systems* 92 (March): 479–489. https://doi.org/10.1016/j.future.2018.03.038.

[6] Dohse, Heidi (2021), "Patient Perspective: Wearable and Digital Health Tools to Support Managing Our Health During the COVID-19 Pandemic and Beyond." *Cardiovascular Digital Health Journal* 2 (1): 88–90. https://doi.org/10.1016/j.cvdhj.2020.12.002.

[7] Edward, Karen Leigh, Loretta Garvey, and Muhammad Aziz Rahman (2020), "Wearable Activity Trackers and Health Awareness: Nursing Implications." *International Journal of Nursing Sciences* 7 (2): 179–183. https://doi.org/10.1016/j.ijnss.2020.03.006.

[8] Guo, Yao, Xiangyu Liu, Shun Peng, Xinyu Jiang, Ke Xu, Chen Chen, Zeyu Wang, Chenyun Dai, and Wei Chen (2021), "A Review of Wearable and Unobtrusive Sensing Technologies for Chronic Disease Management." *Computers in Biology and Medicine* 129 (November 2020). https://doi.org/10.1016/j.compbiomed.2020.104163.

[9] Khandelwal, Gaurav, Nirmal Prashanth, Maria Joseph, and Sang-jae Kim (2020), "Nano Today Triboelectric Nanogenerator for Healthcare and Biomedical Applications." *Nano Today* 33: 100882. https://doi.org/10.1016/j.nantod.2020.100882.

[10] Krzemin, Michał (2021), "Risk Assessment on the Construction Site with the Use of Wearable Technologies." *Ain Shams Engineering Journal* 12 (4): 3411–3417. https://doi.org/10.1016/j.asej.2021.04.006.

[11] Legner, Christopher, Upender Kalwa, Vishal Patel, Austin Chesmore, and Santosh Pandey (2019), "Sweat Sensing in the Smart Wearables Era: Towards Integrative, Multifunctional and Body-Compliant Perspiration Analysis." *Sensors and Actuators, A: Physical.* Elsevier B.V. https://doi.org/10.1016/j.sna.2019.07.020.

[12] Lin, Pei-heng, Lun Chang, Chen Sheu, Pei-heng Lin, Wei-Lun Chang, Sian-Chen Sheu, and Bor-Ran Li (2020), "IScience A Noninvasive Wearable Device for Real-Time Monitoring of Secretion Sweat Pressure by Digital Display a Noninvasive Wearable Device for Real-Time Monitoring of Secretion Sweat Pressure by Digital Display." *Science* 23 (11): 101658. https://doi.org/10.1016/j.isci.2020.101658.

[13] Lukowicz, Paul, Urs Anliker, Jamie Ward, Gerhard Troster, Etienne Hirt, and Christopher Neufelt (2002), "AMON: A Wearable Medical Computer for High Risk Patients." *Proceedings—International Symposium on Wearable Computers, ISWC 2002-Janua* (pp. 133–134). https://doi.org/10.1109/ISWC.2002.1167230.

[14] Maggi, Stefania, Domenico Rogoli, and Fiona Ecarnot (2021), "Healthy Aging in the Context of the Mediterranean Diet—Health -Environment Trilemma." *Aging and Health Research*: 100015. https://doi.org/10.1016/j.ahr.2021.100015.

[15] Mahmud, Md Shaad, Hua Fang, Stephanie Carreiro, Honggang Wang, and Edward W. Boyer (2019), "Wearables Technology for Drug Abuse Detection: A Survey of Recent Advancement." *Smart Health* 13 (August). https://doi.org/10.1016/j.smhl.2018.09.002.

[16] Martin, Thomas, Emil Jovanov, and Dejan Raskovic (2000), "Issues in Wearable Computing for Medical Monitoring Applications: A Case Study of a Wearable ECG Monitoring Device." *International Symposium on Wearable Computers, Digest of Papers*: 43–49. https://doi.org/10.1109/iswc.2000.888463.

[17] Mathew, Ammu Anna, Arunkumar Chandrasekhar, and S. Vivekanandan (2021), "A Review on Real-Time Implantable and Wearable Health Monitoring Sensors Based on Triboelectric Nanogenerator Approach." *Nano Energy.* Elsevier Ltd. https://doi.org/10.1016/j.nanoen.2020.105566.

[18] Ming, Damien K., Sorawat Sangkaew, Ho Q. Chanh, Phung T. H. Nhat, Sophie Yacoub, Pantelis Georgiou, and Alison H. Holmes (2020), "Continuous Physiological Monitoring Using Wearable Technology to Inform Individual Management of Infectious Diseases, Public Health and Outbreak Responses." *International Journal of Infectious Diseases.* Elsevier B.V. https://doi.org/10.1016/j.ijid.2020.05.086.

[19] Ometov, Aleksandr, Viktoriia Shubina, Lucie Klus, Justyna Skibińska, Salwa Saafi, Pavel Pascacio, Laura Flueratoru, et al. (2021), "A Survey on Wearable Technology: History, State-of-the-Art and Current Challenges." *Computer Networks* 193 (March): 108074. https://doi.org/10.1016/j.comnet.2021.108074.

[20] Panicker, Suja Sreeith, and Prakasam Gayathri (2019), "A Survey of Machine Learning Techniques in Physiology Based Mental Stress Detection Systems." *Biocybernetics and Biomedical Engineering*. Elsevier Sp. z o.o. https://doi.org/10.1016/j.bbe.2019.01.004.
[21] Pu, Xianjie, Shanshan An, Qian Tang, Hengyu Guo, and Chenguo Hu (2021), "IScience Ll Wearable Triboelectric Sensors for Biomedical Monitoring and Human-Machine Interface." *Science* 24 (1): 102027. https://doi.org/10.1016/j.isci.2020.102027.
[22] Sana, Furrukh, Eric M. Isselbacher, Jagmeet P. Singh, E. Kevin Heist, Bhupesh Pathik, Antonis A. Armoundas (2020), "Wearable Devices for Ambulatory Cardiac Monitoring: JACC State-of-the-Art Review." *Journal of the American College of Cardiology* 75 (13). https://doi.org/10.1016/j.jacc.2020.01.046.
[23] See, Vincent Y., and Helaine J. Kwong (2020), "Wearing Your Heart on Your Wrist." *JACC: Case Reports* 2 (3): 434–436. https://doi.org/10.1016/j.jaccas.2019.12.037.
[24] Surantha, Nico, Prabadinata Atmaja, and Maulana Wicaksono (2021), "A Review of Wearable Internet-of-Things Device for Healthcare." *Procedia Computer Science* 179 (2020): 936–943. https://doi.org/10.1016/j.procs.2021.01.083.
[25] Taiwo, Olutosin, and Absalom E. Ezugwu (2020), "Informatics in Medicine Unlocked Smart Healthcare Support for Remote Patient Monitoring during Covid-19 Quarantine." *Informatics in Medicine Unlocked* 20: 100428. https://doi.org/10.1016/j.imu.2020.100428.
[26] Tao, Zhiwen, Zhiyong Zhang, Xiaoli Wang, Yongqiang Shi, and Jeffrey Soar (2019), "Comparison of Subsidy Schemes for Reducing Waiting Time: Special Focus on Smart Home Care for Elderly People." *Procedia CIRP* 83: 716–721. https://doi.org/10.1016/j.procir.2019.04.102.
[27] Tipparaju, Vishal Varun, Di Wang, Jingjing Yu, Fang Chen, Francis Tsow, Erica Forzani, Nongjian Tao, and Xiaojun Xian (2020), "Biosensors and Bioelectronics Respiration Pattern Recognition by Wearable Mask Device." *Biosensors and Bioelectronics* 169 (September): 112590. https://doi.org/10.1016/j.bios.2020.112590.
[28] Wang, Haobin, Mengdi Han, Yu Song, and Haixia Zhang (2020), "Design, Manufacturing and Applications of Wearable Triboelectric Nanogenerators." *Nano Energy*: 105627. https://doi.org/10.1016/j.nanoen.2020.105627.
[29] Wang, Si, Yadong Jiang, Huiling Tai, Bohao Liu, Zaihua Duan, Zhen Yuan, Hong Pan, Guangzhong Xie, Xiaosong Du, and Yuanjie Su (2019), "An Integrated Flexible Self-Powered Wearable Respiration Sensor." *Nano Energy* 63 (May): 103829. https://doi.org/10.1016/j.nanoen.2019.06.025.
[30] Uganya, G., Radhika, and N. Vijayaraj (2021), "A Survey on Internet of Things: Applications, Recent Issues, Attacks, and Security Mechanisms." *Journal of Circuits, Systems and Computers* 30 (5): 2130006.
[31] Charan, S., G. Uganya, and M. N. Kumar (2022), "Comparison of Accuracy and Sensitivity in Liver Cancer Segmentation of Magnetic Resonance Images Using Convolutional Neural Network in Comparison with Support Vector Machine." In *2022 14th International Conference on Mathematics, Actuarial Science, Computer Science and Statistics (MACS)* (pp. 1–9). IEEE.
[32] Jhansi, R., and G. Uganya (2023), "Improved Accuracy in Speech Recognition System for Detection of Covid-19 Using K Nearest Neighbour and Comparing with Artificial Neural Network." In *2023 International Conference on Artificial Intelligence and Knowledge Discovery in Concurrent Engineering (ICECONF)* (pp. 1–5). IEEE.
[33] Reddy, P. R. K., and G. Uganya (2022), "Comparative Analysis of Accuracy of Face Recognition System Using CNN and SVM." In *2022 2nd International Conference on Technological Advancements in Computational Sciences (ICTACS)* (pp. 1–5). IEEE.

[34] Jhansi, R., and G. Uganya (2023), "Detection of COVID-19 Patients Using Speech Recognition with Support Vector Machine and Comparing with K Nearest Neighbour Algorithm." In *2023 International Conference on Artificial Intelligence and Knowledge Discovery in Concurrent Engineering (ICECONF)* (pp. 1–5). IEEE.
[35] Ramachandran, B., and Y. C. Liao (2022), "Microfluidic Wearable Electrochemical Sweat Sensors for Health Monitoring." *Biomicrofluidics* 16 (5).
[36] Sumathi, M., N. Vijayaraj, S. P. R. Raja, and M. Rajkamal (2022), "Internet of Thing Based Confidential Healthcare Data Storage, Access Control and Monitoring Using Blockchain Technique." *Computing and Informatics* 41 (5): 1207–1239.
[37] Sumathi, M., S. P. Raja, N. Vijayaraj, and M. Rajkamal (2022), "A Decentralized Medical Network for Maintaining Patient Records Using Blockchain Technology." *Cybernetics and Information Technologies* 22 (4): 129–141.
[38] Tamilvizhi, T., R. Surendran, and R. M. Bommi (2020), "Radio Frequency Identification (RFID) Based Ubiquitous Health Care Data Handling." In *IOP Conference Series: Materials Science and Engineering* (Vol. 994, No. 1, p. 012021). IOP Publishing.
[39] Bommi, R. M., S. V. S. Rajeev, S. Navya, V. S. Teja, and U. Supriya (2023), "Smart Health Care Waste Segregation and Safe Disposal." *Mathematics and Computer Science*, 2 (205).
[40] Singh, A. (2016). "A Modified Signcryption Scheme Using Elliptic Curve Cryptography." In *Proceedings of the National Conference on Recent Innovations in Science, Technology and Management* (pp. 26–27).

6 Securing the Future of Healthcare

Ensuring Privacy and Security in Wearable and Implantable Medical Devices

Gagandeep and Anjali Sharma

6.1 INTRODUCTION

There has been a significant transformation in the healthcare field due to the rapid advancements of wearable and implantable medical devices. These remarkable technological innovations have completely changed how healthcare is provided, offering real-time monitoring, personalized treatment plans, and improved patient engagement. Given the changing nature of healthcare, it is crucial to recognize the significance of security and privacy as both consumers and healthcare providers embrace these devices' possibilities. Regarding wearable and implanted devices, ensuring security and maintaining confidentiality is vital. This section will further explore why this importance cannot be overlooked. These devices, which range from smartwatches to pacemakers, represent the cutting edge of modern healthcare. They hold the promise of improving health outcomes by detecting diseases early and delivering patient-centered care. However, their widespread use has also raised concerns about security and privacy that needs to be carefully addressed. Let's delve into the world of wearable medical technology together, exploring its revolutionary potential while emphasizing the critical need to safeguard patient privacy and data in this new era of healthcare.

6.1.1 Background on Medical Devices That Are Wearable and Implantable

Wearable and implanted devices have become essential to modern healthcare, transforming how we monitor and treat medical conditions. These devices, such as smartwatches, fitness trackers, pacemakers, and insulin pumps, provide data collection and real-time insights that enable personalized and proactive healthcare management. To truly understand the significance of security and privacy concerns in this field, it's important to appreciate the history of wearable and implanted devices. According to a study by PricewaterhouseCoopers (PwC) [1], the global market

DOI: 10.1201/9781032635101-8

for medical devices is expected to reach billions of dollars by 2025. The appeal of these gadgets has grown due to their ability to track activity, monitor vital signs, and administer treatments. Likewise, there have been advancements in implantable medical devices in recent years. These sophisticated devices are designed to monitor, control, or enhance functions [2]. Their purpose is to reduce healthcare costs, improve quality of life, and enhance outcomes. The rising prevalence of diseases, the demand for personalized healthcare solutions, and technological advancements are key factors driving the adoption of wearable and implanted medical devices. By integrating sensors with connectivity and data analytics capabilities, doctors can now easily monitor patients' health conditions. This enables disease detection as well as personalized treatment plans. These devices can revolutionize healthcare by allowing intervention and reducing hospital stays [3]. However, there are concerns about the security and privacy of implantable medical devices due to the sensitive data they collect. Given the nature of physiological and personal health information, wearable and implantable medical devices are highly susceptible to illegal access and cyberattacks. As a result, it becomes crucial to grasp the underlying concepts of these devices to appreciate the necessity, for security and privacy measures [4].

6.1.2 Significance of Healthcare Security and Privacy

Ensuring security and privacy in the healthcare sector is important, particularly with advancements in technologies like wearable and implantable medical devices. Protecting patient data, maintaining the trust between patients and healthcare providers, and preserving the integrity and confidentiality of medical information are all key reasons why security and privacy are vital within the healthcare industry [5]. One critical aspect involves safeguarding data as healthcare systems increasingly digitize records and rely on interconnected devices. With the growth in healthcare data volume and sensitivity, it becomes crucial to uphold data security to prevent unauthorized access or exploitation while preserving patient privacy. Research emphasizes data storage, encryption methods, and access controls as essential to ensure patient information remains confidential [6]. Moreover, privacy holds significance alongside data security. Patients rightly expect their health information to be treated with discretion while retaining control. Healthcare wearable devices that collect and transmit health data present unique challenges regarding privacy protection. Techniques such as anonymization, pseudonymization, and obtaining consent play a pivotal role in safeguarding patient privacy [7]. Apart from protecting patient data, ensuring security and confidentiality within the healthcare industry has implications for patient safety as well as maintaining the integrity of healthcare systems. Ensuring the safety of medical devices is crucial to avoid unauthorized access or tampering that may pose a risk to patients. Security measures, like authentication protocols and encryption, are necessary to reduce potential vulnerabilities [8].

6.1.3 Research Objective and Chapter Overview

The following are the research objectives of this chapter:

- Provide a comprehensive understanding of the purpose of privacy and security in relation to implanted and wearable medical devices.

- Emphasize the importance of protecting patient data, maintaining patient trust in healthcare systems and ensuring patient privacy and security.

To accomplish this goal, the chapter will be organized in this manner:

- First, we will provide an overview of the research topic, emphasizing the increasing importance of privacy and security in the medical field, especially with the rise of wearable and implanted technology.
- Next, we will explore the challenges and risks associated with safeguarding information and privacy when it comes to wearable and implanted medical devices.
- We will then discuss technologies and security measures that can protect patient data, such as data encryption, access controls, and authentication protocols.
- In addition, we will review the regulatory frameworks governing security and privacy in healthcare, including relevant laws, rules, and standards.
- Furthermore, we will examine the vulnerabilities and weaknesses associated with wearables. Implanted medical devices by presenting case studies and real-world examples that highlight the consequences of security breaches and privacy concerns within the healthcare industry.
- To conclude this chapter, we will summarize our findings and contributions while underscoring the importance of robust security protocols in preserving privacy within healthcare.
- Finally, we will offer recommendations for improving security measures for wearable and implanted devices to legislators as well as medical professionals and researchers.
- In addition, we will identify areas for research aimed at advancing security practices in healthcare.

6.2 OVERVIEW OF MEDICAL DEVICES THAT ARE WEARABLE AND IMPLANTABLE

In this section, we will delve into the world of implantable medical devices, exploring their definitions, providing examples, and highlighting their significant importance in modern healthcare. Wearable medical devices have become tools for monitoring physiological parameters and gathering health-related data. These devices include smartwatches, fitness trackers, and biosensors. They are noninvasive and can collect a wide range of data, such as heart rate and sleep patterns, offering valuable real-time insights into our health. On the contrary, implantable medical devices go beyond the surface by delving beneath the skin to offer diagnostic, therapeutic, and monitoring capabilities. By interacting with bodily tissues, these devices provide precise treatment options as well as continuous health information. This section aims to shed light on the definitions and examples of both implantable medical devices while highlighting their roles in shaping the ever-evolving landscape of healthcare.

6.2.1 Definition and Examples of Medical Devices That Are Wearable and Implantable

Devices that can be worn on the body or attached to clothing are commonly referred to as medical devices (see Figure 6.1). These innovative technologies allow for the monitoring of important health data and physiological indicators. Examples of medical technology include biosensors, fitness trackers, and smartwatches. The primary goal of these devices is to provide a noninvasive way to collect various types of data such as heart rate, sleep patterns, activity levels, and even electrocardiogram (ECG) signals. An excellent example is the Apple Watch Series 6, which features sensors like an optical heart rate monitor and a blood oxygen level sensor. This allows users to stay informed about their health in time [9].

On the side, implantable medical devices are specifically created to be placed inside the body for purposes like diagnosis, therapy, or monitoring. They are usually intended for term or permanent use and may involve invasive procedures. Examples of medical devices include pacemakers, artificial joints, and neurostimulators. These devices directly interact with tissues or organs to provide targeted treatment or monitoring. For example, cochlear implants are devices that help individuals with hearing loss regain their ability to hear by stimulating the auditory nerve [10].

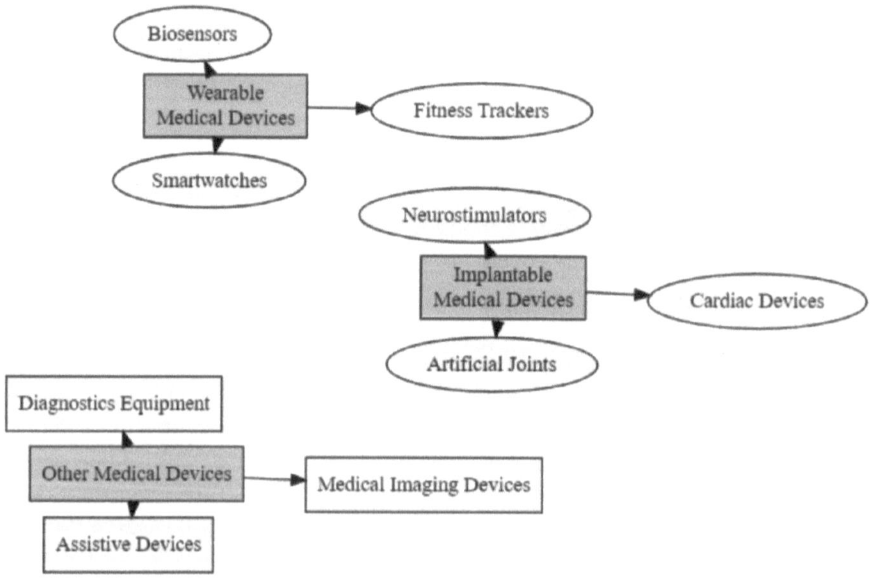

FIGURE 6.1 Types of wearable and implantable medical devices.

Source: Author's self-compilation.

6.2.2 Importance of These Devices in Healthcare Monitoring and Treatment

Wearable technology, like fitness trackers and smartwatches, has the ability to monitor various health parameters in real-time. This includes tracking heart rate, physical activity, sleep patterns, and more. By monitoring these variables, people can gain valuable insights into their overall health and make informed decisions about their lifestyle. An interesting study even demonstrates how wearable technology can help individuals improve their fitness levels and effectively manage health [11]. On the other hand, implantable medical devices offer advanced monitoring and treatment capabilities. They are designed to monitor physiological parameters, deliver targeted therapies, and provide crucial data for clinical decision making. Take implantable cardioverter defibrillators (ICDs) and pacemakers as examples. They constantly monitor heart rhythm and administer appropriate treatments when necessary. Research highlights the effectiveness of these devices in managing arrhythmias and improving patient outcomes [12]. Moreover, these devices enable monitoring and telehealth services, which prove particularly beneficial for patients with chronic conditions or those residing in remote areas. Overall, both wearable technology and implantable medical devices play a role in enhancing healthcare by providing continuous monitoring of health parameters while empowering individuals to take control of their well-being. Healthcare professionals have the advantage of accessing data from the wearable. Implanted devices simplify the detection of potential problems at an early stage and reduce the necessity for frequent hospital visits. Studies suggest that utilizing technologies for remote monitoring can efficiently handle long-term illnesses leading to improved patient results and reduced healthcare costs [13].

6.2.3 Challenges and Vulnerabilities Associated with Their Privacy and Security

One of the obstacles is the possibility of unauthorized access to sensitive patient information [14]. When wearable and implantable devices collect and transmit health data, they become attractive targets for malicious individuals who want to exploit vulnerabilities in the device or communication channels. These security concerns can put privacy at risk and result in negative consequences. Another challenge is the potential for data tampering or manipulation [15]. Hackers could potentially modify the data transmitted by implantable devices, leading to inaccurate medical readings or incorrect treatment decisions. This issue can have implications for patient safety and care. Moreover, the absence of security protocols and regulations creates a significant vulnerability [16]. The widespread use of implantable devices, along with their varying operating systems and communication protocols, makes it difficult to establish a unified security framework. This diversity opens up opportunities for attackers to exploit loopholes. In addition, maintaining privacy becomes challenging due to concerns about data collection and sharing [17]. Wearable and implantable devices often gather personal health data, raising questions about how this information is kept confidential and used with consent. The extensive data collection capabilities of these devices introduce privacy risks that require policies and practices to

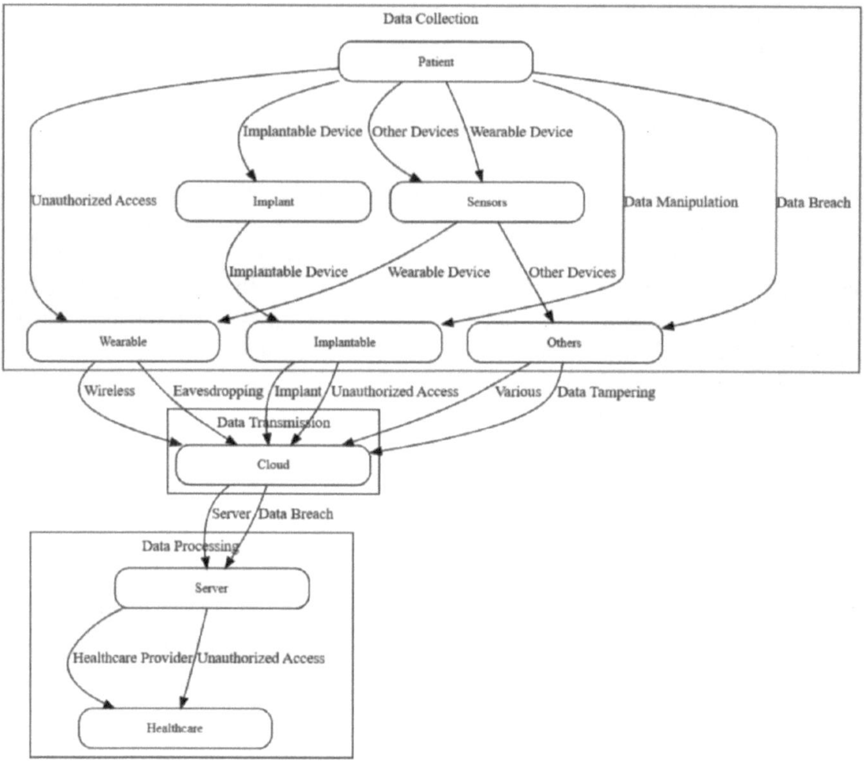

FIGURE 6.2 Data flow in medical devices.

Source: Author's self-compilation.

protect patient information. It is important to incorporate strong security measures, like encryption, authentication techniques, and secure communication channels, to effectively address these concerns and vulnerabilities. In addition, regulatory bodies and healthcare organizations must establish standards and protocols to ensure the privacy and security of wearable devices and implants (see Figure 6.2) [18].

6.3 SECURITY RISKS AND THREATS IN MEDICAL DEVICES THAT ARE WEARABLE AND IMPLANTABLE

This section focuses on the security aspects related to implantable medical devices. These innovative technologies have brought about advancements in healthcare, but they also come with potential security risks and threats. Concerns such as access, interception of data tampering with devices, vulnerabilities to malware, and data breaches are substantial. In this section, we explore the security concerns and threats that these devices may face, emphasizing the crucial importance of robust security measures. We also provide real-world examples of security breaches that highlight their serious consequences and underscore the need for proactive vulnerability management. Understanding these security risks is vital for ensuring patient well-being and maintaining the integrity of healthcare systems.

6.3.1 Types of Security Risks and Potential Threats

One security concern that healthcare professionals need to be aware of is the access to sensitive patient information [19]. This can happen when individuals with intent gain unauthorized entry to the device or its communication channels, which could compromise patient privacy and confidentiality. Another potential threat is data interception or eavesdropping [20]. Attackers might intercept the communication between the implantable device and external systems, allowing them to capture sensitive medical data or manipulate the communication. Device tampering also poses a security risk [21]. Attackers may try to exploit the device's functionality, modify data readings, or disable critical features, potentially endangering patients. In addition, malware and software vulnerabilities introduce security risks (see Figure 6.3) [22].

These devices are vulnerable to malware attacks like viruses or ransomware, which can disrupt their functioning and put data at risk. There is also a concern regarding data breaches and unauthorized disclosure of information [23]. Unauthorized access or accidental exposure of data can lead to identity theft, violations of privacy, and potential misuse of sensitive information. To address these security risks and potential threats, it is crucial to implement security measures such as encryption, authentication procedures, secure communication protocols, and regular software updates [24].

6.3.2 Examples of Past Security Breaches and Their Consequences

A team of cybersecurity researchers recently conducted a study that highlighted the risks associated with insulin pumps. They demonstrated how these pumps could be manipulated remotely, resulting in doses of insulin being administered. Such unauthorized interference could have implications for patients, potentially causing hypoglycemia or hyperglycemia and posing a threat to their lives [25]. In another security incident, a cardiac implantable electronic device was found to have vulnerabilities in its wireless communication protocol. This flaw made it

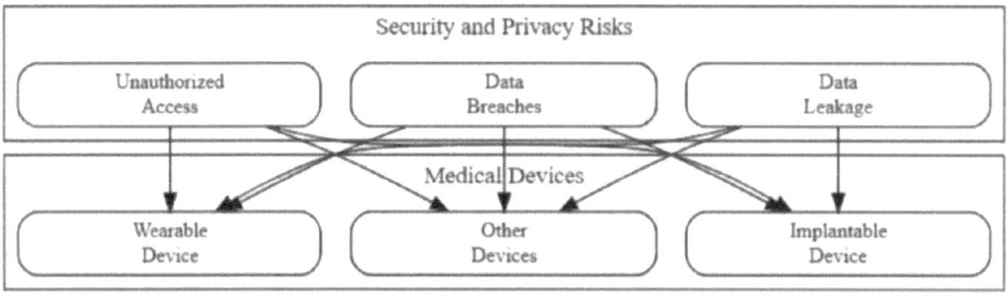

FIGURE 6.3 Security and privacy risks.

Source: Author's self-compilation.

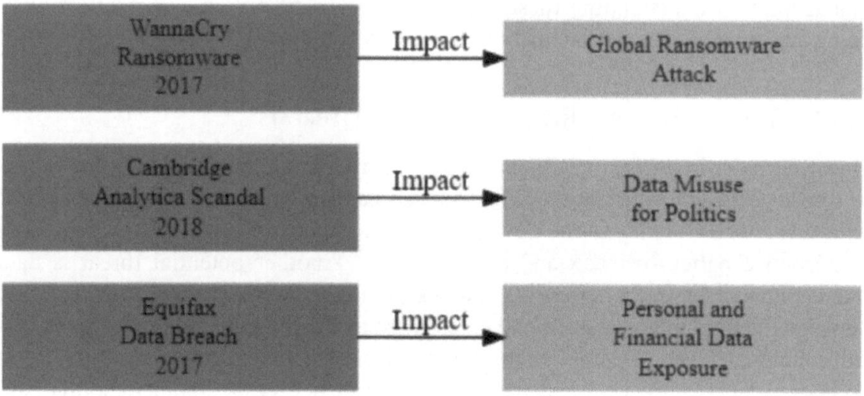

FIGURE 6.4 Examples of security breaches.

Source: Author's self-compilation.

susceptible to attacks granting individuals access to manipulate the device's settings. This breach raised concerns about safety and the potential for life-threatening consequences [26]. In addition, researchers also uncovered vulnerabilities in a used pacemaker system that could potentially allow unauthorized access and control (see Figure 6.4).

Exploiting these vulnerabilities may result in alterations to device settings or manipulation of pacing rhythms, which could pose substantial risks to a patient's health and overall well-being [27]. An impactful incident serves as an example, where a telehealth platform fell victim to a cybersecurity attack resulting in entry and the exposure of patient data. This breach compromised the privacy and confidentiality of health information, undermining the credibility of telehealth services and emphasizing the crucial need for rigorous security protocols [28].

6.4 PRIVACY CONCERNS IN MEDICAL DEVICES THAT ARE WEARABLE AND IMPLANTABLE

In this section, we will explore the topic of privacy concerns related to wearable and implantable medical devices. These advanced technologies provide us with personal health data, but they also raise significant issues regarding the privacy and confidentiality of that data. We will delve into the heart of these concerns, starting with the importance of keeping information confidential and ensuring its protection. The collection, storage, and sharing of medical data present complex challenges, which we will discuss in subsequent sections. We will address topics such as obtaining consent, implementing robust security measures, and adhering to ethical practices. In addition, we will examine the ethical considerations that govern patient privacy, emphasizing the moral obligation to respect patient autonomy and maintain confidentiality. It is crucial to understand and address these privacy concerns in order to establish trust, prioritize patient-centered care, and safeguard health information.

6.4.1 Data Privacy Issues and the Importance of Patient Confidentiality

The collection of health data through wearable and implanted medical devices has raised concerns about data privacy [29]. These devices often capture information like physiological readings, medical history, and even location details. It is crucial to protect the privacy of this data to prevent access, misuse, or disclosure. Patient confidentiality is an aspect of healthcare that ensures individuals have control over their medical information [30]. Keeping data confidential is essential for maintaining trust between patients and healthcare providers. It allows patients to share their health information without worrying about it being revealed to parties. Breaching data privacy and patient confidentiality can have consequences, such as identity theft, financial fraud, discrimination, or compromised medical care. If patients don't trust the privacy and confidentiality of their data, they may hesitate to share information or seek necessary healthcare [31]. Healthcare organizations and regulatory bodies have implemented measures to address these concerns. For example, the General Data Protection Regulation (GDPR) in the European Union provides a framework for safeguarding data, including health-related information [32]. The GDPR emphasizes the importance of informed consent, anonymization of data, and security measures in protecting patient information. To ensure the security of patient data, healthcare providers also need to implement security procedures. To prevent access or the risk of data breaches, it is important to implement encryption, access controls, and secure communication channels [33].

6.4.2 Collection, Storage, and Sharing of Sensitive Medical Data

When it comes to collecting medical information, it's crucial to prioritize privacy concerns. To ensure transparency and trust, healthcare providers and researchers must obtain consent from patients and maintain clear communication about how data is collected (see Figure 6.5) [34]. Robust security measures, such as encryption, access controls, and secure storage systems, are necessary to safeguard medical data against unauthorized access or breaches [35]. It's essential to have protocols in place to prevent unauthorized disclosure or misuse of this information. Sharing medical data requires adherence to privacy regulations and ethical guidelines. Techniques like de-identification and aggregation can help protect privacy while enabling research and analysis [36]. In addition, secure platforms and protocols should be used for data sharing purposes, ensuring that authorized individuals or entities have access. Safeguarding medical data is not only about individual privacy but also about driving advancements in the field of medicine and enhancing patient care through the responsible use of aggregated datasets [37].

6.4.3 Legal and Ethical Considerations Regarding Patient Privacy

Protecting privacy is crucial both from a legal standpoint and as an ethical responsibility in the healthcare field. It is important to abide by privacy and data protection regulations, such as the Health Insurance Portability and Accountability Act (HIPAA) in the United States. Respecting autonomy and confidentiality is a

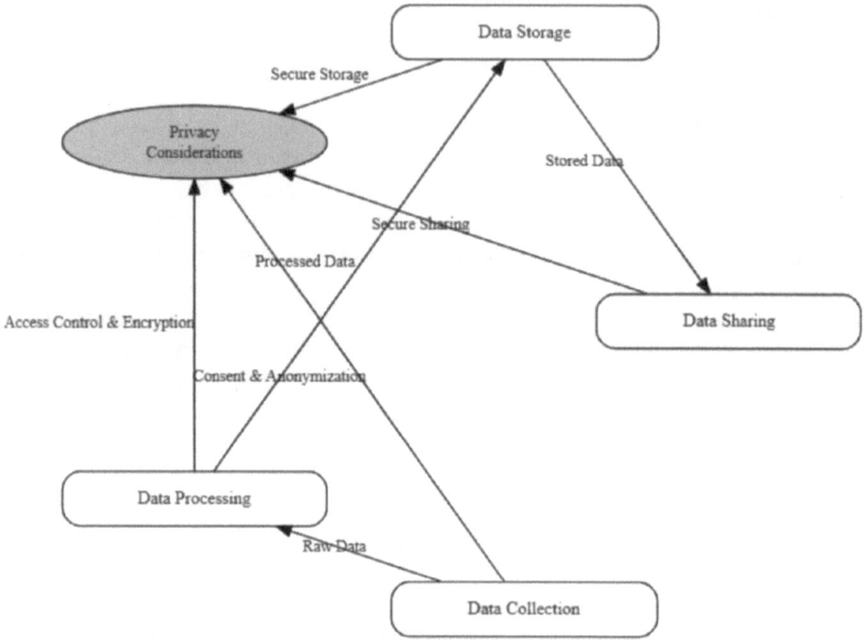

FIGURE 6.5 Patient data life cycle.

Source: Author's self-compilation.

fundamental ethical consideration. In order to gather data, informed consent must be obtained. Healthcare providers have a responsibility to ensure secure transmission and storage of patient information [38]. In addition, it is essential to limit access to data only to authorized individuals to implement robust security measures and maintain strict confidentiality. Upholding ethical standards regarding patient privacy plays a vital role, in building trust, promoting patient-centered care, and safeguarding sensitive health information [39].

6.5 CURRENT PRACTICAL AND SECURITY MEASURES FOR IMPLANTABLE AND WEARABLE MEDICAL DEVICES

Ensuring privacy and security measures is of utmost importance in the realm of wearable and implanted medical devices. To safeguard data during transmission and storage, encryption methods like Advanced Encryption Standard (AES) and Rivest–Shamir–Adleman (RSA) as well as secure communication protocols like Transport Layer Security (TLS) are employed. Access control mechanisms such as role-based access control (RBAC) and multi-factor authentication (MFA) are in place to ensure authorized access only. Compliance with frameworks like HIPAA and GDPR is crucial for protecting patient data, while international standards like ISO 13485 guide manufacturers in meeting quality and security requirements. Healthcare providers establish protocols for secure device usage while providing staff training.

Collaborative efforts with experts, regulatory bodies, and device manufacturers help keep industry standards current, ensuring an environment for healthcare data and preserving patient confidentiality.

6.5.1 Overview of Existing Security Protocols and Encryption Methods

In the healthcare field, it is extremely important to safeguard the confidentiality and integrity of medical information. Various security measures and encryption techniques are employed to prevent access and protect patient privacy (see Figure 6.6). Encryption methods such as AES and RSA are commonly used to transform text into unintelligible cipher text, which can only be understood with the correct decryption key [40]. Alongside encryption, secure communication techniques play a role in ensuring data integrity and confidentiality during transmission. TLS protocol is frequently utilized to establish connections between devices and servers, encrypting data while it's being transmitted and verifying entities using cryptographic methods and certificates [41]. In addition, there are two access control technologies, MFA and RBAC, that assist in enforcing authorized access to medical systems and data [42]. RBAC assigns permissions based on predefined roles, while MFA requires users to provide forms of authentication for enhanced security against unauthorized entry. Furthermore, the Health Level 7 (HL7) framework provides guidelines for the secure exchange of healthcare information. The specifications encompass aspects such as data formats, messaging, and security protocols. These elements are in place to promote compatibility between systems and safeguard information [43].

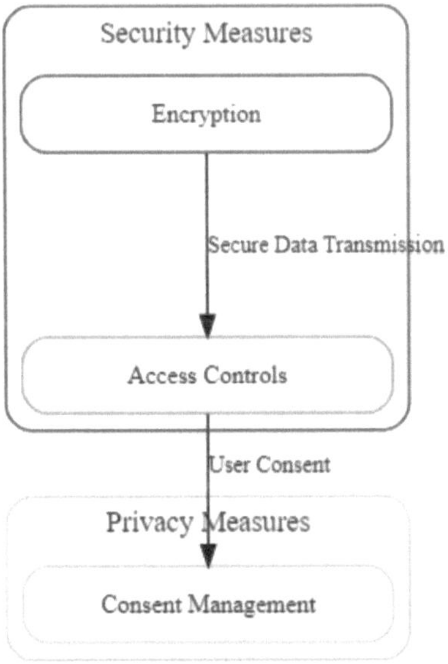

FIGURE 6.6 Security and privacy measures.

Source: Author's self-compilation.

6.5.2 Privacy Frameworks and Regulations in the Healthcare Industry

In the field of healthcare, it is crucial to have privacy frameworks and regulations in place to protect patient information. The United States has HIPAA, which provides guidelines for safeguarding and preserving protected health data. Similarly, the European Union implemented the GDPR to ensure people's data rights and privacy are protected [44]. In addition, many countries have their privacy laws. For instance, Canada has the Personal Information Protection and Electronic Documents Act (PIPEDA). These frameworks establish rules for data collection, storage, access, and disclosure with the aim of guaranteeing patient privacy, informed consent, and secure management of health information. Healthcare organizations must adhere to these regulations in order to protect confidentiality and maintain trust when handling sensitive data [45].

6.5.3 Industry Standards and Best Practices for Device Manufacturers and Healthcare Providers

Adhering to industry standards and best practices is of importance when it comes to ensuring the safety, security, and reliability of wearable and implantable medical devices. Device manufacturers must prioritize compliance with these standards throughout the process, including design, manufacturing, and testing. By following quality management systems like ISO 13485, manufacturers can guarantee that their devices meet requirements and undergo thorough quality control procedures [46]. To minimize vulnerabilities and protect patient data, manufacturers should also strictly adhere to cybersecurity and privacy guidelines set forth by reputable organizations such as the National Institute of Standards and Technology (NIST) and the U.S. Food and Drug Administration (FDA) [47]. On the other hand, healthcare providers need to establish best practices for securely implementing and utilizing these devices. They should have protocols in place for device registration, authentication, and access control to prevent usage or tampering [48]. In addition, providing training and education for healthcare staff will ensure they are well-informed about potential risks while effectively addressing security and privacy concerns. Collaborating with device manufacturers, cybersecurity experts, and regulatory bodies is vital for staying up to date on emerging industry standards as well as best practices [49].

6.6 EMERGING TECHNOLOGIES FOR IMPROVING PRIVACY AND SECURITY

In the changing world of healthcare technology, new advancements are being made to enhance security and privacy measures. Encryption techniques like RSA and Elliptic Curve Cryptography (ECC), along with algorithms like AES, are strengthening data protection. Biometric-based authentication methods, such as fingerprint recognition and voice authentication, are improving access control. Moreover, blockchain technology is transforming data management and consent tracking by ensuring data integrity and transparency while empowering patients with control. The tamper-proof nature of securing sensitive information reduces the risk of unauthorized data access and fosters trust between patients and healthcare providers.

6.6.1 Advances in Encryption and Authentication Techniques

Asymmetric encryption techniques like ECC and RSA have made advancements in the field of encryption. These algorithms are widely used to ensure the transmission of data between devices and healthcare networks due to their strong encryption capabilities [50]. In addition, symmetric encryption techniques, such as the AES, have also evolved to enhance data security and confidentiality [51]. In years, biometric-based authentication methods have gained popularity. Compared with authentication methods like passwords, biometrics such as voice recognition, iris scanning, and fingerprint recognition offer a higher level of security [52]. These techniques ensure that authorized individuals can access sensitive healthcare data and systems. Furthermore, blockchain technology has introduced possibilities for secure data sharing and storage. Blockchain offers improved data integrity and privacy protection by being decentralized and tamper-resistant [53].

6.6.2 Blockchain Technology for Consent Tracking and Safe Data Management

Blockchain, initially developed for Bitcoin transactions, is an unchangeable ledger system. Its purpose is to ensure the reliability and openness of information. Within the healthcare sector, blockchain has the potential to revolutionize data management by providing a secure record of patient data [54]. By utilizing methods and consensus algorithms, blockchain minimizes the requirement for intermediaries while also mitigating the risks of unauthorized data breaches or manipulation (see Figure 6.7) [55].

One of the uses of blockchain in the healthcare industry is keeping track of patient consent. By recording patient consent on the blockchain, healthcare providers and researchers can have a record of permissions given for accessing and using data [56]. This system allows patients to monitor and withdraw their consent whenever necessary, thereby enhancing their privacy and control over health information. Blockchain-based consent management systems also promote transparency and trust among patients' healthcare providers and other stakeholders [57]. In addition, blockchain technology enhances data security by encrypting information and ensuring

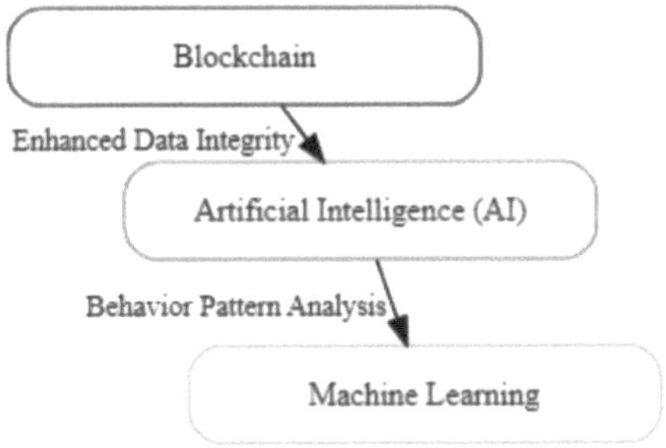

FIGURE 6.7 Emerging technologies.

Source: Author's self-compilation.

secure storage and transfer [58]. The decentralized nature of blockchain reduces the risk of data leaks or vulnerabilities while providing security against unauthorized access [59].

6.7 CHALLENGES AND FUTURE DIRECTIONS

Ensuring the security and privacy of data is paramount as wearable and implanted medical devices become increasingly important in healthcare. In this section, we delve into the obstacles and potential future paths regarding security and confidentiality. Finding the balance between user convenience and robust security measures while also aligning with evolving regulatory demands poses considerable challenges. Nevertheless, by adopting approaches fostering collaboration among stakeholders and actively contributing to ethical and regulatory frameworks, we can pave a promising path for tackling these challenges in the ever-changing landscape of healthcare technology.

6.7.1 Addressing the Trade-Off between Security and Usability

One way to tackle this trade-off is to implement security measures that prioritize user-friendliness. By incorporating design principles that focus on the user, we can seamlessly integrate security features into technology systems without compromising usability [60]. For instance, techniques such as fingerprint or facial recognition authentication offer consumers a quick and secure method to access their devices or accounts [61]. Another approach is to provide clear and easily understandable user education and guidance on security practices. By equipping users with knowledge about risks and recommended security measures, they can make informed choices while navigating technology systems. This helps bridge the gap between security requirements and user expectations. In addition, advancements in technology, such as intelligence and machine learning, can be leveraged to automate security processes and reduce the burden on users. Intelligent algorithms can analyze patterns of user behavior and detect potential security threats, resulting in a more streamlined and user-friendly security experience [62].

6.7.2 Balancing Technological Advancements with Regulatory Requirements

One way to address this issue is by promoting collaboration between technology developers and regulatory bodies. By involving regulators in the development process, potential risks and compliance requirements can be identified and addressed right from the start [63]. This collaboration aids in the creation of guidelines and standards that promote innovation while protecting public interests. In addition, organizations should adopt a privacy-focused and security-conscious approach, integrating requirements into the design and development of technologies [64]. By considering privacy and security implications from the beginning, organizations can avoid compliance problems. Ensure that their innovations align with legal and ethical standards. Furthermore, it is important to monitor and adapt to evolving regulatory landscapes. Regular assessments and audits can help identify any gaps between advancements and regulatory requirements, enabling organizations to make necessary adjustments for compliance [65]. Moreover, active engagement in discussions on ethics is essential for shaping regulatory frameworks. By participating in industry associations, standards bodies, and public consultations, organizations can contribute to the development of policies that strike the right balance between innovation and societal interests [66].

6.8 CASE STUDIES AND EXAMPLES

In this section, we delve into real-life examples and scenarios that shed light on how security and privacy protocols can be used in the realm of implantable and wearable medical devices along with their associated technology. These instances showcase the impact of regulations, industry practices, and technological solutions in safeguarding the confidentiality of medical data. We also explore insights gained from past incidents demonstrating how breaches and controversies have influenced the development of devices and privacy policies, underscoring the crucial role of security in healthcare technology. The case studies and examples mentioned earlier provide perspectives on the practical implementation of security and privacy protocols within the ever-evolving field of medical technology.

6.8.1 Real-World Examples of Successful Security and Privacy Implementations

GDPR is a known example of data protection rules implemented by the European Union. It emphasizes transparency, user consent, and accountability [67]. The successful implementation of GDPR has had an impact on data privacy practices, encouraging organizations worldwide to adopt similar principles to safeguard user data. Another example is HIPAA in the United States. HIPAA establishes security and privacy guidelines for protected health information (PHI), ensuring that authorized individuals handle health information securely and with limited access [68]. In the technology realm, messaging platforms like Signal and WhatsApp have played a role by implementing end-to-end encryption to protect user communications. This encryption ensures that only the sender and recipient can view the content of their messages, enhancing security and privacy. In addition, the Zero Trust architecture framework has proven effective in implementing security measures. With verification and strict access controls, Zero Trust limits user privileges and prevents unauthorized access to sensitive data [69]. Organizations adopting Zero Trust principles have shown security posture and resilience against cyber threats.

6.8.2 Lessons Learned from Past Incidents and Their Impact on Device Design and Policies

The Equifax data breach in 2017 was an event that shed light on security and privacy concerns. This incident exposed the financial information of millions of individuals, emphasizing the need for strong security measures when handling sensitive data [70].

As a result, organizations were prompted to reassess their data protection strategies and prioritize security measures such as encryption and access controls. Another notable event was the Cambridge Analytica scandal, where Facebook user data was unlawfully accessed and manipulated for purposes [71]. This incident raised concerns about data privacy and the ethical use of information. Consequently, there has been increased attention on data practices, transparency, and obtaining user consent in device design and regulations. In 2017, the global cybersecurity incident known as the WannaCry attack affected numerous organizations, highlighting the importance of patch management and timely software updates [72]. This incident showcased the criticality of security measures, like vulnerability management and regular system updates, to prevent and mitigate potential cyber threats.

6.9 CONCLUSION

This section concludes by highlighting the importance of security and privacy in relation to devices that are worn or implanted. While the rapid advancements in healthcare technology offer the potential for enhancing patient care and monitoring, they also introduce new vulnerabilities that need to be addressed. Striking a balance between security and usability is a challenge requiring user-focused design, seamless authentication methods, and education to bridge this gap. It is crucial for all stakeholders, including device manufacturers, healthcare providers, policymakers, and regulators, to prioritize and invest in security measures that protect privacy. Collaboration and knowledge sharing are essential for establishing industry best practices. Furthermore, comprehensive policies and regulations must be advocated to tackle the challenges faced in this field. By doing collectively, we can ensure the highest standards of data protection, and patient safety, and foster trust within the healthcare ecosystem. This sets the foundation for a future where medical implants and wearable gadgetry are more secure while respecting individuals' privacy.

REFERENCES

[1] Chen, K., Ren, J., Chen, C., Xu, W., & Zhang, S. (2020). Safety and effectiveness evaluation of flexible electronic materials for next-generation wearable and implantable medical devices. *Nano Today*, 35, 100939. https://doi.org/10.1016/j.nantod.2020.100939

[2] Choi, S., Han, S. I., Jung, D., Hwang, H. J., Lim, C., Bae, S., Park, O. K., Tschabrunn, C. M., Lee, M., Bae, S. Y., Yu, J. W., Ryu, J. H., Lee, S.-W., Park, K., Kang, P. M., Lee, W. B., Nezafat, R., Hyeon, T., & Kim, D.-H. (2018). Highly conductive, stretchable, and biocompatible Ag–Au core–sheath nanowire composite for wearable and implantable bioelectronics. *Nature Nanotechnology*, 13(11), 1048–1056. https://doi.org/10.1038/s41565-018-0226-8

[3] Rushanan, M., Rubin, A. D., Kune, D. F., & Swanson, C. M. (2014). SoK: Security and privacy in implantable medical devices and body area networks. *2014 IEEE Symposium on Security and Privacy*. https://doi.org/10.1109/sp.2014.40

[4] Long, Y., Li, J., Yang, F., Wang, J., & Wang, X. (2021). Wearable and implantable electroceuticals for therapeutic electrostimulations. *Advanced Science*, 8(8), 2004023. Portico. https://doi.org/10.1002/advs.202004023

[5] Amalraj, J. R., & Lourdusamy, R. (2022). Security and privacy issues in federated healthcare—An overview. *Open Computer Science*, 12(1), 57–65. https://doi.org/10.1515/comp-2022-0230
[6] Burns, A. J., Young, J., Roberts, T., Courtney, J., & Ellis, T. S. (2015). Exploring the role of contextual integrity in electronic medical record (EMR) system workaround decisions: An information security and privacy perspective. *AIS Transactions on Human-Computer Interaction*, 7(3), 142–165. https://doi.org/10.17705/1thci.00070
[7] Shoniregun, C. A., Dube, K., & Mtenzi, F. (2010). Introduction to e-healthcare information security. *Electronic Healthcare Information Security*, 1–27. https://doi.org/10.1007/978-0-387-84919-5_1
[8] Alanazi, R. (2022). Analysis of privacy and security challenges in e-health clouds. *International Journal of Advanced Computer Science and Applications*, 13(9). https://doi.org/10.14569/ijacsa.2022.0130955
[9] Rushanan, M., Rubin, A. D., Kune, D. F., & Swanson, C. M. (2014). SoK: Security and privacy in implantable medical devices and body area networks. *2014 IEEE Symposium on Security and Privacy*. https://doi.org/10.1109/sp.2014.40
[10] Wazid, M., Das, A. K., Kumar, N., Conti, M., & Vasilakos, A. V. (2018). A novel authentication and key agreement scheme for implantable medical devices deployment. *IEEE Journal of Biomedical and Health Informatics*, 22(4), 1299–1309. https://doi.org/10.1109/jbhi.2017.2721545
[11] Song, Y., Min, J., & Gao, W. (2019). Wearable and implantable electronics: Moving toward precision therapy. *ACS Nano*, 13(11), 12280–12286. https://doi.org/10.1021/acsnano.9b08323
[12] Guk, K., Han, G., Lim, J., Jeong, K., Kang, T., Lim, E.-K., & Jung, J. (2019). Evolution of wearable devices with real-time disease monitoring for personalized healthcare. *Nanomaterials*, 9(6), 813. https://doi.org/10.3390/nano9060813
[13] He, T., & Lee, C. (2021). Evolving flexible sensors, wearable and implantable technologies towards bodyNET for advanced healthcare and reinforced life quality. *IEEE Open Journal of Circuits and Systems*, 2, 702–720. https://doi.org/10.1109/ojcas.2021.3123272
[14] Javadi, S. S., & Razzaque, M. A. (2013). Security and privacy in wireless body area networks for health care applications. *Wireless Networks and Security: Issues, Challenges and Research Trends* (pp. 165–187). Springer.
[15] Al Ameen, M., Liu, J., & Kwak, K. (2012). Security and privacy issues in wireless sensor networks for healthcare applications. *Journal of Medical Systems*, 36, 93–101.
[16] Bandyopadhyay, D., & Sen, J. (2011). Internet of things: Applications and challenges in technology and standardization. *Wireless Personal Communications*, 58(1), 49–69. http://doi.org/10.1007/s11277-011-0288-5
[17] Jiang, D., & Shi, G. (2021). Research on data security and privacy protection of wearable equipment in healthcare. *Journal of Healthcare Engineering*, 6656204–6656204. https://doi.org/10.1155/2021/6656204
[18] Aledhari, M., Razzak, R., Qolomany, B., Al-Fuqaha, A., & Saeed, F. (2022). Biomedical IoT: Enabling technologies architectural elements challenges and future directions. *IEEE Access*, 31306. https://doi.org/10.1109/ACCESS.2022.3159235
[19] Mosenia, A., & Jha, N. K. (2017). A comprehensive study of security of internet-of-things. *IEEE Transactions on Emerging Topics in Computing*, 586. https://doi.org/10.1109/TETC.2016.2606384
[20] Siddiqi, M. A., Hahn, G., Hamdioui, S., Serdijn, W. A., & Strydis, C. (2022). Improving the security of the IEEE 802.15.6 standard for medical bans. *IEEE Access*, 62953. https://doi.org/10.1109/ACCESS.2022.3181630

[21] Vakhter, V., Soysal, B., Schaumont, P., & Guler, U. (2022). Threat modeling and risk analysis for miniaturized wireless biomedical devices. *IEEE Internet of Things Journal*, 13338. https://doi.org/10.1109/JIOT.2022.3144130

[22] Malamas, V., Chantzis, F., Dasaklis, T. K., Stergiopoulos, G., Kotzanikolaou, P., & Douligeris, C. (2021). Risk assessment methodologies for the internet of medical things: A survey and comparative appraisal. *IEEE Access*, 40049. https://doi.org/10.1109/ACCESS.2021.3064682

[23] Kim, D.-W., Choi, J.-Y., & Han, K.-H. (2020). Medical device safety management using cybersecurity risk analysis. *IEEE Access*, 115370. https://doi.org/10.1109/ACCESS.2020.3003032

[24] Iwaya, L. H., Ahmad, A., & Babar, M. A. (2020). Security and privacy for mhealth and uhealth systems: A systematic mapping study. *IEEE Access*, 150081. https://doi.org/10.1109/ACCESS.2020.3015962

[25] Rathore, H., Fu, C., Mohamed, A., Al-Ali, A., Du, X., Guizani, M., & Yu, Z. (2018). Multi-layer security scheme for implantable medical devices. *Neural Computing and Applications*, 32(9), 4347–4360. https://doi.org/10.1007/s00521-018-3819-0

[26] Choo, K.-K. R., Kermani, M. M., Azarderakhsh, R., & Govindarasu, M. (2017). Emerging embedded and cyber physical system security challenges and innovations. *IEEE Transactions on Dependable and Secure Computing*, 14(3), 235–236. https://doi.org/10.1109/tdsc.2017.2664183

[27] Beer, E. (2022). Googling patients. *Voices in Bioethics*, 8. https://doi.org/10.52214/vib.v8i.10155

[28] Deering, M. J. (2015). Health information technology: Empowering consumers, patients, and caregivers. *Proceedings of ISIS Summit Vienna 2015—The Information Society at the Crossroads*. https://doi.org/10.3390/isis-summit-vienna-2015-i006

[29] Johnson, I. (2016). Privacy and security risks for mobile health devices. *Journal of Law and the Biosciences*, 3(3), 707–721. https://doi.org/10.1093/jlb/lsw037

[30] Beauchamp, T. L. (1977). Patient autonomy in medical ethics: A critical appraisal. *Journal of Medical Ethics*, 3(4), 143–149. https://doi.org/10.1136/jme.3.4.143

[31] Eysenbach, G. (2000). Consumer health informatics. *British Medical Journal*, 320(7251), 1713–1716. https://doi.org/10.1136/bmj.320.7251.1713

[32] European Commission. (2016). Regulation (EU) 2016/679 of the European parliament and of the council of 27 April 2016 on the protection of natural persons with regard to the processing of personal data and on the free movement of such data, and repealing directive 95/46/EC (General Data Protection Regulation). *Official Journal of the European Union*. https://eur-lex.europa.eu/eli/reg/2016/679/oj

[33] Gurses, A. P., et al. (2017). Security of health information systems: A systematic review. *Journal of the American Medical Informatics Association*, 24(2), e1–e12. https://doi.org/10.1093/jamia/ocw145

[34] Chen, M., Qian, Y., Chen, J., Hwang, K., Mao, S., & Hu, L. (2020). Privacy protection and intrusion avoidance for cloudlet-based medical data sharing. *IEEE Transactions on Cloud Computing*, 8(4), 1274–1283. https://doi.org/10.1109/tcc.2016.2617382

[35] Jin, X.-L., Zhang, M., Zhou, Z., & Yu, X. (2019). Application of a blockchain platform to manage and secure personal genomic data: A case study of lifeCODE.ai in China. *Journal of Medical Internet Research*, 21(9), e13587. https://doi.org/10.2196/13587

[36] Caine, K., & Tierney, W. M. (2014). Point and counterpoint: Patient control of access to data in their electronic health records. *Journal of General Internal Medicine*, 30(S1), 38–41. https://doi.org/10.1007/s11606-014-3061-0

[37] Sendhil, R., & Amuthan, A. (2021). Contextual fully homomorphic encryption schemes-based privacy preserving framework for securing fog-assisted healthcare

data exchanging applications. *International Journal of Information Technology*, 13(4), 1545–1553. https://doi.org/10.1007/s41870-021-00704-z

[38] Gardner, J. M., & Allen, T. C. (2018). Keep calm and tweet on: Legal and ethical considerations for pathologists using social media. *Archives of Pathology & Laboratory Medicine*, 143(1), 75–80. https://doi.org/10.5858/arpa.2018-0313-sa

[39] Powers, J. M., Long, J. M., & Mendonca, W. (2017). Psychosocial, ethical, and legal implications for mutation carriers. *Managing BRCA Mutation Carriers*, 205–234. https://doi.org/10.1007/978-3-319-59198-8_12

[40] Ferguson, N., Schneier, B., & Kohno, T. (2010). *Cryptography Engineering: Design Principles and Practical Applications*. Wiley.

[41] Rescorla, E. (2001). *SSL and TLS: Designing and Building Secure Systems*. Addison-Wesley Professional.

[42] Hernández-Álvarez, L., Bullón Pérez, J. J., Batista, F. K., & Queiruga-Dios, A. (2022). Security threats and cryptographic protocols for medical wearables. *Mathematics*, 10(6), 886. https://doi.org/10.3390/math10060886

[43] Rushanan, M., Rubin, A. D., Kune, D. F., & Swanson, C. M. (2014). SoK: Security and privacy in implantable medical devices and body area networks. *2014 IEEE Symposium on Security and Privacy*. https://doi.org/10.1109/sp.2014.40

[44] Alharbi, R., & Almagwashi, H. (2019). The privacy requirements for wearable IoT devices in healthcare domain. *2019 7th International Conference on Future Internet of Things and Cloud Workshops (FiCloudW)*. https://doi.org/10.1109/ficloudw.2019.00017

[45] Ahouanmenou, S., Van Looy, A., & Poels, G. (2022). Information security and privacy in hospitals: A literature mapping and review of research gaps. *Informatics for Health and Social Care*, 48(1), 30–46. https://doi.org/10.1080/17538157.2022.2049274

[46] Vogl, G. W., Weiss, B. A., & Helu, M. (2016). A review of diagnostic and prognostic capabilities and best practices for manufacturing. *Journal of Intelligent Manufacturing*, 30(1), 79–95. https://doi.org/10.1007/s10845-016-1228-8

[47] Morgan, P., Kao, M. B., Kilvington, L., Alexander, R., & Gheraibia, Y. (2020). Review of legal frameworks, standards and best practices in verification and assurance for infrastructure inspection robotics. *SSRN Electronic Journal*. https://doi.org/10.2139/ssrn.3645997

[48] Murphy, R. J., Sukkarieh, M., Haass, J., & Hriljac, P. (2015). *Guidebook on Best Practices for Airport Cybersecurity*. https://doi.org/10.17226/22116

[49] Hartmann, C. J., Fliegen, S., Groiss, S. J., Wojtecki, L., & Schnitzler, A. (2019). An update on best practice of deep brain stimulation in Parkinson's disease. *Therapeutic Advances in Neurological Disorders*, 12, 175628641983809. https://doi.org/10.1177/1756286419838096

[50] Rivest, R. L., Shamir, A., & Adleman, L. (1978). A method for obtaining digital signatures and public-key cryptosystems. *Communications of the ACM*, 21(2), 120–126. https://doi.org/10.1145/359340.359342

[51] National Institute of Standards and Technology. (2001). *Advanced Encryption Standard (AES)*. https://csrc.nist.gov/projects/advanced-encryption-standard

[52] Rane, S., & Yeung, D. (2014). Biometric authentication. In *Encyclopedia of Cryptography and Security* (2nd ed., pp. 116–119). Springer. https://doi.org/10.1007/978-1-4899-7489-2_343

[53] Ekblaw, A., Azaria, A., Halamka, J. D., & Lippman, A. (2016). A case study for blockchain in healthcare: "MedRec" prototype for electronic health records and medical research data. *Proceedings of IEEE Open & Big Data Conference*, 25–30. https://doi.org/10.1109/OBD.2016.8

[54] Nakamoto, S. (2008). *Bitcoin: A peer-to-peer electronic cash system*. https://bitcoin.org/bitcoin.pdf

[55] Swan, M. (2015). *Blockchain: Blueprint for a New Economy*. O'Reilly Media.

[56] Ekblaw, A., Azaria, A., Halamka, J. D., & Lippman, A. (2016). A case study for blockchain in healthcare: "MedRec" prototype for electronic health records and medical research data. *Proceedings of IEEE Open & Big Data Conference*, 25–30. https://doi.org/10.1109/OBD.2016.8

[57] Agbo, C. C., Mahmoud, Q. H., & Eklund, J. M. (2019). Blockchain technology in healthcare: A systematic review. *Healthcare*, 7(2), 56. https://doi.org/10.3390/healthcare7020056

[58] Azaria, A., Ekblaw, A., Vieira, T., & Lippman, A. (2016). MedRec: Using blockchain for medical data access and permission management. In *Proceedings of IEEE International Conference on Open and Big Data* (pp. 25–30). https://doi.org/10.1109/OBD.2016.8

[59] Dagher, G. G., Mohler, J., Milojkovic, M., Marella, P. B., & Ancile Consortium. (2018). Ancile: Privacy-preserving framework for access control and interoperability of electronic health records using blockchain technology. *Sustainable Cities and Society*, 39, 283–297. https://doi.org/10.1016/j.scs.2018.02.036

[60] Sasse, M. A., Brostoff, S., & Weirich, D. (2001). Transforming the "weakest link"—a human/computer interaction approach to usable and effective security. *BT Technology Journal*, 19(3), 122–131. https://doi.org/10.1023/A:1011265715113

[61] Jain, A. K., Ross, A., & Prabhakar, S. (2004). An introduction to biometric recognition. *IEEE Transactions on Circuits and Systems for Video Technology*, 14(1), 4–20. https://doi.org/10.1109/TCSVT.2003.818349

[62] Walker-Roberts, S., Hammoudeh, M., Aldabbas, O., Aydin, M., & Dehghantanha, A. (2019). Threats on the horizon: Understanding security threats in the era of cyber-physical systems. *The Journal of Supercomputing*, 76(4), 2643–2664. https://doi.org/10.1007/s11227-019-03028-9

[63] Ng, W. L., Lee, J. M., Zhou, M., Chen, Y.-W., Lee, K.-X. A., Yeong, W. Y., & Shen, Y.-F. (2020). Vat polymerization-based bioprinting—process, materials, applications and regulatory challenges. *Biofabrication*, 12(2), 022001. https://doi.org/10.1088/1758-5090/ab6034

[64] Hernandez-Ramos, J. L., Geneiatakis, D., Kounelis, I., Steri, G., & Nai Fovino, I. (2020). Toward a data-driven society: A technological perspective on the development of cybersecurity and data-protection policies. *IEEE Security & Privacy*, 18(1), 28–38. https://doi.org/10.1109/msec.2019.2939728

[65] Lee, S.-W., Gandhi, R., Muthurajan, D., Yavagal, D., & Ahn, G.-J. (2006). Building problem domain ontology from security requirements in regulatory documents. *Proceedings of the 2006 International Workshop on Software Engineering for Secure Systems*. https://doi.org/10.1145/1137627.1137635

[66] Thakkar, S., Anklam, E., Xu, A., Ulberth, F., Li, J., Li, B., Hugas, M., Sarma, N., Crerar, S., Swift, S., Hakamatsuka, T., Curtui, V., Yan, W., Geng, X., Slikker, W., & Tong, W. (2020). Regulatory landscape of dietary supplements and herbal medicines from a global perspective. *Regulatory Toxicology and Pharmacology*, 114, 104647. https://doi.org/10.1016/j.yrtph.2020.104647

[67] European Union. (2018). *General Data Protection Regulation (GDPR)*. https://eur-lex.europa.eu/eli/reg/2016/679/oj

[68] U.S. Department of Health & Human Services. (n.d.). *Summary of the HIPAA Privacy Rule*. www.hhs.gov/hipaa/for-professionals/privacy/laws-regulations/index.html

[69] Signal Messenger. (n.d.). *Technology Overview*. https://signal.org/docs/

[70] Al-Tawil, K., & Papadaki, M. (2018). Lessons learned from the equifax data breach: Legal, regulatory, and ethical implications. *Journal of Business Ethics*, 161(2), 457–471. http://doi.org/10.1007/s10551-018-4069-y

[71] Cadwalladr, C. (2018). The Cambridge analytica files: 'I made Steve Bannon's psychological warfare tool'. *The Guardian*. www.theguardian.com/news/2018/mar/17/data-war-whistleblower-christopher-wylie-faceook-nix-bannon-trump

[72] Ollmann, G. (2017). WannaCry: Lessons learned from the world's biggest ransomware attack. *Information Security Technical Report*, 22(4), 62–67. http://doi.org/10.1016/j.istr.2017.10.001

7 Robust Security and Privacy of Wearable and Implantable Medical Devices

Protecting Lives and Data

Ashwini A., Banu Priya Prathaban, and D. Ferlin Deva Shahila

7.1 INTRODUCTION

The Internet of Things (IoT) integration has improved quality of life by fostering insightful discoveries, increasing productivity, and lowering costs. Implantable medical devices (IMDs), including artificial pacemakers, drug delivery systems, and neuro-stimulators, utilize embedded miniaturized computer systems to autonomously carry out a variety of therapeutic and health monitoring duties [1]. The use of big data [2], blockchain [3], and the IoT [4] has led to the rise of health 2.0 and medicine 4.0. Wireless Body Area Networks (WBANs), also known as e-healthcare, have recently emerged as a promising approach that will revolutionize how healthcare is obtained. This is due to the rapid development of wearable medical devices and wireless communication. For instance, a new development in technology called "Tactile Internet" using 5G scenarios allows surgeons to undertake surgery with patients in faraway locations [5]. This has made it possible to develop brand-new approaches to asset management, medical equipment maintenance, remote monitoring, autonomous assistive solutions, medicine administration, early warning, and proactive treatment. Healthcare professionals, researchers, government organizations, and insurance firms are just a few of the users who have access to and exchange this medical information. While its expanding use has increased its usefulness and the ability to monitor patients' health, it has also greatly increased the environment's complexity. This will considerably boost the efficiency of healthcare by facilitating and speeding up processes like clinical examinations and emergency medical response.

Ubiquitous health monitoring (UHM), emergency medical response systems (EMRS), computer-assisted rehabilitation (CARF), and the promotion of healthy lifestyles are just a few of the unique applications that can be made possible thanks

 DOI: 10.1201/9781032635101-9

to the WBAN [6–8]. Due to this diversity, technology requirements are much more complex in this situation, but security and privacy are now more important than ever. Therefore, it is of prime importance to create cost-effective, systematic procedures for healthcare, especially for nations that lack medical facilities and qualified personnel. Critical security flaws and privacy issues with sensitive personal data are the result of this. Therefore, before widespread implementation of these IMDs is feasible, the underlying security vulnerabilities must be resolved. A tiny wireless device called an IMD is inserted into a patient's body for medical reasons [9]. The key to leveraging IMD technology and protecting privacy is to make security a facilitator of secure and protected data transit, exchange, and use. Even though the battery for these gadgets is incorporated into the body and is not rechargeable, it must have a lengthy lifespan. The patient safety aspect of healthcare, on the contrary, is another factor that makes it more difficult to meet the situation's current requirements.

The security methods must essentially be as small as feasible because of strict resource limits on devices inside a WBAN, particularly the sensor nodes [10]. In this regard, several official rules adopted by various nations and continents, including the Medical Device Directive [11], the Federal Drug Administration [12], the European Medicines Agency [13], and the Medical Device Regulation [14], have established a legal framework for classifying medical procedures and devices according to varying levels of risk, enabling the classification of the entities involved. However, no additional controls exist to guarantee the privacy and security of the data that has been stored. Various security plans have been researched recently in terms of security and usability in an emergency. A strong security solution needs to support both emergency and everyday use scenarios to be practical. The difficult practical concerns that lie beneath these topics are then examined. We then survey relevant work in both WBANs and Wireless Sensor Networks to further investigate the problem space. Evaluation is done on how well WBANs work and make some suggestions for the future.

7.2 NEED FOR WEARABLE AND IMPLANTABLE MEDICAL DEVICES

The need for portable medical devices and IMDs is brought on by the changing healthcare environment, the desire to advance the field of medicine, and a number of other issues in contemporary healthcare [15]. Through continuous tracking, real-time data investigation, and personalized interventions, these devices make use of technology to improve patient care and make healthcare systems more effective (Figure 7.1).

The specific needs of medical devices that can be worn or implanted are given below:

a) Permanent tracking and timely detection: To detect irregularities or changes early, many medical illnesses need constant monitoring [16]. Live information on activity levels, vital signs, and other health parameters is provided via wearable technology, including medical patches, fitness trackers, and smart watches.

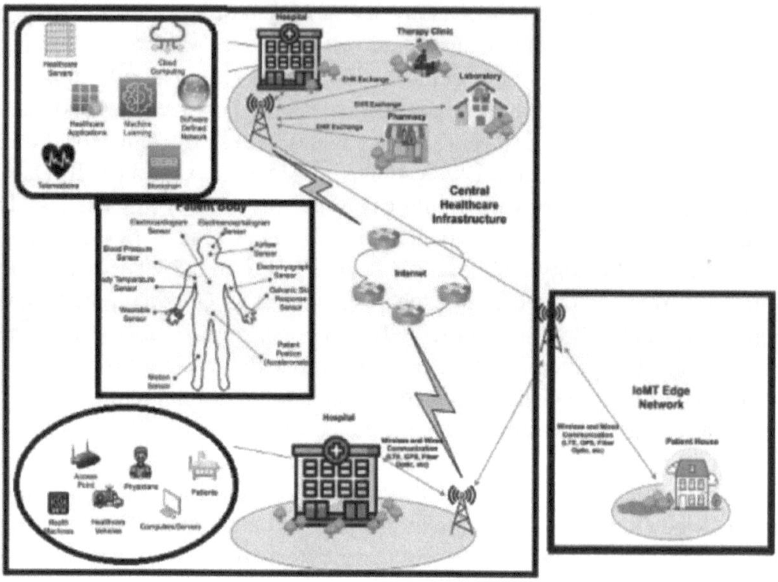

FIGURE 7.1 Healthcare systems—overall architecture.

b) Disease management: Hypertension, diabetes, and heart disease are examples of chronic diseases that need constant management and observation. Sensor-equipped wearables can monitor a variety of functions, including glucose levels, heart rate, and blood pressure [17]. Patients can offer their health records, enabling them to make detailed treatment decisions and modifications that enhance managing their illnesses.
c) Monitoring of patients: Remote patient monitoring is made possible by wearable devices, which are essential for providing healthcare, particularly for people who live in remote locations or have limited mobility [18,19]. The necessity for regular in-person visits can be reduced because patients can be followed remotely. This is especially helpful for treating chronic illnesses, caring for the elderly, and providing postoperative care.
d) Healthcare-driven systems: Huge volumes of health-related data are produced by wearable technology. To get insights into patient disease patterns, behavior, and effectiveness, data has to be analyzed [20]. This data can be used by healthcare professionals to personalize treatment programs, forecast health trends, and make wise choices.
e) Personalized medical care: Every individual's health needs and responses are unique. Wearable devices provide a wealth of personalized data that can guide healthcare decisions [21]. By analyzing a patient's physiological data, lifestyle patterns, and genetic markers, healthcare providers can create personalized treatment plans that are more effective and aligned with the patient's needs.
f) Patient engagement and empowerment: Health requirements and responses vary from person to person. Wearable technology offers a wealth of

personalized data that can provide medical choices [22,23]. Healthcare professionals can develop individualized treatment regimens that are more efficient and in line with the patient's demands by analyzing a patient's lifestyle choices, physiological data, and genetic markers.

g) Research innovation: Wearable technology data is used to advance medical innovation and research. Researchers can examine illness progression, treatment outcomes, and general health trends using large-scale datasets [24].
h) Independent living with aging: Healthcare services are in greater demand as the world's population ages. In order to maintain their freedom and quality of life, older people need the assistance of wearable technology [25]. These gadgets can keep an eye on falls, medication compliance, and general health.
i) Response in emergency cases: Pacemakers and other IMDs, such as heart monitors, provide early intervention in life-threatening situations [26]. These devices can provide lifesaving shocks, detect abnormal heartbeats, and communicate data to doctors and nurses, ensuring prompt action and perhaps saving lives.
j) Healthcare cost reduction and efficiency: Wearable and implantable medical technology can improve the effectiveness of the healthcare system by lowering readmission rates to hospitals, cutting down on pointless visits, and enabling better patient management [27]. In order to offer proactive, individualized, and data-driven healthcare solutions, portable medical implants are required.

7.3 SECURITY RISKS AND THREATS IN MEDICAL DEVICES

As healthcare becomes increasingly interconnected and dependent on technology, security concerns and attacks in medical equipment have grown to be a significant concern. These dangers could have detrimental effects, from compromising patient privacy to possible injury from interfering with health care equipment [28–30] (Figure 7.2). Some of the security problems related to medical devices are given below:

a) A breach of data: Sensitive patient data is frequently transmitted and collected by medical devices. Data breaches caused by insufficient safety protocols can jeopardize patient privacy and expose sensitive personal health data to criminals.
b) Unauthorized and illegal access: Medical equipment that is linked to networks or internet connections is susceptible to intrusion. When hackers gain unauthorized access to a system, they may alter its operation, take confidential medical information, or even introduce malicious code or viruses.
c) Viruses and malware: Malware or viruses can interfere with a medical device's normal operation, possibly causing inaccurate diagnoses, inappropriate treatments, or even failure of the device. Within a hospital network, malicious malware might also spread to further devices.
d) Ransomware: Attacks using ransomware encrypt data or functionality on a device and hold it hostage until the ransom is paid [31]. If a medical gadget is impacted, it might be impossible to access crucial patient data or carry out lifesaving operations until the ransom charge is paid.

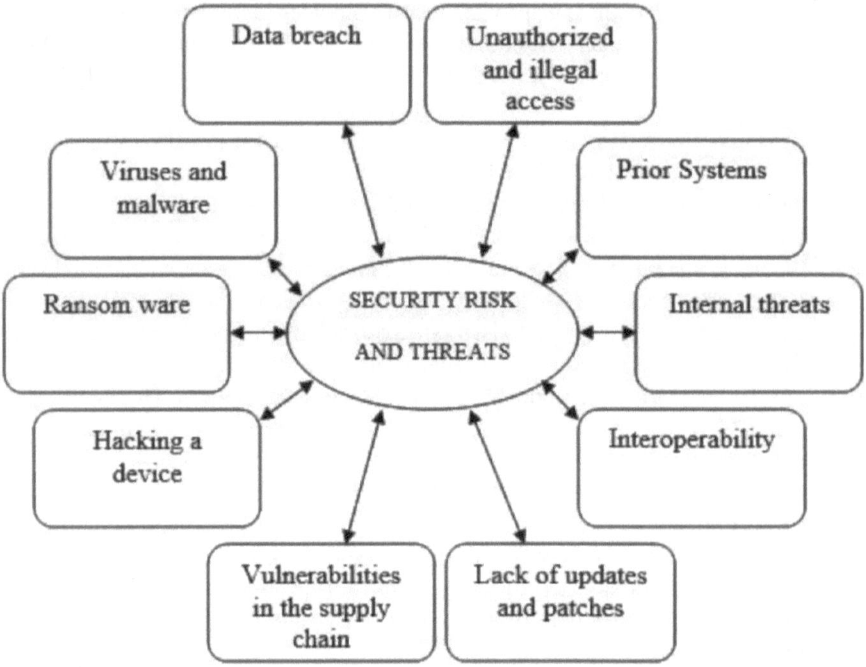

FIGURE 7.2 Security risks and threats in implantable medical devices.

e) Hacking a device: Malicious actors may actively tamper with medical equipment to alter settings, influence operation, or jeopardize patient safety. This can result in improper medicine dosages, modifications to the treatment plan, or other negative effects.
f) Vulnerabilities in the supply chain: Supply chain issues might potentially result in security risks. Medical instruments themselves could be affected if parts utilized in those devices are infected with malicious software during production or distribution.
g) Lack of updates and patches: Medical equipment frequently uses software that could be vulnerable. Devices may continue to be vulnerable to known security concerns if producers do not regularly release security patches or if medical professionals do not install them right away.
h) Risks of integration and interoperability: Medical equipment is increasingly communicating with one another and other systems, which increases the risk of a security compromise by potentially affecting other devices in the network.
i) Internal threats: Intentional or unintentional misuse of medical equipment by healthcare staff could endanger the safety of patients and data security. Negligence, data theft, and unauthorized access are some examples of insider dangers [32].
j) Prior systems: Due to outdated technology or poor design practices, older medical devices could not have strong security protections. These devices may be more vulnerable to attacks and more difficult to adequately update or secure.

7.4 SECURE DESIGN APPROACHES

For wearable and implantable medical equipment to be secure and private, an extensive structure with secure design methodologies must be developed. Here is a well-organized framework using secure design techniques:

- Risk evaluation and specifications gathering: Identify any potential hazards, weaknesses, or threats connected to the wearable or implanted device. Regulatory laws and industry best practices should be utilized to determine security and privacy needs.
- Security by design: Include privacy issues from the beginning of the design process. Limit the acquisition of individual data to what is required for device operation. To ensure access, strong authentication methods like biometrics or multi-factor authorization should be utilized.
- Data-encrypted, secure communication protocols: Select novel services for sending information between the gadget and the outside systems if security has to be ensured. For both data in transit and at rest, encryption is used.
- Secure software development: It complies with accepted software development lifecycle (SDLC) processes and secure coding standards. Updating of libraries and software components frequently to fix security flaws is of prime concern. Rigorous vulnerability evaluations and penetration examinations to detect and address any potential security flaws have to be conducted enabled with the device's defenses against different physical attack types.
- User education and knowledge: It informs the users of potential danger, privacy options, and recommended procedures. Users should have access to settings for privacy and consent management so that they can manage data sharing. It offers reliable consent management systems that let consumers control their data preferences. Through exercises and simulations, the incident management plan is tested.
- Regulatory compliance: Observe how regulatory requirements change and adjust the framework as necessary. Continuously monitor the performance and security of your devices, and update or patch them as necessary. Utilize proactive monitoring and intelligence on threats to deal with developing risks.
- Integration with third parties and app security: Before being integrated with the device, third-party apps as well as services should be checked for security and privacy. Impose security rules and secure code standards on third-party developers.

Manufacturers may make sure that their products meet the highest possible levels of security and privacy by including these safe design principles at every stage of the development and lifecycle of wearable or implantable health devices [33,34]. This framework preserves patient health information while also advancing ethical and responsible medical technologies.

7.5 KEY STANDARDS USED IN THE DESIGN OF WEARABLE AND IMPLANTABLE MEDICAL DEVICES

Adherence to numerous standards is necessary when designing wearable and implanted medical devices to guarantee their safety, efficacy, and compatibility. These standards include instructions for the design, production, evaluation, and compliance with regulations of devices. The standards for a system of quality control particular to the medical equipment business are outlined in the ISO 13485 standard. To guarantee constant product quality and legal compliance, it covers a variety of device planning, development, manufacturing, and post-market operations. Medical device risk management is applied to medical devices according to ISO 14971. This standard serves as a roadmap for manufacturers as they identify, evaluate, and reduce hazards related to medical devices at every stage of their lifecycle. In order to guarantee the security and efficiency of devices, it emphasizes the significance of risk management [35]. Although it is not standard, the U.S. Food and Drug Administration (FDA) advisory explains the design control requirements that manufacturers of medical products should follow. It offers an organized approach to the process of creating and developing them to create safe and effective gadgets. The ISO 14155 standard specifies acceptable clinical procedures for the clinical testing of medical equipment for human subjects. When wearable and implantable technologies undergo clinical trials, the standards and prerequisites for appropriate clinical practices are outlined, including guidelines for conduct, monitoring, reporting outcomes, and study design. IEEE 11073 acts as medical device communication standard for health informatics.

The seamless integration of medical devices and their data transfer with medical information systems of some standards are depicted in Figure 7.3. It enables smooth communication between implantable and wearable medical devices as well as other medical technology. Device specialization on glucose meters with ISO/IEEE 11073–10404 entirely depends on Health Informatics—Personal Health Device Communication addresses the communication needs for glucose meters and

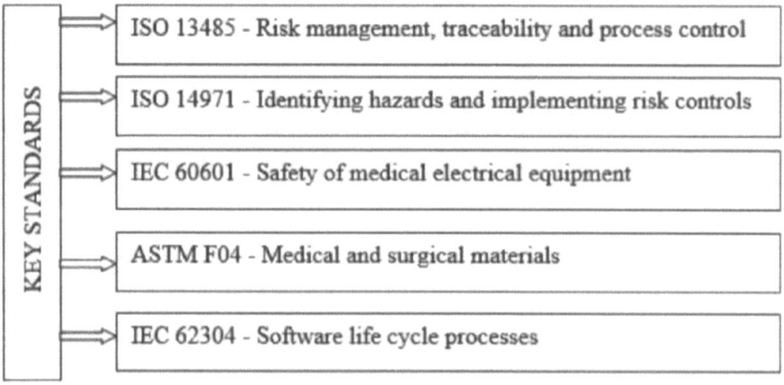

FIGURE 7.3 Key standards used in implantable medical devices.

other diabetes-related devices. This particular standard from the IEEE 11073 series ensures reliable data interchange in this area. For network-connected equipment, including medical devices, UL 2900 provides cybersecurity guidelines. To safeguard patient data and device operation, it aids manufacturers in evaluating and addressing potential cybersecurity issues [36]. The creation of safe, efficient, and legal medical devices that satisfy the requirements of both patients and medical professionals is made possible by observing these guidelines.

7.5.1 Authentication and Access Control

The safety and confidentiality of user data as well as the operation of wearable devices depend heavily on authentication and access control. To prevent unauthorized access and data breaches, it is critical to develop strong authentication systems and access restrictions as fashionable gadgets become increasingly integrated into everyday life and healthcare [37,38].

- Authentication is the procedure of confirming a user's or device's identity before allowing access to particular features or data. Following are a few authentication techniques frequently used in wearable technology:
  - To unlock and use the wearable gadget, users must enter a login or personal authentication number. Although straightforward, this approach is susceptible to speculation or social engineering.
  - Modern wearables frequently include biometric identification features like heart rate variability analysis, fingerprint recognition, and facial recognition. These techniques offer a more practical and safe way to confirm a user's identification.
  - Multi-factor authentication (MFA) integrates two or more identification elements, such as something the user is (their fingerprint), something they have (their smartphone), and something they know (a password). This increases security by another level.
  - Wearable gadgets can make use of proximity-based authentication, in which case a smartphone or other reliable gadget verifies the user's presence and authorizes access to the gadget [39].
- Access control refers to the process of controlling which characteristics, functions, and data are available to distinct users or roles in the system. Here is an example of how control of access is used in wearable technology:
  - Various user roles are established, each with different levels of access to data and device features. For instance, a healthcare expert may have access to settings that have become more sophisticated than those of a casual user.
  - Permission-based users must authorize particular rights to programs or features on the device they are wearing to gain access. For instance, access to the device's senses and data may be required by a fitness app.
  - Remote device locking and wiping capabilities can assist in preventing unauthorized access to a wearable device and its data if it is lost or stolen.

Data on the device that is worn should be encrypted before being stored or sent to other systems. This makes sure that even in the event of unauthorized access, the data will still be secure and unreadable. A few wearable users establish time restrictions for the use of particular features or data access [40]. For example, a user could grant a physician with temporary access for monitoring purposes. In some circumstances, this aids in preventing unauthorized access or use. Personal profiles and host modes are features that wearable devices can provide, enabling different people to have customized preferences and access permissions. Usability should be taken into consideration while designing identification and access control to prevent user experience degradation. Manufacturers of wearable technology should also release firmware and software updates regularly to fix security flaws and enhance overall security posture. Effective authorization and access management mechanisms are necessary to retain user confidence and protect user privacy as wearable devices manage critical health data and link to numerous networks.

7.5.2 Data Protection and Encryption

In order to guarantee the confidentiality and safety of personally identifiable information stored and transferred by wearable devices, data protection and encrypting are crucial components. Strong data protection mechanisms must be put in place since wearable devices manage user preferences, personal health data, and potentially sensitive communications [41–43]. This is done to avoid unauthorized access, data breaches, and additional security issues. The following is an explanation of data encryption and protection in wearable technology:

- Protecting data from unauthorized access, alterations, or disclosure is known as data protection. Data protection in the setting of wearable technology includes several tactics and procedures to guarantee the privacy, availability, and accuracy of user data, including:
  - Access controls use usernames and passwords, fingerprints, or MFA techniques to make sure that only those with authorization can access the device and its contents.
  - Client authorization requires consumers to provide particular permissions to wearable device applications or services, limiting the data that each program can access. Ensure that the gadget boots up with trustworthy and validated software components to avoid the installation of unauthorized or malicious applications.
  - To prevent unauthorized access in the event if a wearable device is lost or stolen, encrypt data saved on the internal hard drive or external memory of the device. Giving consumers the option to remotely secure or delete their devices in the event of loss or theft prevents unauthorized access to the data on the device.
  - Data minimization reduces the risk of data breaches by collecting and storing just the data required for the device's specified operation. Maintain the wearable device's firmware and operating system up to date regularly to fix security flaws and improve data security.

- Data is encrypted when it is changed into a fragmented form that can only be unlocked using a key for decryption. As a result, even if unauthorized parties obtain the encrypted material, they will be unable to decrypt it without the right key [44]. Data is encrypted both when it is at rest (on the device) and while it is in transit between the device and the systems:
  - Encrypt communication between the device being worn and external systems (e.g., mobile apps or cloud servers) using end-to-end encryption. This guarantees that data is transmitted privately and securely.
  - To guarantee the security of encrypted data, use strong encryption techniques like Advanced Encryption Standard (AES), which implement methods to identify any unauthorized alterations to encrypted data. This makes it easier to verify that the data wasn't altered during storage or transport.
  - To prevent unauthorized access to encrypted data, properly handle encryption keys, including their development, storage, distribution, revocation, and rotation.

Data shared between the wearable gadget and external systems can be encrypted by using safe communication protocols, like Transport Layer Security (TLS), which are used in this instance. Hardware-based encryption mechanisms are utilized where they are available because they can offer more security and performance than software-based ones.

7.5.3 Secure Communication Protocols

The integrity, privacy, and data transmitted between wearable devices and external systems, such as mobile devices, cloud servers, or healthcare providers, depend on secure communication protocols in wearable devices [45]. To prevent unauthorized access, interceptions, and manipulation of sensitive information, these protocols employ encryption and other safety features. In wearable technology, secure communication protocols are described as follows:

a) TLS: TLS is a popular technology that guarantees safe communication over networks like the Internet. It creates an encrypted connection to stop data modification and eavesdropping between the wearable gadget and a server or other device. Digital certificates are used by TLS to confirm the legitimacy of the server and guarantee that data is transferred to the appropriate receiver.
b) Datagram Transport Layer Security (DTLS): This TLS variant was created primarily for real-time or low-latency communication applications, such as wearable technology. It has comparable encryption and security capabilities to TLS but is tailored for datagram (packet) transmission, making it appropriate for situations like streaming information from sensors from wearable.
c) Wireless Secure Communications: For handheld gadgets that use Bluetooth to communicate, the device's Bluetooth Secure Communications feature

improves security by creating secure connections between devices using elliptic curve encryption. This shields the information sent between wearable and linked devices like smartphones from eavesdropping and unauthorized access [46].

d) Zigbee Security: Zigbee is a wireless transmission protocol that is frequently used in home automation systems and wearable technology. To enable safe communication between devices within a Zigbee network, it provides several security capabilities, including the use of encryption, user authentication, and data integrity checks.
e) Secure Message Queuing Telemetry Transport (MQTT): MQTT is a lightweight communication technology that is frequently used for Internet and wearable device connectivity. This variant includes encryption and authentication procedures to assure security, making it appropriate for data transmission between devices and platforms in the cloud.
f) CoAP (Constrained Application Protocol) with DTLS: CoAP was created for devices with restricted computational resources, making it appropriate for wearable technology. CoAP offers authentication and encryption for IoT and wearable devices when combined with DTLS, enabling secure connection.
g) A group of technologies called IPSec (Internet Protocol Security) are used to protect IP connections. IPSec is a networking protocol that is frequently used, but it can also be modified to allow for secure communication among wearable technology and additional systems, especially when those other systems are a component of larger networks.
h) OpenID Connect: Although they are not communication protocols, OpenID Connect (for authentication) is frequently used when combined with safe communication protocols to manage access and the authentication process for wearable devices accessing third-party services and application programming interfaces (APIs).

To implement secure communication methods in wearable devices, it is necessary to carefully choose the best protocol for the device's intended use case, configure encrypted parameters, manage keys, and make sure that the device receives regular updates to fix security flaws.

7.5.4 Vulnerability Assessment and Penetration

To find and fix any flaws or vulnerabilities in wearable devices, penetration testing and vulnerability assessment are essential security procedures [47]. These procedures assist designers, programmers, and security experts in proactively identifying security holes and weaknesses in the hardware, software, and infrastructure of the devices. An explanation of vulnerability evaluation and penetration testing in technological devices is provided below:

- The process of vulnerability assessment is methodically locating potential weak points, security holes, and vulnerabilities in wearable technology.

It checks wearable gear for known vulnerabilities, out-of-date software, configuration errors, and lax security settings using automated tools and procedures. To make sure that nothing is missed during the assessment, a thorough inventory of all wearable device parts, software, and systems is given [48]. Risk avoidance is followed by evaluating the potential consequences and likelihood of exposure for detected vulnerabilities to determine which of them should be fixed first. Vulnerabilities are categorized according to their types, such as software bugs, unsafe setups, shoddy authentication, or holes in physical protection.

- Penetration testing simulates actual attacks to evaluate the security features of wearable technology. It outlines the penetration test's objectives, methodology, and attack types that will be replicated. It also defines the penetration test's scope, collecting data about the architecture, software, network connections, and possible entry points for attackers of the wearable gadget. It makes an effort to take advantage of flaws discovered during the vulnerability evaluation to obtain access to or control over the devices. The potential for an attacker to get access to and steal private information from the wearable device is simulated. It assesses the degree that an attacker can keep control of the device, switch to other structures, or consistently undermine the security of the device [49]. By ensuring that wearable technology is built with strong security features, these practices help to lower the risk of hacking, unauthorized access, and other incidents related to security. To further improve the overall safety and security of wearable technology, regulatory frameworks frequently call for these security evaluations to be performed as part of the device's design and certification processes.

7.5.5 Testing Incident Response and Management

A critical component of verifying the security and efficacy of a device's reaction to security breaches, breaches of data, and other incidents is testing response in wearable devices. To assess the readiness, efficacy, and efficiency of a firm's incident response strategy, incident response testing simulates a variety of security incidents. These situations can involve device theft physically, unauthorized access, malware infections, and data breaches. The existing emergency preparedness plan is examined to make sure it is thorough, current, and specifically suited to the vulnerabilities and risks related to wearable technology [50]. Test objectives specify specific goals for the testing, such as quantifying the time needed to identify and address events, gauging the efficiency of communication channels, and also the coordination of reaction teams.

Security event simulations based on pre-established scenarios could entail testing the device's security features and the response strategy using simulated attacks, dummy data, or controlled breaches. Response execution enacts the incident response strategy as though an actual occurrence had taken place. This involves alerting the appropriate parties, isolating the impacted systems, gathering information, and taking the appropriate preventative measures. Communication assesses how successfully customers, stakeholders, partners, and regulatory authorities are communicated within the incident response team. Communication needs to be prompt, precise, and consistent. Every step of the procedure for responding to an incident must be documented.

Evaluation measures how well the incident response strategy was carried out. In addition to assisting organizations in detecting and responding to security problems effectively, testing response to incidents and management in wearable devices indicates readiness to regulators, clients, and stakeholders [51]. Conducting emergency response testing regularly makes ensuring that wearable technology is prepared to address security problems and lessen their effect on customer information, device operation, and user confidence.

7.5.6 User Education and Awareness

In the context of wearable technology, user awareness, the fact of education, refers to the act of giving users the information and abilities they need to operate their devices efficiently, safely, and ethically. Through education, users will be better equipped to comprehend the features, capabilities, potential risks, and recommended practices related to their wearable technology. Manufacturers may improve privacy, device security, and user experience by educating users. The characteristics, operations, and capacities of wearable technology should be explained to users. This includes being able to use the device's features, navigate its user interface, and set it up. Users can provide their data with confidence to this education, which also allays privacy worries. Regular updates frequently include security updates that fix flaws [52].

Users need to be informed about typical cybercriminal strategies, such as misleading emails and text messages. They ought to understand how to spot and reject dubious communications. They need to exercise caution while providing unneeded personal information to third-party applications. Users should be aware of what to do if their wearable is stolen or misplaced. To stop unauthorized access, the gadget might be remotely locked or erased.

Users of wearable technology should be conscious of the settings related to confidentiality that are accessible to them and how to modify them to suit their preferences [53]. From the moment a user purchases a wearable gadget until the end of its life cycle, user awareness and education should be a continuous endeavor. To support customer education and foster a high level of privacy and safety awareness among wearable device users, manufacturers should offer concise and straightforward user documentation, intuitive user interfaces, and regular updates.

7.6 ETHICAL CONSIDERATIONS IN MEDICAL DEVICES

To make sure that patient safety and privacy are prioritized during the design, development, and use of medical devices, ethical concerns are essential. The invention, use, and ongoing management of medical devices are all governed by ethical considerations since they have the potential to have a substantial impact on patients' lives. The following are some crucial medical device ethics to keep in mind:

- The primary consideration in the construction and usage of medical equipment should be patient safety. Devices should undergo extensive testing to make

sure they benefit patients without doing any damage. Before giving their consent to use a medical device, individuals have the opportunity to be fully aware of the risks, advantages, and possible outcomes of doing so [54,55]. Patients' decisions are respected, and autonomy is ensured by informed consent.
- Strong privacy and data security protections must be implemented in healthcare equipment that collect and send patient data. To avoid unauthorized access or breaches, patient health information must be protected.

Accessibility should be taken into account while designing medical equipment so that everybody, irrespective of their physical or mental abilities, can utilize and benefit from them.

Manufacturers and developers ought to be open and honest about the capabilities, potential restrictions, and potential conflicts of interest of their products [56–58]. Accountability guarantees that users of the equipment accept responsibility for its results.

- All patients, irrespective of socioeconomic class, color, ethnicity, or region, should have access to medical devices. Ensuring equal access reduces healthcare inequities.
- Regulations and policies established by the appropriate health authorities must be followed by medical equipment. Devices must comply with basic standards for safety and quality to be compliant.
- Even after medical equipment is put on the market, the manufacturer should keep an eye on its performance. A dedication to the well-being of patients can be seen in the timely identification and resolution of problems.
- Manufacturers should make every effort to reduce any potential risks or damages related to the use of healthcare devices, including any hazards that may result from unintended consequences or design faults. Healthcare practitioners should uphold the highest standards of ethics and prioritize the needs of their patients when designing, developing, prescribing, or administering medical devices.

Through study, creative thinking, and user input, manufacturers should constantly look for ways to increase the security, efficacy, and usefulness of their medical equipment. Legal, societal, and professional obligations are linked with moral issues in medical devices [59]. To guarantee that medical devices favorably impact patient care and general well-being as technology develops, everyone in the medical device ecosystem, including makers, medical professionals, regulators, and patients, must work together to address these ethical concerns.

7.7 SECURITY AND PRIVACY—A KEY DEVELOPMENT

Specifically in the context of quickly advancing technology and digital transformation, security and privacy are crucial development considerations in many different industries. The importance of ensuring strong security measures and safeguarding people's privacy has increased as hacking, cyberattacks, and unauthorized access to personal data are happening more frequently (Table 7.1).

TABLE 7.1
Various Mechanisms for Security and Privacy in Wearable Implantable Medical Devices

Attacks	Schemes	Description	Limitation with Future Works
Traffic Analysis, Eavesdropping, Sniffing, Compromised Node, Data Breach	Schemes for physical layer security	It depends on exploiting communication channel characteristics to secure the edge network.	Improve Internet of Medical Things (IoMT) security performance by implementing intelligent reflective surface (IRS) techniques.
Message Disclosure, Eavesdropping, Sniffing	Data combination	To safeguard sensitive patient data, deploy secure and private data aggregation from various distributed medical equipment.	There are still unanswered questions in the healthcare industry, particularly in federated learning, data quality, model precision, incentive mechanisms, and so on.
Sniffing, Data Breach, Denial-of-Service (DoS), Tampering, Man-in-the-Middle (MitM)	Blockchain	Utilizing blockchain technology in the healthcare industry to provide features like decentralized recording and sharing with any third party.	Because of its high level of innovation, blockchain has to be enhanced in certain areas, including storage, interoperability, data sharing, and scalability.
MitM, Password Intrusion, Weak Authentication, Side Channel	Biometric-based mutual authentication protocols	This method aims to securely verify patients and medical professionals.	Cost-effective adoption with improved performance and efficiency and adaptation to mobile healthcare contexts.
Routing Attack, Traffic Analysis, Sniffing, DoS, Impersonation	Methodical key management	Used to safeguard data exchanges made in the healthcare context	• Complexity and high power consumption of traditional methods • Need for adaptation to newly emerging scenarios and technology
Routing Attack, Battery, Depletion, Wormhole, Flooding, Gray Hole	Mechanisms for routing like Software-Defined Network (SDN), congestion, and the like	Use of secure collection and routing techniques to reduce communication expenses and transmission costs.	• Using these routing techniques in mobile healthcare situations • Enhanced security and energy efficiency
Eavesdropping, Replying Attack, Tampering	Proxy-based	Medical device security measures placed above the devices.	The necessity of increasing efficiency and decreasing expenses thus increasing the number of security and privacy-enhancing measures.
Malware, Compromised Node, Sensor Spoofing, Ransomware, MitM, Sybil	Anomaly detection	Using various machine learning methods, such as Support Vector Machines and Random Forest to identify unusual behavior.	Ongoing research and development into real-time, autonomic detection of abnormal behaviors in healthcare settings.

- Data leaks and security breaches reduce consumer confidence. Trust among users can be fostered by creating safe systems and preserving privacy, which is essential for the survival and expansion of businesses [60].
- Expensive fines, court fees, and reputational harm. These potential expenses can be avoided by organizations by investing in security measures.
- Cybersecurity threats must be continuously monitored since they are ever-evolving. Strong security protocols must be incorporated into development processes to protect against vulnerabilities and assaults.
- As wearable technology, IoT, and healthcare innovations evolve, it is essential to protect patient and medical data from unauthorized access and other risks.
- Privacy by style and focused on users' principles of design makes sure that safety and confidentiality considerations are built into the development process from the beginning rather than being introduced as an afterthought.
- Businesses are implementing digital technology to improve user experiences and streamline operations. To prevent exposing private information throughout this shift, security and privacy issues must be addressed.
- As cloud-based systems and remote work become more popular, it has become increasingly difficult to secure data and communications. To safeguard against breaches of information and online dangers, robust safety safeguards are necessary [61].
- Companies that gather user data for a variety of objectives, including monetization, must strike a balance between the ethical use of the data and the rights of individuals to privacy. Data practices that are transparent are crucial.
- While innovations like edge computing, blockchain, and artificial intelligence (AI) have the potential to alter industries, they also pose new security risks. It is essential to provide secure and moral implementations.
- To lessen the effects of security breaches, development plans should incorporate well-defined incident response protocols. Resilience-building techniques aid organizations in quickly recovering from such catastrophes.

7.8 FUTURE TRENDS AND CHALLENGES

Wearable device security and privacy in the future will face both huge hurdles and great improvements. Wearable technology is influencing a variety of industries, including consumer electronics and healthcare, as it continues to advance in technology. Advanced biometric identification techniques, such as vein pattern recognition, gait analysis, and even brainwave authentication, will likely be incorporated into wearable technology to improve user identity verification [62]. To protect data at several levels, wearable devices will use multi-layered security strategies that include hardware security modules, secure enclaves, and cutting-edge encryption algorithms. Privacy laws will require more stringent data protection procedures as wearable technology collects more health-related information. The trend toward zero-trust security, in which people and gadgets are never taken at their word, will be followed by wearable technology. The important elements will be continuous authentication, behavior analysis, and risk assessment.

To improve privacy, wearables may use federated learning, a method that trains machine learning models directly on user devices while keeping data localized [63,64]. By enabling users to manage who has access to their data, blockchain technology may be used to secure the integrity of health-related data obtained through wearable technologies. To better protect against physical attacks, many wearable will incorporate secure components or tamper-resistant chips to store critical data. Wearable technology risks the exposure of personal medical information since it collects and transmits sensitive data, making it a tempting target for hackers and data breaches. A bigger attack surface is created by wearable's frequent connections to smartphones, cloud services, and other devices. Security lapses can result from flaws in any connected component. Risks might arise from the security of third-party apps that interface with wearables. Best practices for app development must be followed; however, it might be difficult to guarantee that each software upholds strict security requirements.

Users need to be informed about possible security threats and the best ways to use wearable technology. It is essential to make sure consumers are aware of how to adjust privacy preferences and spot phishing scams. It can be difficult for manufacturers to make sure that wearable gadgets receive security updates and fixes during their entire lifecycle because they can be used for several years [65,66]. It can be challenging to implement strong security measures while keeping wearable devices' energy economy and battery life. Wearable technology has a bright future, but solving security and privacy issues is essential to ethically reaping these benefits. In order to address these issues and trends, manufacturers, developers, regulators, and users must cooperate. This will help to maintain the privacy and security of wearable technology users.

7.9 CONCLUSION

Wearable medical device and IMD security and privacy are at the forefront of both technological development and ethical considerations. These gadgets have the potential to revolutionize healthcare, improve patient monitoring, and enhance general well-being. These possibilities come with important obligations, though, to guarantee the security, privacy, and moral handling of private health information. The development of implanted and wearable technology calls for a comprehensive approach to security and privacy. To solve the complex problems these technologies present, manufacturers, developers, healthcare professionals, regulatory agencies, and users must work together. It takes constant dedication to best practices, developing technology, and legal compliance to strike a balance between innovations and protecting patient information. The importance of strong authentication systems, secure communication protocols, data encryption, and vulnerability assessments cannot be stressed as the interconnected healthcare ecosystem keeps growing. In addition, user education and awareness are crucial for enabling people to make wise choices regarding their health data and device usage.

Exciting technologies like better biometrics, multi-layered security, and blockchain integration are predicted for the future and will help wearable and implantable devices be more secure and private. These breakthroughs will, however, also be accompanied by enduring difficulties, such as regulatory complications, data breaches, and the requirement for ongoing updates to handle new risks. An ethical commitment is still crucial in this quickly changing environment. The extent to which wearable medical devices and IMDs improve patient care, boost medical research, and enhance individual well-being will ultimately depend on how successfully security and privacy features are incorporated into these devices. We can create a future where wearable and implantable devices not only improve health outcomes but also sustain the trust and confidence of patients and stakeholders by giving priority to these factors.

REFERENCES

[1] Wahid Khan, Eameema Muntimadugu, Michael Jaffe, and Abraham J. Domb (2014), Implantable medical devices. In *Focal Controlled Drug Delivery*, pp. 33–59. IEEE.

[2] Naeem Abbas, Abdul Basir, Amjad Iqbal, Muhammad Yousaf, Adeel Akram, and Hyoungsuk Yoo (2022), Ultra-miniaturized antenna for deeply implanted biomedical devices. *IEEE Access*, vol. 10, pp. 54563–54571.

[3] Abdulatif Alabdulatif, Ibrahim Khalil, and Mohammad Saidur Rahman (2022), Security of blockchain and AI-empowered smart healthcare: Application-based analysis. *Applied Sciences*, vol. 12, no. 21, p. 11039.

[4] J. C. Tellez Gaytan (2022), A literature survey of security and privacy issues in internet of medical things. *International Journal of Computations, Information and Manufacturing (IJCIM)*, vol. 2, no. 2.

[5] Vincent Omollo Nyangaresi (2023), Privacy preserving three-factor authentication protocol for secure message forwarding in wireless body area networks. *Ad Hoc Networks*, vol. 142, no. 103–117.

[6] Lubna Hashlamoun (2022), *Quality of Service Position Based Routing Protocol for Wireless Body Area Networks.* Palestine Polytechnic University.

[7] Lisha Zhong, Shuling He, Jinzhao Lin, Jia Wu, Xi Li, Yu Pang, and Zhangyong Li (2022), Technological requirements and challenges in wireless body area networks for health monitoring: A comprehensive survey. *Sensors*, vol. 22, no. 9, p. 3539.

[8] Yasmeen Al-Saeed, Eman Eldaydamony, Ahmed Atwan, Mohammed Elmogy, and Osama Ouda (2021), Efficient key agreement algorithm for wireless body area networks using reusable ECG-based features. *Electronics*, vol. 10, no. 4, p. 404.

[9] Longfei Wu and John Du (2023), Designing novel proxy-based access control scheme for implantable medical devices. *Computer Standards & Interfaces*, vol. 87, p. 103754.

[10] Sayon Duttagupta, Eduard Marin, Dave Singele, and Bart Preneel (2023), HAT: Secure and practical key establishment for implantable medical devices. In *Proceedings of the Thirteenth ACM Conference on Data and Application Security and Privacy*, pp. 213–224. ACM.

[11] Rajveer Bhaskar, Monika Ola, Subhash Singh Saharan, and Vishal Nikam (2023), Regulatory requirements for Medical Device Registration in Kingdom of Saudi Arabia, US and European Union. *International Journal of Drug Regulatory Affairs*, vol. 11, no. 2, pp. 75–85.

[12] Antonio Lopez Martinez, Manuel Gil Perez, and Antonio Ruiz-Martínez (2023), A comprehensive review of the state-of-the-art on security and privacy issues in healthcare, *ACM Computing Surveys*, vol. 55, no. 12, pp. 1–38.

[13] Ana Longras, Teresa Pereira, and Antonio Amaral (2023), Cybersecurity challenges in healthcare medical devices. In *International Conference on Internet of Everything*, pp. 66–75. Springer.

[14] Abhinanda Kar, Nadim Ahamad, Mahima Dewani, Lisha Awasthi, Runali Patil, and Rinti Banerjee (2022), Wearable and implantable devices for drug delivery: Applications and challenges. *Biomaterials*, vol. 283, p. 121435.
[15] Kourosh Kakhi, Roohallah Alizadehsani, H. M. Dipu Kabir, Abbas Khosravi, Saeid Nahavandi, and U. Rajendra Acharya (2022), The internet of medical things and artificial intelligence: Trends, challenges, and opportunities. *Biocybernetics and Biomedical Engineering*, vol. 42, no. 3, pp. 749–771.
[16] Shouvik Mukherjee, Shariq Suleman, Roberto Pilloton, Jagriti Narang, and Kirti Rani (2022), State of the art in smart portable, wearable, ingestible and implantable devices for health status monitoring and disease management, *Sensors*, vol. 22, no. 11, p. 4228.
[17] Daniel A. Domingo-Lopez, Giulia Lattanzi, Lucien H. J. Schreiber, Eimear J. Wallace, Robert Wylie, Janice Sullivan, Eimear B. Dolan, and Garry P. Duffy (2022), Medical devices, smart drug delivery, wearables and technology for the treatment of Diabetes Mellitus. *Advanced Drug Delivery Reviews*, vol. 185, p. 114280.
[18] Shruti Suhas Kute, Amit Kumar Tyagi, and S. U. Aswathy (2022), Industry 4.0 challenges in e-healthcare applications and emerging technologies. In *Intelligent Interactive Multimedia Systems for e-Healthcare Applications*, pp. 265–290. Springer.
[19] Kristine M. Alpi, Christie L. Martin, Joseph M. Plasek, Scott Sittig, Catherine Arnott Smith, Elizabeth V. Weinfurter, Jennifer K. Wells, Rachel Wong, and Robin R. Austin (2023), Characterizing terminology applied by authors and database producers to informatics literature on consumer engagement with wearable devices. *Journal of the American Medical Informatics Association*, vol. 30, no. 7, pp. 1284–1292.
[20] Ravishankar S. Ulle (2023), Challenges and opportunities for IoT deployment in India's healthcare sector. *Computational Intelligence for Clinical Diagnosis*, pp. 561–571.
[21] Wesley Mcloughlin and Ian Mcloughlin (2023), Wearable and implantable medical devices: A fantastic voyage, In *Medical Innovation*, pp. 195–206. CRC Press.
[22] Zhuofu Liu, Vincenzo Cascioli, and Peter W. Mccarthy (2023), Healthcare monitoring using Low-Cost sensors to supplement and replace human sensation: Does it have potential to increase independent living and prevent disease? *Sensors*, vol. 23, no. 4, p. 2139.
[23] Minhong Tan, Yang Xu, Ziqi Gao, Tiejun Yuan, Qingjun Liu, Rusen Yang, Bin Zhang, and Lihua Peng (2022), Recent advances in intelligent wearable medical devices integrating biosensing and drug delivery. *Advanced Materials*, vol. 34, no. 27, p. 2108491.
[24] Lida Kouhalvandi, Ladislau Matekovits, and Ildiko Peter (2022), Magic of 5G technology and optimization methods applied to biomedical devices: A survey. *Applied Sciences*, vol. 12, no. 14, p. 7096.
[25] Vladimir Vakhter, Betul Soysal, Patrick Schaumont, and Ulkuhan Guler (2022), Threat modeling and risk analysis for miniaturized wireless biomedical devices. *IEEE Internet of Things Journal*, vol. 9, no. 15, pp. 13338–13352
[26] Zankhana Patel, Yash Velankar, Chandan Trivedi, and Parita Oza (2022), Wireless implantable medical devices security and privacy: A survey. *Smart Energy and Advancement in Power Technologies: Select Proceedings of ICSEAPT 2021*, vol. 2, pp. 69–87.
[27] W. A. E. L. Toghuj and N. I. D. A. L. Turab (2022), A survey on security threats in the internet of medical things (IoMT). *Journal of Theoretical and Applied Information Technology*, vol. 100, no. 10, pp. 3361–3371.
[28] A. Ashwini and S. Murugan (2020), Automatic skin tumour segmentation using prioritized patch based region–A novel comparative technique. *IETE Journal of Research*, vol. 66, pp. 1–12.

[29] Mohammad Kamrul Hasan, Taher M. Ghazal, Rashid A. Saeed, Bishwajeet Pandey, Hardik Gohel, Ala'A. Eshmawi, Sayed Abdel-Khalek, and Hula Mahmoud Alkhassawneh (2022), A review on security threats, vulnerabilities, and counter measures of 5G enabled Internet-of-Medical-Things. *IET Communications*, vol. 16, no. 5, pp. 421–432.

[30] F. Alsubaei and S. Shiva (2017), Security and privacy in the internet of medical things: Taxonomy and risk assessment. In *2017 IEEE 42nd Conference on Local Computer Networks Workshops (LCN Workshops)*, pp. 112–120. IEEE.

[31] H. M. Alzoubi (2022), Digital transformation and SMART-the analytics factor. In *2022 International Conference on Business Analytics for Technology and Security*, ICBATS, pp. 1–11.

[32] H. M. Alzoubi, A. Ali, A. W. Septyanto, I. Chaudhary, H. A. Hamadi, and Z. F. Khan (2022), Applied artificial intelligence as event horizon of cyber security. In *2022 International Conference on Business Analytics for Technology and Security*, pp. 1–7. IEEE.

[33] H. M. Alzoubi and R. Yanamandra (2020), Investigating the mediating role of information sharing strategy on agile supply chain in supply chain performance. *Uncertain Supply Chain Management*, vol. 8, no. 2, pp. 273–284.

[34] A. Ashwini, K. E. Purushothaman, Banu Priya, Prathaban M. Jenath, and R. Prasanna (2023), Automatic traffic sign board detection from camera images using deep learning and binarization search algorithm. In *2023 International Conference in Recent Advances in Electrical, Electronics, Ubiquitous Communication and Computational Intelligence (RAEEUCCI)*. IEEE.

[35] H. M. Alzoubi, J. R. Hanaysha, M. E. Al-Shaikh, and S. Joghee (2022), Impact of innovation capabilities on business sustainability in small and medium enterprises. *FIIB Business Review*, vol. 11, no. 1, pp. 67–78.

[36] M. El Khatib, A. Al Mulla, and W. Al Ketbi (2022), The role of blockchain in e-governance and decision-making in project and program management. *Advanced Internet of Things*, vol. 12, no. 3, pp. 88–109.

[37] H. M. Alzoubi, S. Joghee, and A. R. Dubey (2020), Decisions effectiveness of FDI investment biases at real estate industry: Empirical evidence from Dubai smart city projects. *International Journal of Scientific and Technology Research*, vol. 9, no. 3, pp. 3499–3503.

[38] A. K. Singh, A. Solanki, A. Nayyar, and B. Qureshi (2020), Elliptic curve signcryption-based mutual authentication protocol for smart cards. *Applied Sciences*, vol. 10, no. 22, p. 8291.

[39] H. M. Alzoubi, M. A. Nuaimi, and N. A. Alnazer (2017), Analyzing the appropriate cognitive styles and its effect on strategic innovation in Jordanian Universities. *International Journal of Business Excellence*, vol. 13, no. 1, pp. 127–140.

[40] S. R. Guntur, R. R. Gorrepati, and V. R. Dirisala (2018), Internet of Medical Things, *Medical Big Data and Internet of Medical Things*, vol. 4, no. 6, pp. 271–297.

[41] A. A. Kashif, B. Bakhtawar, A. Akhtar, S. Akhtar, N. Aziz, and M. S. Javeid (2021), Treatment response prediction in hepatitis c patients using machine learning techniques. *International Journal of Innovation and Technology Management*, vol. 1, no. 2, pp. 79–89.

[42] F. Ma, T. Sun, L. Liu, and H. Jing (2020), Detection and diagnosis of chronic kidney disease using deep learning-based heterogeneous modified artificial neural network. *Future Generation Computer Systems*, vol. 111, pp. 17–26.

[43] N. Alsharari (2021), Integrating blockchain technology with internet of things to efficiency. *International Journal of Innovation and Technology Management*, vol. 1, no. 2, pp. 1–13.

[44] A. Ashwini and V. Kavitha (2021), Automatic skin tumor detection using online tiger claw region based segmentation–A novel comparative technique. *IETE Journal of Research*, vol. 69, pp. 1–9.
[45] T. M. Ghazal (2022), AI-based prediction of capital structure: Performance comparison of ANN SVM and LR models. *Computational Intelligence and Neuroscience*, vol. 2022, pp. 1–13, 2022.
[46] G. M. Qasaimeh and H. E. Jaradeh (2022), The impact of artificial intelligence on the effective applying of cyber governance in Jordanian commercial banks. *International Journal of Innovation and Technology Management*, vol. 2, no. 1.
[47] H. M. Alzoubi, M. In'airat, and G. Ahmed (2022), Investigating the impact of total quality management practices and Six Sigma processes to enhance the quality and reduce the cost of quality: The case of Dubai. *International Journal of Business Excellence*, vol. 27, no. 1, pp. 94–109.
[48] A. K. Singh and S. Kumar (2023, June). Performance analysis of ECC-based security solutions for internet of medical things. In *International Conference on Data Analytics & Management*, pp. 337–347. Springer Nature Singapore.
[49] Asem Alzoubi (2022), Machine learning for intelligent energy consumption in smart homes. *International Journal of Computer Aided Manufacturing*, vol. 2, no. 1, p. 1.
[50] B. Amrani, A. Z., Urquia, and I. Vallespir (2022), Industry 4.0 technologies and lean production combination: A strategic methodology based on links quantification. *International Journal of Innovation and Technology Management (IJTIM)*, vol. 2, no. 2, pp. 33–51.
[51] T. M. Ghazal (2021), IoMT cloud-based intelligent prediction of breast cancer stages empowered with deep learning. *IEEE Access*, vol. 9, pp. 146478–146491.
[52] F. Del and G. Solfa (2022), Impacts of cyber security and supply chain risk on digital operations: Evidence from the UAE pharmaceutical industry Federico Del Giorgio Solfa. *International Journal of Innovation and Technology Management (IJTIM)*, vol. 2, no. 2, pp. 18–32.
[53] H. M. Alzoubi (2021), Modelling supply chain information collaboration empowered with machine learning technique. *Intelligent Automation and Soft Computing*, vol. 30, no. 1, pp. 243–257.
[54] R. Kumar and R. Tripathi (2021), Towards design and implementation of security and privacy framework for Internet of Medical Things (IoMT) by leveraging blockchain and IPFS technology. *The Journal of Supercomputing*, vol. 77, no. 8, pp. 7916–7955.
[55] H. M. Alzoubi and Y. Ramakrishna (2022), Empirical investigation of mediating role of six sigma approach in rationalizing the COQ in service organizations. *Operations and Supply Chain Management*, vol. 15, no. 1, pp. 122–135.
[56] Maged Farouk (2022), Studying human robot interaction and its characteristics. *International Journal of Computations, Information and Manufacturing*, vol. 2, no. 1, p. 1.
[57] H. M. Alzoubi, G. Ahmed, A. Al-Gasaymeh, and B. Al Kurdi (2020), Empirical study on sustainable supply chain strategies and its impact on competitive priorities: The mediating role of supply chain collaboration. *Management Science Letters*, vol. 10, no. 3, pp. 703–708.
[58] H. M. Alzoubi, H. Elrehail, J. R. Hanaysha, A. Al-Gasaymeh, and R. Al-Adaileh (2022), The role of supply chain integration and agile practices in improving lead time during the COVID-19 crisis. *The International Journal of Service Science, Management, Engineering, and Technology*, vol. 13, no. 1, pp. 1–11.
[59] Nada Ratkovic (2022), Improving home security using blockchain. *International Journal of Computations, Information and Manufacturing*, vol. 2, no. 1, pp. 1.

[60] M. Abdel-Basset, G. Manogaran, and M. Mohamed (2018), Internet of Things (IoT) and its impact on supply chain: A framework for building smart, secure and efficient systems. *Future Generation Computer Systems*, vol. 86, pp. 614–628.

[61] Edward Probir Mondol (2022), The role of VR games to minimize the obesity of video gamers. *International Journal of Computer Aided Manufacturing*, vol. 2, no. 1, p. 1.

[62] Saad Masood Butt (2022), Management and treatment of type 2 diabetes. *International Journal of Computations, Information and Manufacturing*, vol. 2, no. 1, pp. 1.

[63] S. F. Nasim, M. R. Ali, and U. Kulsoom (2022), Artificial intelligence incidents & ethics a narrative review. *International Journal of Technology, Innovation and Management*, vol. 2, no. 2, pp. 52–64.

[64] H. M. Alzoubi (2022), Securing smart cities using blockchain technology. In *2022 1st International Conference on AI in Cybersecurity*, pp. 1–4. IEEE.

[65] H. M. Alzoubi (2022), Fusion-based supply chain collaboration using machine learning techniques. *Intelligent Automation & Soft Computing*, vol. 31, no. 3, pp. 1671–1687.

[66] S. Akhtar, A. Bakhtawar, and B. Akhtar (2022), Extreme programming vs scrum: A comparison of agile models. *International Journal of Innovation, Management and Technology* (IJTIM), vol. 2, no. 2, pp. 80–96, 2022.

8 Unleashing the Leap toward Secured IoMT Environment in WBAN

A Comprehensive Review

Banu Priya Prathaban, R. Subash, M. Sathesh, and Ashwini A.

8.1 BACKGROUND

The Internet of Things (IoT) refers to the interconnected network of everyday physical objects and devices embedded with sensors, software, and connectivity capabilities. These objects can collect and exchange data with each other and with central systems over the internet, enabling them to make informed decisions and perform actions without direct human intervention [1]. The networking infrastructure allows IoT devices to communicate efficiently and securely. Security is critical to IoT due to the sensitive data collected and transmitted by devices. Encryption, authentication, authorization, and regular software updates are essential to safeguard IoT systems from cyber threats. Blockchain technology could enhance security and trust in IoT ecosystems by ensuring the integrity of data transactions [2].

The Internet of Medical Things (IoMT) is a subcategory of the IoT that specifically emphasizes the integration of medical devices, healthcare systems, and patient data through connected technology [3]. IoT is revolutionizing the healthcare industry by enhancing patient care, improving operational efficiency, and enabling better data-driven decision-making for healthcare providers [4]. Protecting patient data from unauthorized access and breaches is a primary concern due to the sensitive nature of health information. Ensuring compatibility between different IoMT devices and healthcare systems is crucial for seamless data exchange. IoMT devices need to be reliable and accurate, as medical decisions may be based on the data they provide [5]. Developing ethical guidelines and regulations for IoMT will help address concerns related to data privacy, patient consent, and responsible use of technology [6].

A Wireless Body Area Network (WBAN) is a specialized type of wireless network designed to connect wearable and implantable medical devices placed on or inside the human body [7]. WBANs enable real-time monitoring of physiological

DOI: 10.1201/9781032635101-10

data, allowing for continuous health tracking, early disease detection, and personalized healthcare [8]. These networks are a crucial component of the broader IoT ecosystem, focusing specifically on healthcare applications [9]. WBAN devices are designed to be small, lightweight, and unobtrusive, allowing them to be comfortably worn on the body or even implanted. Miniaturization is essential for user comfort and acceptance [10].

WBAN devices are equipped with various sensors to monitor physiological parameters such as heart rate, temperature, blood pressure, and oxygen levels. Actuators can also provide interventions, such as drug delivery [11]. WBAN devices communicate wirelessly using short-range technologies such as Bluetooth, Zigbee, or other low-power protocols. This allows seamless data transfer to a central system or healthcare provider. Data collected from WBAN devices is aggregated and analyzed to provide insights into the wearer's health status. Advanced data analytics and machine learning can identify patterns, anomalies, and potential health issues [12]. WBANs find applications in healthcare, including remote patient monitoring, disease management, assisted living, postoperative care, and early intervention [13]. Following are the major challenges and considerations in WBAN: WBAN devices need to be energy-efficient to extend battery life, especially when worn for prolonged periods. The human body can cause signal attenuation and interference, affecting the reliability of data transmission. Reliable communication protocols are crucial [14]. Medical data collected by WBAN devices is sensitive and requires robust security measures to protect against unauthorized access and breaches. Ensuring accurate and consistent data from wearable sensors is essential for making informed medical decisions [15].

8.2 IoMT SYSTEM ARCHITECTURE

The architecture of an IoMT system defines how various components, devices, and services interact to deliver connected healthcare solutions. IoMT system architecture is designed to seamlessly integrate medical devices, data processing, communication, and analytics, ensuring that healthcare providers can collect, analyze, and act upon patient data effectively while maintaining security and privacy [16]. Incorporating artificial intelligence (AI) capabilities at the edge will enable real-time data analysis and decision-making directly on wearable devices, reducing latency and enhancing responsiveness [17]. Blockchain technology might be integrated to improve data security, transparency, and patient control over their medical data [18]. Collaboration between healthcare institutions could involve federated learning, allowing AI models to be trained collectively without sharing raw patient data. Developing standardized communication protocols and data formats will improve interoperability between different IoMT devices, systems, and platforms. The IoMT architecture will increasingly integrate with telemedicine platforms, enabling remote patient monitoring and telehealth consultations [19]. An effective IoMT system architecture allows the seamless flow of data, insights, and interventions, contributing to improved patient care, early disease detection, and personalized treatment plans. As technology advances, IoMT architectures will evolve to accommodate new trends and challenges in the dynamic field of connected healthcare [20].

1. **Perception Layer:** This layer comprises medical devices and sensors that collect data from patients, their environment, and various healthcare parameters. These devices include wearable health trackers, smart medical sensors, implantable devices, and other monitoring equipment. The perception layer is responsible for capturing real-time health data and transmitting it to the subsequent layers for further processing.
2. **Communication Layer:** The communication layer facilitates data transmission between devices, sensors, and the central system. Various wireless communication protocols, such as Bluetooth, Wi-Fi, Zigbee, and cellular networks, establish connectivity. This layer ensures that data collected from medical devices can be reliably transferred to the data processing components.
3. **Data Processing Layer:** In this layer, collected data undergoes preprocessing, aggregation, and transformation. Edge computing involves processing tasks closer to the data source, reducing latency, and optimizing bandwidth usage. Cloud computing allows for extensive data analysis, storage, and access from various locations.
4. **Data Analytics and Insights Layer:** The data analytics layer processes the preprocessed data to extract meaningful insights. Advanced analytics, machine learning algorithms, and AI are applied to identify patient data patterns, anomalies, and trends. This layer is crucial in enabling early disease detection, predictive analytics, and personalized treatment recommendations.
5. **Integration Layer:** The integration layer connects the IoMT system with other healthcare systems and applications, such as electronic health records (EHR), hospital information systems (HIS), and telemedicine platforms. Integration ensures that relevant patient data is shared with healthcare professionals and accessible within existing healthcare workflows.
6. **Decision Support Layer:** The insights generated by the data analytics layer are used in the decision support layer to assist healthcare providers in making informed decisions. Clinical decision support systems (CDSS) can provide alerts, recommendations, and guidelines based on the analyzed patient data.
8. **User Interface Layer:** The user interface layer provides user-friendly interfaces for healthcare providers, patients, and caregivers to interact with the IoMT system. This can include web dashboards, mobile apps, and wearable device interfaces that display real-time health data, notifications, and insights [21].

8.3 SECURITY AND PRIVACY REQUIREMENTS FOR IOMT

As IoMT endures progress, it should ensure the security and privacy of sensitive healthcare data; this becomes paramount and important to society. IoMT systems collect and transmit personal health information, making them potential targets for cyberattacks and breaches [22]. To mitigate these risks and build trust in IoMT applications, robust security and privacy measures must be implemented. By prioritizing security and privacy, the healthcare industry can harness the benefits of IoMT while minimizing the risks associated with handling sensitive medical data [23]. The key security and privacy requirements for IoMT are the following:

1. **Data Encryption:** All sensitive data, including patient health information, must be encrypted during transmission and storage.
2. **Secure Device Identity and Onboarding:** IoMT devices should have unique identities and be securely onboarded onto the network.
3. **Regular Software Updates:** Firmware and software on IoMT devices should be regularly updated to patch vulnerabilities and address security flaws. Attackers can exploit outdated software.
4. **Data Minimization and Anonymization:** Only essential patient data should be collected, and data minimization principles should be followed to reduce the amount of sensitive information stored. Anonymization techniques can protect patient identities while retaining useful data for analysis.
5. **Data Retention Policies:** IoMT systems should have clear data retention policies, specifying how long data will be stored and when it will be securely deleted. Retaining data only for as long as necessary reduces the risk of data exposure.
6. **Regulatory Compliance:** IoMT systems must adhere to relevant regulations and standards, such as HIPAA (Health Insurance Portability and Accountability Act) in the United States or GDPR (General Data Protection Regulation) in the European Union. Noncompliance can lead to severe penalties.
7. **Incident Response Plan:** IoMT systems should have a well-defined incident response plan to address and mitigate the impact of security breaches [24]. Figure 8.1 depicts the security and privacy requirements for IoMT devices.

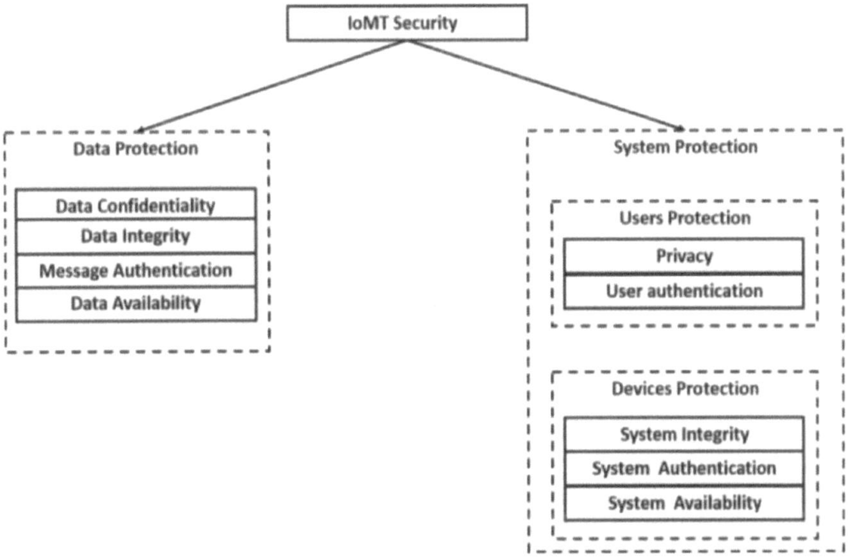

FIGURE 8.1 IoMT—security and privacy requirements.

8.4 WBAN COMMUNICATION ARCHITECTURE

WBANs are specialized networks that facilitate the seamless and wireless communication of medical sensors and devices placed on or inside the human body [25]. WBANs are critical in enabling real-time health monitoring, remote patient care, and personalized healthcare interventions. The communication architecture of a WBAN is designed to ensure reliable, low-power, and secure data transmission within the unique context of wearable and implantable medical devices [26].

WBANs consist of various sensors and devices that capture physiological data such as heart rate, temperature, and blood pressure. These sensors can be integrated into wearable, implantable, or even ingestible devices [27]. The Body Area Network Coordinator (BANC) is a central WBAN controller. It manages the communication between sensors and external devices, aggregating data and facilitating wireless connectivity. Wireless communication modules or transceivers are embedded within the sensors and devices [28]. These transceivers use short-range wireless technologies to transmit data to the BANC or other authorized external devices [29]. Personal Area Network (PAN) interface connects the BANC to external devices, such as smartphones, tablets, or gateways. This interface transmits data to remote healthcare providers or storage platforms [30]. The BANC collects data from different sensors, aggregates it, and performs initial data processing. This layer may include algorithms for data fusion, noise reduction, and basic analysis. Low-power and short-range communication protocols establish reliable connections within the WBAN [31].

Common protocols include Bluetooth Low Energy (BLE), Zigbee, and IEEE 802.15.6. WBANs support various data transmission modes, including real-time streaming, periodic reporting, event-triggered transmission, and on-demand historical data retrieval [32]. The following are the four communication modes in WBAN: In continuous monitoring, the real-time data streams from sensors to the BANC allow for constant health tracking and immediate alerts in case of anomalies. In periodic reporting, the sensors transmit data regularly, conserving energy while providing updated health insights. In event-triggered transmission, the sensors transmit data only when specific events or thresholds are met. This minimizes unnecessary data transfer and saves power. In on-demand retrieval, the authorized external devices can request historical health data from the WBAN when needed. The key characteristics of WBAN Communication Architecture are the following: WBAN devices operate on limited battery power, necessitating communication protocols and strategies that minimize energy consumption to extend the device lifespan. WBANs are designed for short-range communication, typically within a few meters or centimeters. This ensures minimal interference and optimized data transmission. The human body can cause signal attenuation and interference. WBAN communication architecture must account for these challenges and ensure reliable data transmission. WBANs enable real-time health monitoring by transmitting timely data from sensors to the BANC and external devices for immediate analysis and intervention. WBAN communication architecture must ensure that critical health data is transmitted with high reliability, minimal latency, and optimal bandwidth utilization [33].

8.4.1 WBAN Challenges

WBANs offer immense potential for healthcare applications by enabling real-time health monitoring, remote patient care, and personalized interventions. Though WBANs provide transformative possibilities for healthcare, they also pose a range of challenges that need to be addressed through technological innovation, careful design, and comprehensive security measures. The exploration of the difficulties associated with WBANs is detailed below [34]: WBAN devices are often constrained by limited battery capacity due to their small form factors. Ensuring energy-efficient communication and data processing is crucial to extending the device lifespan and reducing the need for frequent battery replacements. The reliability and correctness of information composed by sensors in WBANs are paramount. Factors such as signal attenuation, noise interference, and the placement of sensors on the body can affect data quality.

Techniques for noise reduction, error correction, and data fusion are necessary to ensure reliable health data. The human body and the surrounding environment can interfere with wireless communication within WBANs. This interference can degrade data transmission quality and reliability. Coexistence with other wireless devices and networks, such as Wi-Fi and Bluetooth, poses additional challenges that must be managed effectively. WBANs are designed for short-range communication, typically within a few meters. Ensuring reliable data transmission without signal loss or interruption, especially in dynamic environments, is challenging. Radio signals can be attenuated as they pass through the human body's tissues. This attenuation can affect the strength and quality of signals between sensors and devices. Strategies to mitigate signal loss and optimize signal propagation are necessary. WBANs deal with sensitive health records. Ensuring robust encryption, authentication, and access control mechanisms are in place to protect patient privacy and prevent unauthorized access to health data. As the number of devices in a WBAN grows, managing communication, coordination, and synchronization among devices becomes complex.

Developing scalable network management strategies is essential to maintain network performance and efficiency. WBANs must support real-time data transmission for applications like remote monitoring and emergency alerts. Ensuring timely data transfer without introducing significant latency is a challenge, especially considering the limited bandwidth of some communication protocols. The variety of WBAN strategies and announcement protocols can lead to interoperability issues. Standardization efforts are needed to ensure seamless communication and compatibility among devices from different manufacturers. Wearable and implantable devices collect intimate health data. Complementing the assistance of real-time monitoring using patient confidentiality disquiets obliges for clear communication, informed consent, and transparency in data usage. Healthcare regulations and data protection laws (e.g., HIPAA, GDPR) apply to collecting and transmitting patient health data through WBANs. Ensuring compliance with these regulations is crucial to avoid legal and financial consequences. Developing and implementing WBAN technologies can be costly, making them inaccessible to certain patient populations or healthcare facilities. Balancing cost-effectiveness with quality is a challenge. The challenges of WBAN are depicted in Figure 8.2. WBANs raise ethical questions regarding data ownership, informed consensus, and accountable usage of health data for research and diagnosis.

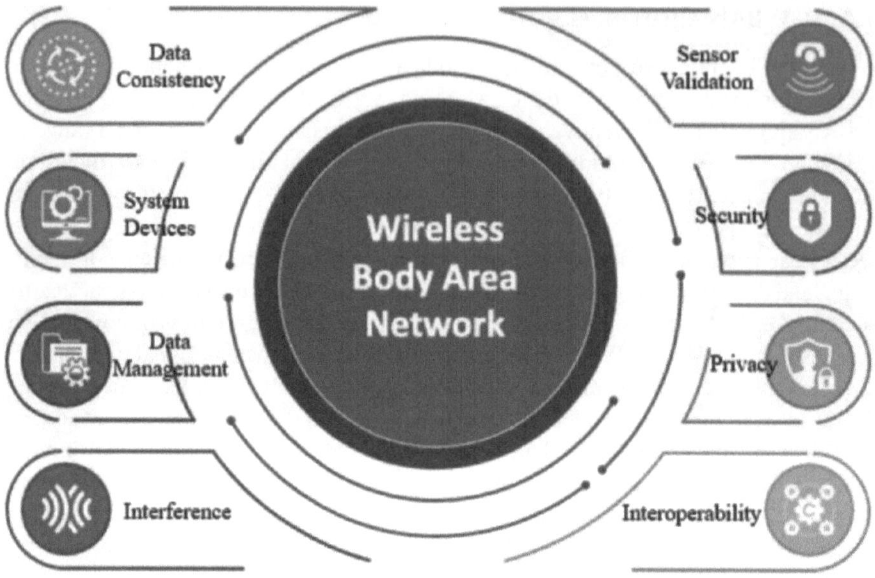

FIGURE 8.2 Challenges in WBAN.

8.5 SECURITY AND PRIVACY REQUIREMENTS FOR WBAN

WBANs present unique security and privacy challenges due to the nature of the data they handle—sensitive health information collected from wearable and implantable medical devices. Protecting patient data and ensuring consumer confidentiality are acute to fostering confidence in WBANs and the broader healthcare ecosystem. The security and privacy requirements for WBANs are paramount to ensuring patient trust, data integrity, and the overall success of connected healthcare. By implementing robust security measures and adhering to stringent privacy guidelines, WBANs can realize their potential to revolutionize healthcare while maintaining the highest data protection standards.

The security and privacy requirements specific to WBANs are as follows: Integrity checks should be applied to data to detect unauthorized tampering or alterations. Hashing algorithms and digital signatures can be used to ensure the data's integrity during transmission and storage. Clearly define the ownership of the data collected by WBAN devices. Consumers ought to regulate their records and the ability to manage how it's used and shared. Guarantee that merely approved devices are allowed to join the WBAN. This includes implementing secure device registration and validation processes. Segment the WBAN from extra systems to decrease the threats. Implement isolation mechanisms to prevent unauthorized access between devices within the network. Implement intrusion detection and prevention systems to monitor for suspicious activities within the WBAN and take appropriate actions to prevent security breaches. Choose communication protocols prioritizing security and privacy, such as protocols with strong encryption and authentication mechanisms.

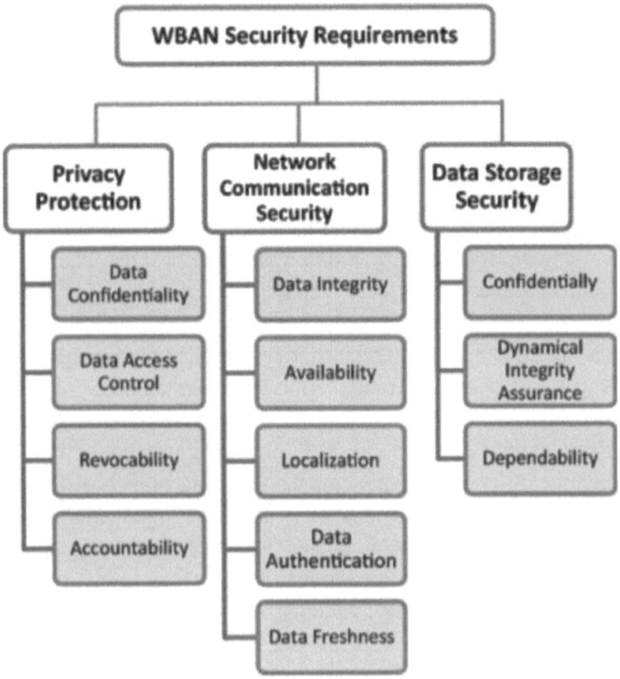

FIGURE 8.3 WBAN—security and privacy requirements.

Ensure compliance with relevant healthcare regulations such as HIPAA, GDPR, and U.S. Food and Drug Administration (FDA) guidelines to protect patient rights and data privacy. Develop a comprehensive incident response plan to handle security breaches effectively, minimize damage, and communicate with affected parties [35]. The security requirements of WBAN are shown in Figure 8.3.

8.6 THREATS AND ATTACKS IN WBAN

WBANs introduce transformative healthcare possibilities, but their unique characteristics expose them to various security threats and attacks. These threats can compromise patient privacy, disrupt healthcare operations, and compromise the integrity of medical data. Attackers gaining unauthorized access to WBAN devices or networks can compromise patient data, manipulate sensor readings, and disrupt healthcare services. Intercepting wireless communication between devices within a WBAN can expose sensitive health data, leading to privacy breaches and unauthorized access to patient information, which is called as eavesdropping. Attackers altering the data transmitted by WBAN sensors can lead to incorrect diagnosis, treatment, and decisions based on manipulated health information [36]. This is termed "data tampering and manipulation." Attackers can record and replay legitimate data transmissions to trick the system into accepting unauthorized commands or transactions, compromising data integrity, called replay attacks.

Denial-of-service (DoS) attacks flood WBAN devices or network components with excessive traffic, causing service disruption and delayed data transmission and impacting patient care. Jamming attacks involve emitting interference signals to disrupt WBAN communication, potentially interfering with critical medical monitoring and patient care. Physical access to WBAN devices can lead to tampering, unauthorized access, and data leakage. WBAN devices can be infected with malware or ransomware, leading to device malfunction, data loss, or hijacking. Man-in-the-Middle (MitM) attacks involve intercepting and altering communication between devices, allowing attackers to gain unauthorized access, manipulate data, or inject malicious commands. Attackers can impersonate authorized WBAN devices or users to gain unapproved admittance to the link and collect sensitive information, called identity spoofing. Cloning WBAN devices can lead to unauthorized replication of patient data, unauthorized device usage, and potential fraud, termed "device cloning." Attackers intercept data transmissions between WBAN devices and backend systems, potentially gaining access to sensitive patient information. Malicious or negligent actions by authorized users within healthcare organizations can lead to data breaches or system compromises. Inadequate data encryption and security measures can lead to data leakage from stored patient information on WBAN devices. Attackers can manipulate or counterfeit sensor readings, leading to false medical data that can result in incorrect diagnoses and treatments, termed "sensor spoofing." Attackers targeting WBAN devices with battery depletion attacks can render the devices useless, impacting patient monitoring and care, called battery depletion attacks. Unsecured physical storage of WBAN devices can lead to theft, unauthorized access, and potential data breaches [37]. Primary threats and attacks happening in the WBAN system are depicted in Figure 8.4.

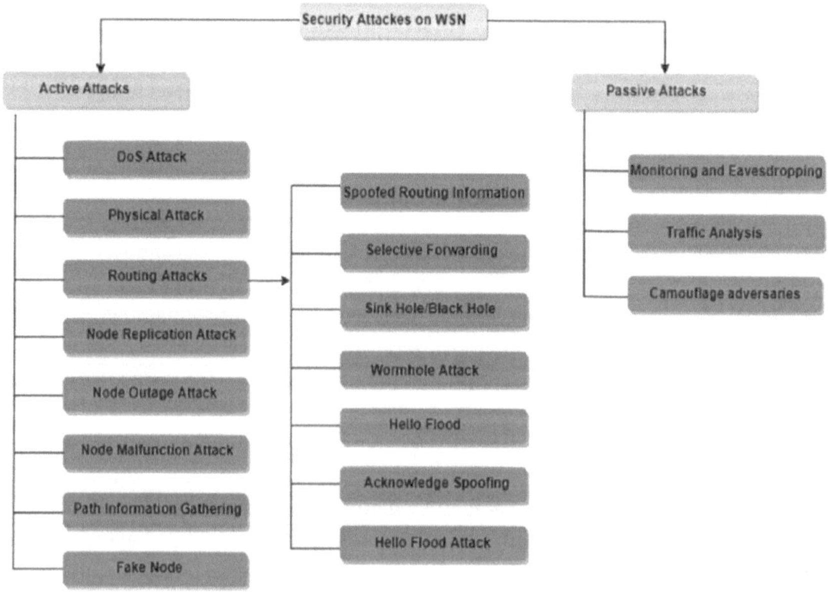

FIGURE 8.4 WBAN—threats and attacks.

8.7 SECURITY SOLUTIONS FOR WBAN

As WBANs become integral to healthcare, robust security solutions are important to guard patient records, ensure device integrity, and preserve the trust of both patients and healthcare providers. Strong encryption procedures to secure data can be constructed. Encryption confirms that even if data is captured, it is incomprehensible to unlicensed persons. Always communication protocols prioritizing security, such as protocols with strong encryption, mutual authentication, and data integrity verification, should be chosen. Intrusion Detection and Prevention Systems (IDS/IPS) solutions to monitor for abnormal activities, unauthorized access attempts, and potential security breaches within the WBAN should be deployed. Major security schemes for WBAN are depicted in Figure 8.5.

The authenticity and integrity of devices within the WBAN through digital signatures, device certificates, and integrity checks should be carried out. Secure key management practices should be conducted to protect encryption keys and ensure proper distribution and rotation. WBAN devices should be protected from physical tampering and theft by utilizing biological security mechanisms, such as tamper-evident packaging and secure storage. Healthcare personnel, patients, and users should be trained about security best practices, potential threats, and how to recognize and respond to security incidents. Always adhere to healthcare regulations, data protection laws, and industry standards relevant to WBANs to ensure legal and ethical data handling. A comprehensive incident response plan should be developed to manage and mitigate security breaches effectively and minimize damage. If WBAN data is transmitted to the cloud or edge for analysis, ensure secure integration with appropriate security measures in place. Working with cybersecurity experts and professionals to assess, design, and implement robust security measures for WBANs is the best option [38].

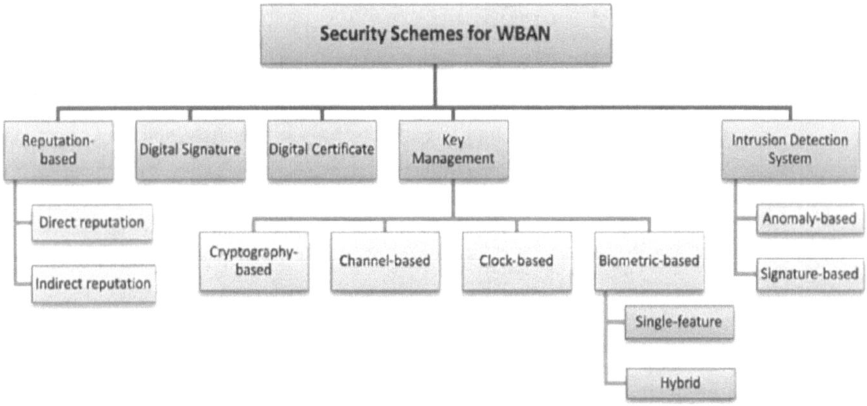

FIGURE 8.5 Security schemes for WBAN.

8.7.1 Recent Security Measures for WBAN

Biometric authentication methods, such as fingerprint or facial recognition, have been explored to provide strong user authentication in WBANs. These methods offer an additional layer of security beyond traditional passwords or PINs. Context-aware security involves adapting security measures based on the specific context of the environment and the user. In WBANs, this could mean adjusting security settings based on the user's location, activity, or health condition. Behavioral biometrics analyze unique patterns in how individuals interact with their devices, such as typing speed or touchscreen gestures. These patterns can be used to identify users and detect anomalies, providing continuous authentication [39].

Implementing secure boot processes and device attestation ensures that only trusted and authenticated software components can run on WBAN devices. This prevents unauthorized or malicious software from compromising the device's security. With the advent of quantum computing, there's growing concern about the potential for current encryption methods to be broken. Researchers are exploring post-quantum cryptographic techniques to ensure long-term data security in WBANs [40]. Homomorphic encryption allows working out on encrypted data deprived of decrypting it, preserving data privacy while allowing for analysis. This technique could be used in WBANs to perform data analytics on encrypted health data. Techniques like federated learning enable machine learning models to be trained on decentralized data sources, such as WBANs, without sharing raw data. This maintains data privacy while deriving useful insights. WBANs can reduce the need for transmitting sensitive data over long distances. AI-powered edge devices can also detect anomalies and potential threats in real time [41]. Table 8.1 depicts the security requirements, threats, and its possible solutions in WBAN systems. Various security criteria for WBAN are shown in Figure 8.6.

Blockchain technology can be used to maintain an immutable and tamper-proof record of health data collected from WBANs. This enhances data integrity and

TABLE 8.1
Security Threats, Requirements, and Solutions

Sl. No.	Security Threats	Security Requirements	Security Solutions
1	Unauthorized access	Trust setup	Random/public key encryption
2	Message disclosure	Confidentiality	Encrypting network layer
3	Message modification	Integrity and authenticity	Hash function
4	DoS	Usability	Redundant routing
5	Information leakage	Confidentiality	Encrypting network layer
6	Message tampering	Integrity	Digital signature
7	Compromised node	Resilience	Node revocation
8	Node capture	Resilience	Tamper proof nodes
9	Routing attacks	Secure routing	Routing protocols
10	Intrusions/malicious actions	Trust set-up	Intrusion detection system

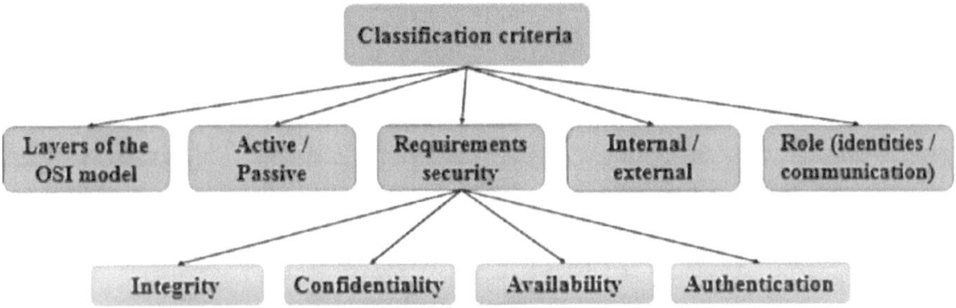

FIGURE 8.6 Security criteria for WBAN.

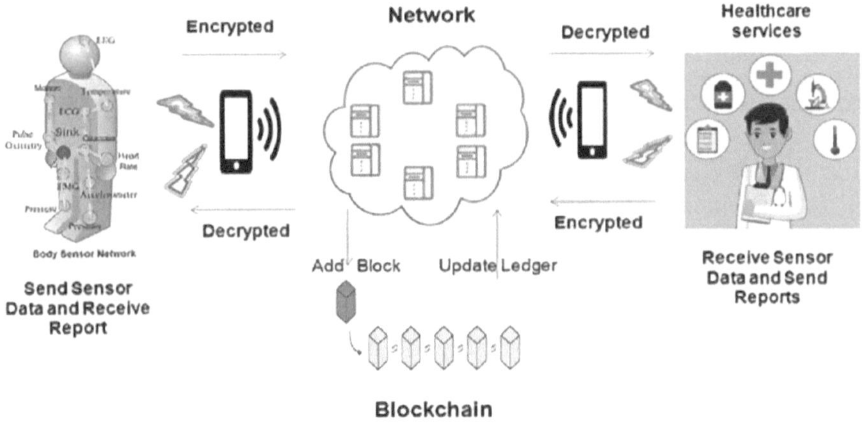

FIGURE 8.7 Idea behind blockchain technology in WBAN.

ensures the accuracy of medical records. The zero trust approach involves treating every device, user, and network connection as untrusted until proven otherwise. Implementing this model in WBANs can enhance security by continuously verifying access requests. In the event of a security breach, incorporating measures for rapid recovery and remediation can minimize the impact of attacks and reduce downtime in WBAN systems [42]. The idea behind blockchain technology in WBAN is shown in Figure 8.7.

8.8 DISCUSSIONS AND RECOMMENDATIONS

The summary of the discussions and recommendations provided in our previous conversation about WBANs, including their architecture, security, privacy, challenges, and solutions, is as follows: The IoMT architecture consists of sensing, network, cloud infrastructure, and application layers. The sensing layer includes wearable devices and sensors that collect patient health data. The network layer enables data transmission and communication between devices and external systems. The cloud infrastructure layer processes and stores data, providing advanced analytics and insights. The application layer includes healthcare applications that utilize the collected data for patient monitoring, diagnostics, and interventions. Blockchain technology can be used to maintain an immutable and tamper-proof record of health data collected from WBANs. This enhances data integrity and ensures the accuracy of medical records. The zero trust approach involves treating every device, user, and network connection as untrusted until proven otherwise. The deployment of blockchain-based WBAN system is depicted in Figure 8.8. Strong security measures are essential due to the sensitive nature of health data in IoMT. Encryption, secure authentication, and data integrity mechanisms are crucial. Data privacy, user consent, and control over data sharing must be prioritized. Compliance with healthcare regulations (e.g., HIPAA, GDPR) is vital for protecting patient rights. The sensing layer encompasses wearable and implantable devices that collect physiological data. Advanced sensors measure parameters like heart rate, temperature, and activity levels. Accurate data collection and reliable sensor calibration are crucial for effective health monitoring. The network layer enables data transmission and communication among IoMT devices. Short-range wireless technologies like BLE and Zigbee are commonly used. Reliable communication protocols and efficient data transfer are essential for real-time monitoring. Deep learning, machine learning, and data fusion provide valuable insights. Scalable and secure cloud infrastructure is necessary to handle diverse healthcare data sources.

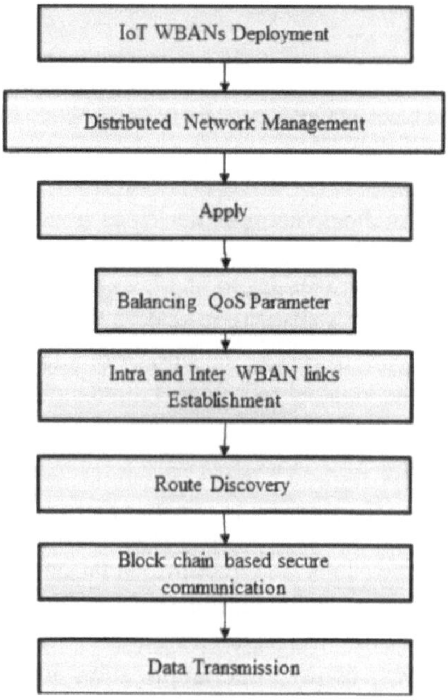

FIGURE 8.8 Deployment of blockchain-based WBAN system.

IoMT systems face threats like unauthorized access, eavesdropping, data tampering, and more. Security measures include encryption, authentication, intrusion detection, and regular updates. Collaborating with security experts, staying informed about emerging threats, and proactive risk assessment are crucial. Similar to IoMT, WBANs require encryption, authentication, data integrity, and compliance with regulations like HIPAA and GDPR. Privacy-preserving techniques, secure device onboarding, and user education are important. Recent measures include biometric authentication, context-aware security, behavioral biometrics, and more. Quantum-resistant cryptography, blockchain, and secure boot processes enhance WBAN security. Privacy-preserving machine learning and secure cloud integration are emerging trends. The following are the major recommendations:

- Implement strong security measures, including encryption, authentication, and intrusion detection.
- Adhere to relevant healthcare regulations and standards to guarantee legal compliance.
- Educate healthcare personnel, patients, and users about security best practices.
- Collaborate with cyber security experts and consider emerging security technologies.
- Perform regular security audits, penetration testing, and privacy impact assessments.
- Stay informed about the evolving security landscape and adapt measures accordingly.

8.9 OPEN RESEARCH DIRECTIONS

The field of WBANs and the broader IoMT presents numerous exciting research directions that can further advance healthcare, enhance patient outcomes, and address the challenges faced in connected healthcare systems. Here are some open research directions in this domain:

1. **Security and Privacy Enhancements:** Develop innovative encryption techniques and secure communication protocols to counter emerging threats and quantum computing challenges. Explore the integration of AI and machine learning for real-time anomaly detection and adaptive security responses in WBANs. Investigate privacy-preserving analytics methods that allow data analysis while maintaining patient privacy.
2. **Cross-Device Integration:** Research ways to seamlessly integrate data from various wearable and implantable devices to create a holistic patient health profile. Develop standardized communication protocols and interoperability structures.
3. **Edge and Fog Computing:** Study how edge and fog computing can alleviate latency and privacy concerns for decreasing the need for cloud transmission. Investigate how AI algorithms can be deployed at the edge to provide real-time insights while respecting data privacy.

4. **AI-Driven Medical Diagnostics:** Explore the potential of AI and deep learning algorithms to analyze data from WBANs for early disease detection, personalized treatment recommendations, and diagnostic accuracy improvement.
5. **Wearable Device Design:** Innovate in wearable device design, considering factors like power efficiency, comfort, and ease of use. Research new materials and miniaturization techniques to improve wearability and user experience.
6. **Ethical Considerations:** Investigate ethical challenges related to data proprietorship and informed approval for using the health records collected by WBANs. Develop ethical guidelines for the responsible use of AI and machine learning in medical diagnostics.
7. **Integration with Telemedicine:** Study how WBANs can enhance telemedicine by enabling real-time patient monitoring, remote consultations, and accurate data transmission to healthcare providers.
8. **Regulatory Compliance and Standards:** Research the impact of evolving regulations (such as GDPR, HIPAA) on WBAN deployment and develop strategies for maintaining compliance. Contribute to the establishment of industry standards for secure and interoperable WBANs.
9. **Energy Harvesting and Power Efficiency:** Investigate energy harvesting techniques to power WBAN devices using body heat, motion, or other ambient sources. Research efficient power management strategies to extend the battery life of wearable devices.
10. **Human-Centric User Interfaces:** Explore innovative user interfaces, such as haptic feedback or brain–computer interfaces, to improve interaction with WBAN devices, especially for users with physical limitations.
11. **Data Fusion and Contextual Understanding:** Develop advanced data fusion techniques to integrate data from multiple sources, such as wearable sensors, environmental sensors, and medical records. Explore AI models that can understand and interpret contextual information to provide more accurate insights.
12. **Patient Engagement and Behavioral Change:** Research ways to leverage WBAN data to encourage healthy behaviors and lifestyle changes among patients, promoting proactive healthcare management.
13. **Data Visualization and Interpretation:** Develop intuitive data visualization tools that allow both healthcare professionals and patients to easily interpret complex health data collected by WBANs. Continuing research in these directions will contribute to the advancement of WBAN technology, improving healthcare delivery, patient outcomes, and overall quality of life.

REFERENCES

[1] Mathers CD, Loncar D. Updated projections of global mortality and burden of disease, 2002–2030: Data sources, methods and results. *PLos Med (World Health Org)* 2006:1–8.

[2] Wang J, Zhang Z, Yang X, Zuo L, Kim JU. Data security and privacy of e-healthcare in electronic medical environment. *ASTL SIA* 2013;22:92–98.
[3] Sharma DA. Wireless health care monitoring system with data security and privacy. *Int J Res Comput Eng Electron* 2013;2(2).
[4] Yazdandoost KY, Sayrafian-Pour K. *Channel Model for Body Area Network (BAN).* IEEE P802, 15, 08-0780; 2009. <http://math.nist.gov/mcsd/savg/papers/15-08-0780-09-0006-tg6-channel-model.pdf> [accessed 6.9.16].
[5] Naranjo PG, Shojafar M, Mostafaei H, Pooranian Z, Baccarelli E. P-SEP: A prolong stable election routing algorithm for energy-limited heterogeneous fog supported wireless sensor networks. *J Supercomput* 2016;1–23.
[6] Rocker C, Ziefle M. E-health, assistive technologies and applications for assisted living: Challenges and solutions. *Med Inform Sci Ref* 2011:23–48.
[7] Fragopoulos AG, Gialelis J, Serpanos D. Imposing holistic privacy and data security on person centric eHealth monitoring infrastructures. In: *12th IEEE International Conference on E-Health Networking Applications and Services (Healthcom).* Lyon, France: IEEE, 2010, pp. 127–134.
[8] Wang J, Zhang Z, Xu K, Yin Y, Guo P. A research on security and privacy issues for patient related data in medical organization system. *Int J Security Appl* 2013;7(4):287–298.
[9] Crosby GV, Ghosh T, Murimi R, Chin CA. Wireless body area networks for healthcare: A survey. *Int J Ad Hoc, Sensor Ubiquitous Comput* 2012;3(3):1–26.
[10] Siddiqui MA, Kamal MB, Moinuddin H. Towards the development of cross layer approach for energy efficiency and mobile wireless body area networks. *Int J Comput Inform Technol* 2013:542–547.
[11] Rocker C, Ziefle M. E-health, assistive technologies and applications for assisted living: Challenges and solutions. *Med Inform Sci Ref* 2011;392. ISBN13: 9781609604691.
[12] Rehman OU, Javaid N, Bibi A, Khan ZA. Performance study of localization techniques in wireless body area sensor networks. In: *11th IEEE International Conference on Trust, Security, and Privacy in Computing and Communications*. Liverpool, UK: IEEE, 2012, pp. 1968–1975.
[13] Pathania S, Bilandi N. Security issues in wireless body area network. *Int J Comput Sci Mobile Comput* 2014;3(4):1171–1178.
[14] Li M, Lou W, Ren K. Data security and privacy in wireless body area networks. *IEEE Wirel Commun* 2010;17(1):51–58.
[15] Al Ameen M, Liu J, Kwak K. Security and privacy issues in wireless sensor networks for healthcare applications. *J Med Syst* 2012;36(1):93–101.
[16] Tachtatzis C, Di Franco F, Tracey DC, Timmons NF, Morrison J. An energy analysis of IEEE 802.15. 6 scheduled access modes for medical applications. In: *International Conference on Ad Hoc Networks*. Berlin Heidelberg: Springer; 2011, pp. 209–222.
[17] Latré B, Braem B, Moerman I, Blondia C, Demeester P. A survey on wireless body area networks. *Wireless Netw* 2011;17(1):1–8.
[18] Kumar R, Mukesh R. State of the art: Security in wireless body area networks. *Int J Comput Sci Eng Technol (IJCSET)* 2013;4(5):622–630.
[19] Kargar MJ, Ghasemi S, Rahimi O. Wireless body area network: From electronic health security perspective. *Int J Reliable Quality E-Healthcare (IJRQEH)* 2013;2(4):38–47.
[20] Ferdous MS, Chowdhury F, Moniruzzaman M. A taxonomy of attack methods on peer-to-peer network. In: *The Proceedings of the 1st Indian Conference on Computational Intelligence and Information Security.* Madurai, India: ICCIIS; 2007, pp. 132–138.
[21] Han ND, Han L, Tuan DM, In HP, Jo M. A scheme for data confidentiality in cloud-assisted wireless body area networks. *Inf Sci* 2014;284:157–166.

[22] Tewari A, Verma P. Security and privacy in e-healthcare monitoring with WBAN: A critical review. *Int J Comput Appl* 2016;136(11).
[23] Fatema N, Brad R. Security requirements, counterattacks and projects in healthcare applications using WSNs—a review. *Int J Comput Network Commun (IJCNAC)* 2014;2(2).
[24] Kavitha T, Sridharan D. Security vulnerabilities in wireless sensor networks: A survey. *J Inf Assurance Sec* 2010;5(1):31–44.
[25] Somasundaram M, Sivakumar R. Security in wireless body area networks: A survey. In: *International Conference on Advancements in Information Technology ICBMG.* Singapore: IPCSIT, 2011, p. 20.
[26] Li J, Ren K, Zhu B, Wan Z. Privacy-aware attribute-based encryption with user accountability. In: *International Conference on Information Security.* Berlin Heidelberg: Springer; 2009, pp. 347–62.
[27] Javadi SS, Razzaque MA. Security and privacy in wireless body area networks for health care applications. In: *Wireless Networks and Security.* Berlin Heidelberg: Springer; 2013, pp. 165–187.
[28] Office for Civil Rights, United State Department of Health and Human Services. *Medical Privacy. National Standards of Protect the Privacy of Personal-Health-Information* <www.hhs.gov/ocr/privacy/hipaa/administrative/privacyrule/index.html> [accessed 20.12.15].
[29] Health Information Technology for Economic and Clinical Health Act HITECH. *Ways and Means and Science Technology* <http://waysandmeans.house.gov/media/pdf/110/hit2.pdf> [accessed 10.11.15].
[30] Kim KJ, Hong SP. Privacy care architecture in wireless sensor networks. *Int J Distrib Sens Netw* 2013:1–8.
[31] Ullah S, Higgins H, Braem B, Latre B, Blondia C, Moerman I, Saleem S, Rahman Z, Kwak KS. A comprehensive survey of wireless body area networks. *J Med Syst* 2012;36(3):1065–1094.
[32] Latif R, Abbas H, Assar S. Distributed denial of service (DDoS) attack in cloud-assisted wireless body area networks: A systematic literature review. *J Med Syst* 2014;38(11):1.
[33] Kifayat K, Merabti M, Shi Q, Llewellyn-Jones D. Security in wireless sensor networks. In: *Handbook of Information and Communication Security.* Berlin Heidelberg: Springer; 2010, pp. 513–552.
[34] Ramli SN, Ahmad R, Abdollah MF, Dutkiewicz E. A biometric-based security for data authentication in wireless body area network (WBAN). In: *The 15th International Conference on Advanced Communication Technology (ICACT).* PyeongChang, Korea: IEEE; 2013, pp. 998–1001.
[35] Zhang GH, Poon CC, Li Y, Zhang YT. A biometric method to secure telemedicine systems. In: *Annual International Conference of the IEEE Engineering in Medicine and Biology Society.* Minneapolis, MN, USA: IEEE; 2009, pp. 701–704.
[36] Saleem S, Ullah S, Kwak KS. A study of IEEE 802.15.4 security framework for wireless body area networks. *Sensors* 2011;11(2):1383–1395.
[37] Crosby GV, Ghosh T, Murimi R, Chin CA. Wireless body area networks for healthcare: A survey. *In J Ad Hoc, Sensor Ubiquitous Comput* 2012;3(3):1.
[38] Ullah S, Mohaisen M, Alnuem MA. A review of IEEE 802.15.6 MAC, PHY, and security specifications. *Int J Distrib Sens Netw* 2013:24.
[39] Misic J. Enforcing patient privacy in healthcare WSNs using ECC implemented on 802.15. 4 beacon enabled clusters. In: *2008 Sixth Annual IEEE International Conference on Pervasive Computing and Communications (PerCom).* Hong Kong: IEEE; 2008, pp. 686–691.

[40] Ahmadi A, Shojafar M, Hajeforosh SF, Dehghan M, Singhal M. An efficient routing algorithm to preserve k-coverage in wireless sensor networks. *J Supercomput* 2014;68(2):599–623.

[41] Zhao Z. An efficient anonymous authentication scheme for wireless body area networks using elliptic curve cryptosystem. *J Med Syst* 2014;38(2):1–7.

[42] Ali SH. Novel approach for generating the key of stream cipher system using random forest data mining algorithm. In: *The Sixth of IEEE International Conference on Developments in eSystemsEngineering (DeSE)*. Abu Dhabi, UAE: IEEE; 2013, pp. 259–269.

Part 3

Security Frameworks and Solutions for WBANs and E-Health Systems

9 Security and Privacy Frameworks for WBANs and E-Health Systems

Anuj Kumar Singh, Sachin Kumar, and Ankit Garg

9.1 INTRODUCTION

Modern technology has made it possible to apply complex solutions that improve people's quality of life. Researchers studying the development of technology have examined health data from various sources to gain understanding and solve issues related to health. Thus, advancements in integrated healthcare technology can potentially improve patient outcomes and increase efficiency at several levels of the healthcare system. The development of innovative electronic health (e-health) application systems can effectively tackle challenges in conventional healthcare systems by implementing strong patient safety protocols, enabling broad access to data, remotely monitoring inpatients, expediting clinical interventions, and utilizing decentralized electronic health records. These systems can handle patient data and health information, raise patient outcomes, encourage collaboration, lower costs, and generally increase the efficacy of e-healthcare services [1]. They can also improve patients' quality of life. A groundbreaking innovation that provides adaptable, scalable, and effective solutions is the integration of cloud computing and Internet of Things (IoT) in patient healthcare systems. Comparing this combination with traditional networks, there are benefits, including quicker record access, more information security, ease of implementation, and energy savings. Continuous innovation may be fostered, and IoT cloud-based e-health systems can considerably improve healthcare services. These systems store medical data in the cloud and employ IoT networks to enable connectivity between users, services, and servers.

The Wireless Body Area Network (WBAN) facilitates the integration of low-power, intelligent, tiny sensor nodes inside, on, or around the body of an individual to monitor physiological processes and the external environment. It has drawn a lot of researchers in the last several years from both academia and industry because of its enormous potential to transform healthcare technology completely. WBANs support numerous consumer electronics (CE) and medical applications. WBANs, for instance, allow patients to have long-term remote health monitoring without limiting their ability to engage in regular activities. The human body is equipped with many nodes, including electrocardiogram (ECG), electromyography (EMG), and electroencephalography (EEG), which gather physiological parameters and transmit them to a medical server located far away for additional examination [2].

DOI: 10.1201/9781032635101-12

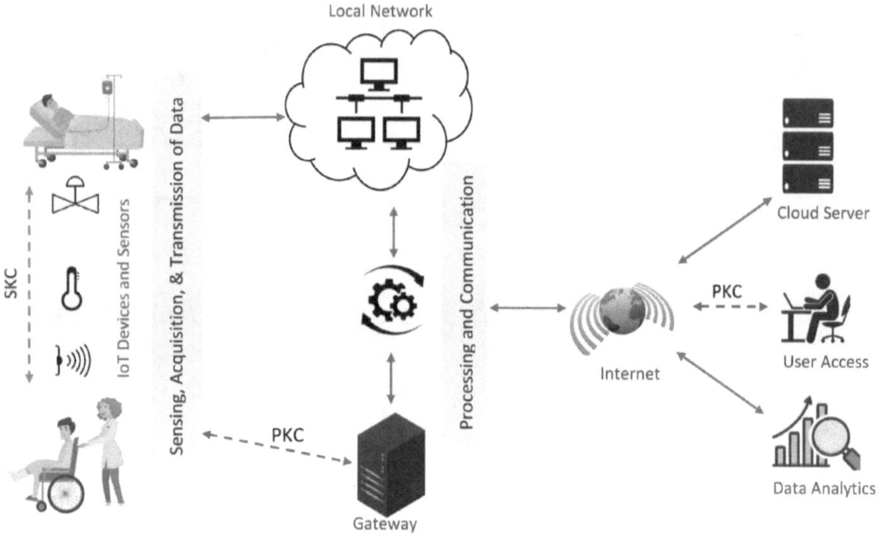

FIGURE 9.1 A privacy-preserving generic healthcare security framework.

Figure 9.1 depicts a privacy-preserving, secure IoT-based generic healthcare architecture with three processing layers: data analytics, processing and communication, and data sensing and acquisition [3]. Cryptographic techniques such as public key cryptography (PKC) and symmetric key cryptography (SKC) are essential to ensure the security of the data kept or transferred in a healthcare system. It is important to remember that public key cryptography techniques are used to secure information at the communication and processing layer. In contrast, SKC techniques are used to protect data at the data sensing and acquisition layer. Furthermore, distant server-based back-end data exchange services also employ public key methods.

9.1.1 Security Requirements of E-health Systems

Digital platforms that store, handle, and transmit health-related data are known as electronic health (e-health) systems. Security is a top priority since medical data is sensitive. WBAN and e-health systems must adhere to stringent security regulations safeguarding patient confidentiality, data integrity, and overall system performance [4].

Access Control: To guarantee that only individuals with permission can access patient data, implementation of strong access control measures is essential. Verification of user identities must be ensured by employing robust authentication techniques like multi-factor authentication.

Data Confidentiality: Protection of data while it is in transit and at rest by using encryption techniques. A secure encryption system should be used to protect sensitive data, including personal and medical records, from unwanted access.

Audit Trails and Logging: To monitor system activity and patient record access, thorough audit logs should be maintained. To find and address any unlawful or questionable activity, periodic examination and evaluation of audit trails is required.

Integrity: There must be procedures in place to guarantee the accuracy of medical data and guard against illegal changes. To confirm the legitimacy and consistency of electronic health records, hash algorithms or digital signatures must be utilized.

Device Security: It must be ensured that the computers, servers, and medical equipment utilized in the e-health system are secure. Security measures, including antivirus software, safe setups, and routine software updates, must be implemented.

User Training and Awareness: Make security best practices more widely known by providing healthcare workers with regular training and informing users on the dangers of social engineering and other forms of fake communication.

Network Security: Firewalls, intrusion detection and prevention systems, and other security measures can be used to safeguard the network architecture. The protection of data being transferred between various e-health system components by using secure communication protocols is required.

Backup and Recovery: Make sure that patient records are accessible during data loss or system failures by implementing regular data backups. Verify and test the efficacy of the backup and recovery processes.

Secure Software Development: When creating and updating software for e-health systems, certain coding techniques must be used. Routine security assessments and code reviews need to be done to find and fix vulnerabilities.

Privacy Protection: Patient privacy and adherence to privacy laws must be ensured by putting privacy-preserving procedures in place.

Secure Authentication and Authorization: It must ensure that the authorization and authentication procedures are safe and adhere to industry standards. Utilize the least privilege concept when reviewing and updating user roles and permissions.

Interoperability Security: To guarantee interoperability without sacrificing security, security safeguards must be in place when transferring data with other healthcare systems.

9.1.2 Security Attacks on E-health Systems

Since e-health systems handle private medical data, they are vulnerable to several security breaches. A summary highlighting the specific attacks on e-health systems [5,6], their description, impact, and mitigation techniques has been presented in Table 9.1.

WBANs and e-health systems require careful consideration of security and privacy issues. Several frameworks and standards have been created to address these issues. The next sections of this chapter have identified and explained a few of the major security and privacy frameworks for WBANs and e-health systems.

TABLE 9.1
Summary of Attacks on E-health Systems

Attack	Description	Impact	Mitigation
Data Breach	Unauthorized access to medical histories or patient records.	Patient privacy breaches, possible identity theft, and improper use of medical records.	Encryption, access controls, and regular security audits.
Device Vulnerabilities	Taking advantage of weaknesses in IoT or medical device connectivity.	Illegal access to medical equipment and compromised patient safety.	Segmenting the network, updating firmware, and conducting regular security audits.
Phishing Attacks	Deceptive emails or messages for disclosing private information, including their login passwords.	Compromised user accounts and unauthorized access to e-health systems.	Multi-factor authentication, email filtering, and user training.
Malware and Ransomware	Malicious software that targets computers with the intention of ransom demands.	Possible financial losses, disruption of the system, and loss of data.	Antivirus software, regular system updates, and employee training on phishing prevention.
Physical Attacks	Illegal physical access to computers, servers, or other vital infrastructure.	Potentially interrupted services, unauthorized system changes, and theft of data.	Security protocols for hardware, surveillance, and physical access controls.
Data Interception	Tapping into data in transit between systems or devices.	Possible data manipulation and unauthorized access to patient information.	Encryption, secure communication protocols, and VPNs.
Denial-of-Service (DoS)	Disrupting regular operation of the e-health system by flooding it with unwanted traffic.	Potential data loss, delays in patient care, and service outages.	Intrusion detection systems, network firewalls, and distributed denial-of-service (DDoS) mitigation services.
Insider Threats	Malice or carelessness of someone inside the medical establishment.	Information theft, unauthorized access, or purposeful information destruction.	Monitoring user activity, stringent access rules, and employee training.
Man-in-the-Middle Attacks	Secretly listening in and influencing conversations between two entities.	Sensitive data is being accessed without authorization and may be modified.	Network monitoring, secure communication protocols, and encryption (SSL/TLS).
SQL Injection	Inserting malicious SQL queries to exploit the weaknesses in the e-health database.	Potential data loss, data modification, and unauthorized access.	Regular security audits, parameterized queries, and input validation.

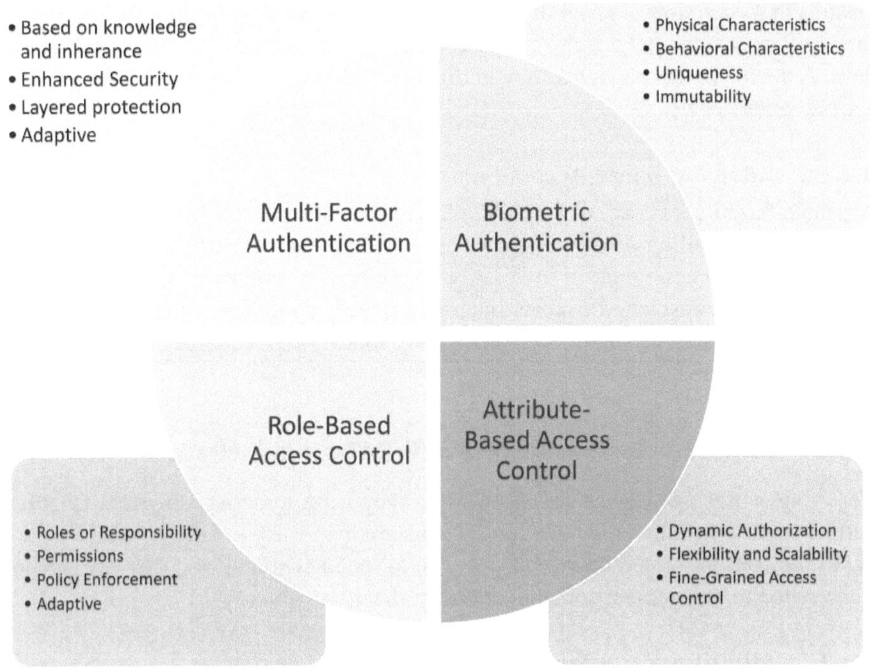

FIGURE 9.2 Attributes of access control and authentication frameworks in e-health systems.

9.2 ACCESS CONTROL AND AUTHENTICATION FRAMEWORKS

Healthcare technology is advancing at a rapid pace. WBANs and e-health systems have transformed patient care and monitoring. This innovation does, however, raise the possibility of data breaches and unwanted access. Strong access control and authentication mechanisms are essential to addressing these issues [7,8]. This section of the chapter investigates the role of authentication and access control in protecting WBANs and e-health systems. Figure 9.2 shows the attributes of access control and authentication frameworks in e-health systems.

9.2.1 Access Control Frameworks in WBANs and E-Health Systems

Limiting who or what can view or use system resources is known as access control, and it is a basic security feature. Access control techniques are crucial in protecting sensitive patient data in the context of WBANs and e-health systems, limiting access to vital information to only those permitted [9]. The two primarily used access control approaches in e-health systems are listed below.

9.2.1.1 Role-Based Access Control

Role-based access control (RBAC) is a popular access control paradigm that allocates permissions to individuals according to their roles in the system. RBAC ensures that each user is given the proper amount of access according to their duties in e-health systems, where different healthcare professionals interact with patient data. A doctor has extra rights to read diagnostic results, whereas a nurse is able to access patient vitals.

9.2.1.2 Attribute-Based Access Control

Attribute-based access control (ABAC) expands access control by considering several factors, including resource properties, environmental factors, and user characteristics. More precise control in WBANs and e-health systems is made possible by ABAC, which considers variables including patient condition, time, and location. This dynamic approach improves security by modifying access rights in response to contextual data.

9.2.2 Authentication Frameworks in WBANs and E-Health Systems

Verifying a user's or device's identity when they want to access a system is called authentication. Strong authentication systems are necessary to keep sensitive health data safe from unwanted access [10,11]. The following explains the two key categories needed to accomplish authentication in e-health systems.

9.2.2.1 Multi-Factor Authentication

Multi-factor authentication (MFA) boosts security by combining several authentication techniques. An additional layer of security is added to e-health systems by requiring users to enter two or more forms of identification, such as a password and a onetime code sent to a mobile device. If one authentication element is compromised, MFA is very good at preventing unwanted access.

9.2.2.2 Biometric Authentication

In e-health systems, biometric identification techniques like voice recognition, retinal or iris recognition, and fingerprint scanning offer a high degree of security. By acting as trustworthy identifiers, these distinct physiological or behavioral traits lower the possibility of unwanted access. Patient data protection is improved at the point of collection when biometric authentication is integrated into WBANs.

Even while access control and authentication frameworks greatly improve WBANs and e-health systems security, problems still exist. Not to mention addressing user acceptance and guaranteeing interoperability across many systems are balancing security and usability. Furthermore, as technology develops, it becomes necessary to create flexible frameworks that may change to meet new threats.

9.3 FRAMEWORKS BASED ON PRIVACY-PRESERVING TECHNIQUES

Patient privacy becomes a critical problem when WBANs and e-health systems gather and transmit real-time health data. Strong privacy-preserving frameworks are necessary given issues including illegal access, data breaches, and the possible abuse of sensitive health information [12]. Differential privacy, homomorphic encryption, and federated learning are the three foremost techniques that can be used in developing privacy-preserving frameworks for e-health systems.

9.3.1 Homomorphic Encryption

Using a homomorphic encryption technique, encrypted data can be computed without needing to be first decrypted. Homomorphic encryption guarantees the confidentiality of sensitive patient data during processing in e-health systems, where data is shared among multiple entities for analysis and decision-making. This preserves anonymity while enabling safe, cooperative research and analysis. Homomorphic encryption provides a secure and privacy-focused method in WBANs, where real-time health data is continuously created. In the rapidly changing world of linked health technology, it is essential to protect sensitive patient data since it enables smooth joint analysis and decision-making across healthcare devices without jeopardizing individual privacy.

9.3.2 Differential Privacy

Differential privacy is a privacy-preserving method that modifies data by adding controlled noise or randomness. In doing so, it preserves personal privacy and permits the extraction of valuable aggregate data. Using differential privacy guarantees that the data shared in WBANs, where private health records are continuously created, is anonymized and cannot be linked to specific people. Anonymity is guaranteed by differential privacy, which also keeps certain people from being identified. This privacy-preserving strategy achieves a critical balance between protecting sensitive patient information and maintaining confidentiality in the networked world of WBAN technology.

9.3.3 Federated Learning

Decentralized machine learning techniques like federated learning enable model training on several servers or devices without requiring the sharing of raw data. Without centralizing sensitive data, federated learning allows the creation of tailored health models in WBANs, where wearable devices continuously provide health data. Patient privacy is preserved, and the risk of data leakage is reduced with this decentralized method. This novel method can create personalized health models, which reduces exposure risk by not centralizing data. Federated learning benefits WBANs because it facilitates safe, effective, and privacy-preserving collaborative analysis. This learning aligns with a contemporary healthcare paradigm that values data-driven insights without sacrificing individual privacy.

9.4 SECURITY MANAGEMENT FRAMEWORKS

To protect e-health systems from potential attacks, it is imperative to implement effective security management frameworks [13,14]. This section discusses the significance of security management and four major frameworks that strengthen e-health networks.

9.4.1 ISO/IEC 27001: A Pillar of Information Security Management

An internationally acknowledged framework for information security management systems (ISMS) is the ISO/IEC 27001 standard. Following ISO/IEC 27001 guarantees the installation of thorough security controls in the context of e-health systems. The framework places significant emphasis on risk assessment and management, guiding companies in identifying vulnerabilities and implementing countermeasures to reduce potential threats. E-health systems create an organized approach to information security by implementing ISO/IEC 27001, which supports patient data availability, confidentiality, and integrity.

9.4.2 HITRUST: Elevating Trust in Healthcare

The Health Information Trust Alliance (HITRUST) offers a customized architecture designed to address the particular difficulties faced by the healthcare sector [15]. Health Insurance Portability and Accountability Act (HIPAA) and Health Information Technology for Economic and Clinical Health (HITECH) are just two examples of the standards and laws harmonized by the certifiable HITRUST CSF (Common Security Framework). Compliance with industry-specific regulations and information risk management are demonstrated by e-health systems that follow HITRUST recommendations. By combining security, privacy, and compliance concerns, this framework provides a comprehensive method for protecting patient data.

9.4.3 COBIT: Bridging IT Governance and Security

A framework called Control Objectives for Information and Related Technologies (COBIT) was created by ISACA to coordinate information technology (IT) governance with business objectives [16]. COBIT offers a governance framework that prioritizes efficient control and risk management in e-health systems, where IT plays a central role. E-health companies can improve their capacity to manage and safeguard data by implementing COBIT principles, which help guarantee the dependability and integrity of vital healthcare systems.

9.5 INTRUSION DETECTION AND PREVENTION FRAMEWORKS

Identifying and reacting to illegal or malevolent behavior entail monitoring and assessing network or system activity using intrusion detection. Alternatively, intrusion prevention takes proactive steps to neutralize or stop possible threats before they have a chance to jeopardize the system's integrity. Consolidated, these elements provide an all-encompassing Intrusion Detection and Prevention (IDP) framework essential to preserving medical records' privacy and accuracy [17]. The security of WBANs and e-health systems is greatly enhanced by IDP frameworks, which serve as watchful gatekeepers against unknown intrusions and possible threats. The

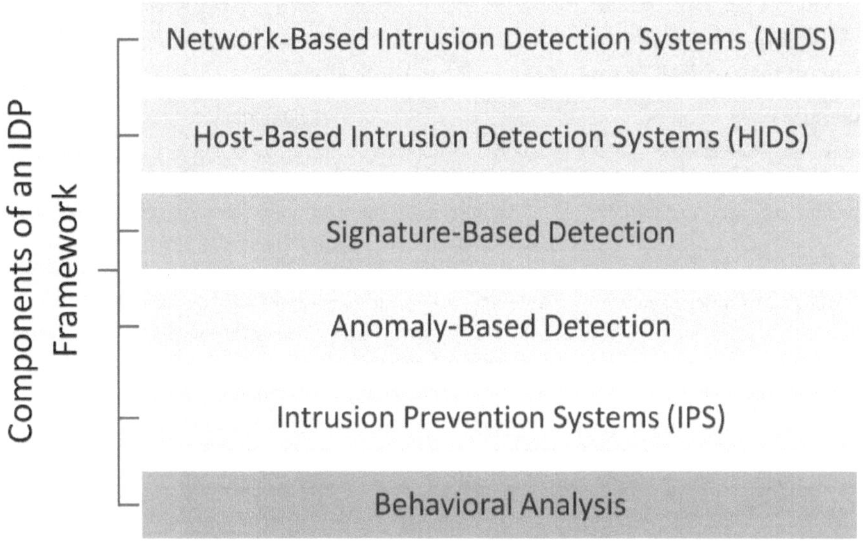

FIGURE 9.3 Components of IDP framework for e-health systems.

significant components of an IDP framework have been listed in Figure 9.3 and are discussed below.

- (i) *Network-Based Intrusion Detection Systems (NIDS):* NIDS examines network data and trends to find anomalies or known attack signatures. To ensure that only authorized parties may access sensitive data and patient records, NIDS can monitor data transmissions in e-health systems.
- (ii) *Host-Based Intrusion Detection Systems (HIDS):* File integrity, user authentication, and system logs are among the tasks and devices that HIDS monitor. HIDS can be used by wearable devices in WBANs to detect any unauthorized access or manipulation of health data.
- (iii) *Signature-Based Detection:* A recognized pattern of harmful activity is the foundation for signature-based detection. WBANs can stop certain assaults or malware signatures from infecting linked devices by using signature-based detection to identify them.
- (iv) *Anomaly-Based Detection:* Deviations from expected behavior are found via anomaly-based detection. Anomaly-based detection can trigger alerts if an abnormally high volume of data requests or access attempts is found in an e-health system.
- (v) *Intrusion Prevention Systems (IPS):* When risks are identified, IPS take preventive action by obstructing malicious activity. IPS in WBANs can actively stop unwanted access to wearable technology and step in to safeguard health information.
- (vi) *Behavioral Analysis:* Errors from usual patterns can be found by analyzing user and device activity.

TABLE 9.2
Advantages of IDP Frameworks in E-health Systems

Advantage	Explanation
Early Threat Detection	IDP frameworks reduce the impact of attacks by enabling early detection of possible security violations.
Data Integrity and Confidentiality	IDP frameworks support the preservation of patient health data's confidentiality and integrity by guarding against unwanted access and manipulation.
Compliance with Regulations	E-health systems can comply with legal mandates like HIPAA and maintain the secure processing of patient data by putting in place strong IDP frameworks.
User Trust and Confidence	Patients, healthcare providers, and other stakeholders have increased trust levels since private health information is secure when a strong IDP framework is in place.

To detect anomalous data retrieval or access behaviors, e-health systems can use behavioral analysis to alert users to potential security risks (see Table 9.2).

Safeguarding patient data and upholding stakeholder confidence will become increasingly important as the healthcare sector embraces digital transformation. This may be achieved by prioritizing cybersecurity through efficient IDP frameworks [18]. WBANs and e-health systems can be fully utilized without jeopardizing the security and privacy of patient information, thanks to the integration of cutting-edge technologies and preemptive security measures.

9.6 COMMUNICATION SECURITY FRAMEWORKS

A strong communication security framework is necessary to smoothly share sensitive health data between the networked WBAN and e-health systems. This section examines the importance of communication security within the framework of WBANs and e-health systems, emphasizing essential elements and tactics for preserving the privacy and accuracy of medical records. To mitigate the vulnerabilities related to the transmission of health data over networks, communication security is crucial in e-health systems and WBANs. Patient privacy and the accuracy of medical information are seriously threatened by unauthorized access, interception, or interference with this data. Information shared across devices, systems, and healthcare providers is safeguarded from potential cyberattacks using a strong communication security architecture [19]. The major components of a communication security framework for e-health systems have been demonstrated in Figure 9.4 and are explained below.

(i) *Confidentiality Ensuring Protocols:* Encrypting health data during transmission using robust encryption techniques like Advanced Encryption Standard (AES) helps prevent unwanted access. Encryption is essential in e-health systems to protect patient, healthcare provider, and database data transfers.

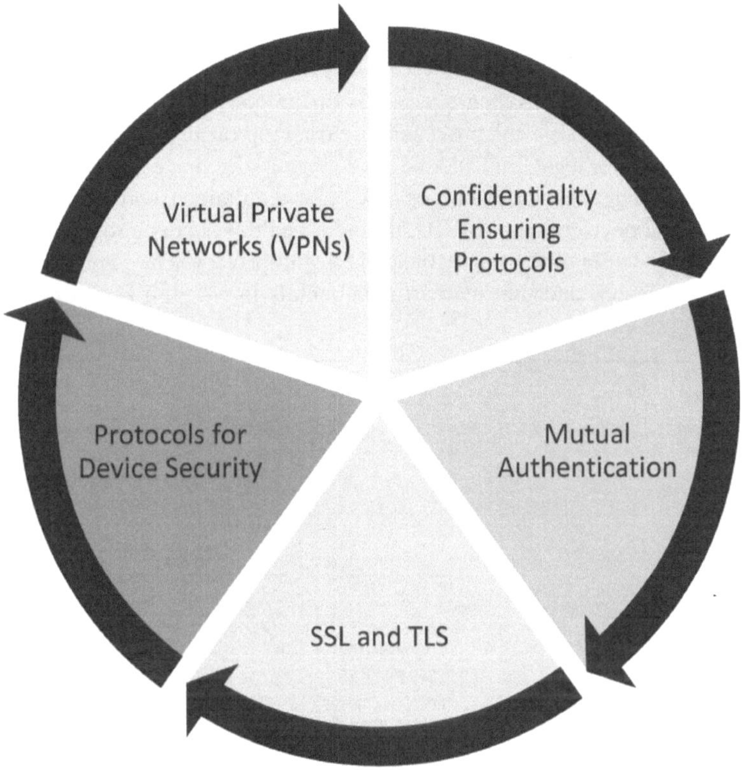

FIGURE 9.4 Components of communication security framework for e-health systems.

(ii) *Mutual Authentication:* By using mutual authentication, it is ensured that both sides of a communication confirm each other's identities before sharing information. Mutual authentication in e-health systems improves the communication channel's credibility between healthcare providers and data repositories [20].

(iii) *Secure Sockets Layer (SSL) and Transport Layer Security (TLS):* By offering encryption and authentication, the SSL and TLS protocols create secure communication channels. To protect patient information during telemedicine and remote consultations, e-health systems rely on SSL/TLS for secure data exchanges over the internet.

(iv) *Virtual Private Networks (VPNs):* VPNs improve the secrecy of data transferred between linked devices by establishing safe, private networks across the internet. VPNs can be used in WBANs to create secure links for data exchange between healthcare servers and wearable devices.

(v) *Protocols for Device Security:* Wearable device communication is standardized and secure when secure communication protocols are used, such as IEEE 802.15.6 for WBANs. These standards help to ensure that health data is transmitted within WBANs in an overall safe manner.

9.7 REGULATORY, ETHICAL, AND LEGAL FRAMEWORKS

E-health systems and WBANs are products of the convergence of technology and healthcare, revolutionizing the provision and tracking of medical care. But there is a complicated web of legal, ethical, and regulatory issues that come with this digital transformation. This chapter examines how laws, regulations, and ethical standards have shaped the development of WBANs and e-health systems, promoting technology responsibly while protecting patient rights and privacy. The regulatory, honest, and legal framework and the related fundamentals in e-health systems have been shown in Figure 9.5.

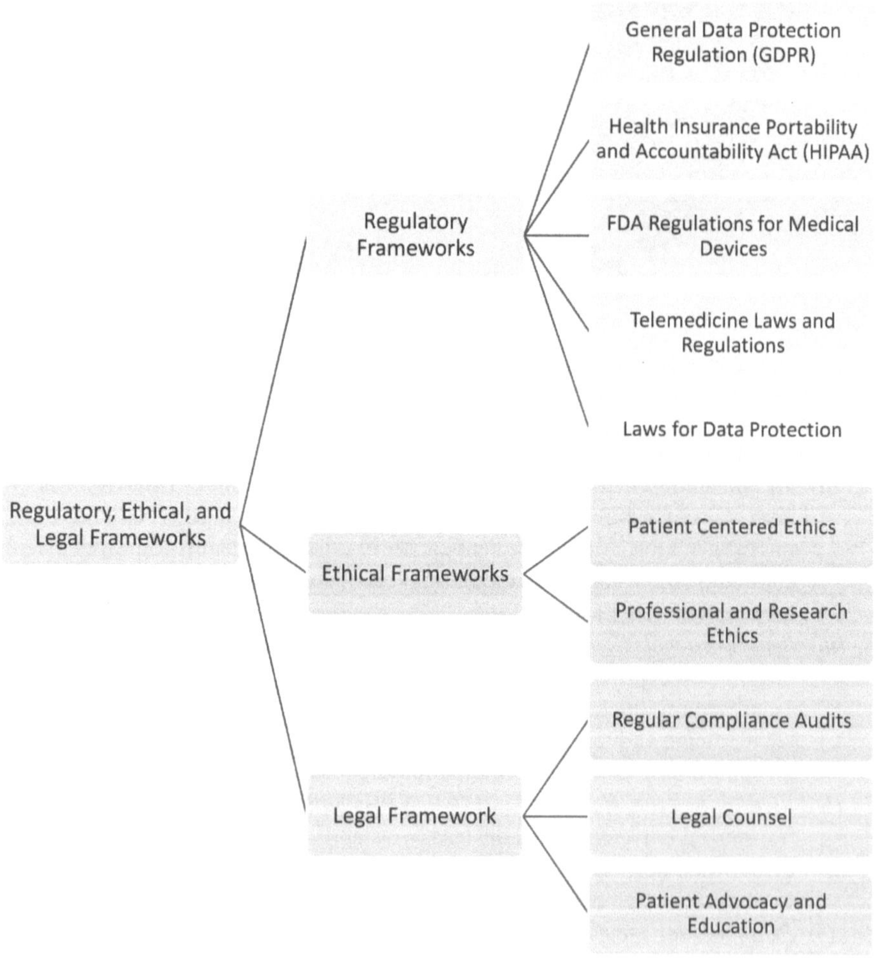

FIGURE 9.5 Regulatory, ethical, and legal framework and the related fundamentals in e-health systems.

9.7.1 Regulatory Frameworks for E-Health Systems

Examination of important regulations and their effects on the creation, application, and management of digital health solutions has been done here, highlighting the vital significance of regulatory frameworks for e-health systems and WBANs.

9.7.1.1 General Data Protection Regulation

Within the European Union (EU), the General Data Protection Regulation (GDPR) provides a comprehensive regulatory framework for data protection and privacy for e-health systems. By stressing the requirement for clear user consent, open data processing procedures, and the right to be forgotten, GDPR expands its application to the processing of health-related data. To acquire and keep consumers' trust while respecting the highest data protection standards, e-health systems must ensure GDPR compliance [21].

9.7.1.2 Health Insurance Portability and Accountability Act

HIPAA is one of the main legislative frameworks in the United States for protecting patient information. HIPAA establishes strict guidelines to guarantee sensitive health data's confidentiality, integrity, and availability. HIPAA regulations, which encompass a wide range of healthcare operations, including telehealth services and electronic health records (EHRs), must be followed by e-health systems. To create a safe and private environment for patient data in electronic health systems, compliance with HIPAA regulations is essential [22].

9.7.1.3 Food and Drug Administration's Regulations for Medical Devices

Medical devices, especially those incorporated into e-health systems, are subject to important regulations by the U.S. Food and Drug Administration (FDA). Through an extensive approval process, the FDA guarantees the efficacy and safety of medical equipment. To ensure the dependability and security of the technologies used in patient care and monitoring, e-health systems that incorporate medical devices must abide by FDA standards.

9.7.1.4 Telemedicine Laws and Regulations

Compliance with telemedicine legislation becomes crucial as telemedicine integrates increasingly into e-health systems. Worldwide variations in these rules cover topics including certification, payback, and care requirements for medical treatments provided remotely. Ensuring legal compliance and smooth virtual healthcare service delivery requires e-health systems to maneuver through the intricate terrain of telemedicine laws.

9.7.1.5 Laws for Data Protection

Many nations have passed unique data protection legislation in addition to GDPR to protect personal information. These legal frameworks control the gathering, handling, and archiving health data in electronic health systems. Compliance with local data protection laws is essential to avoid legal consequences for privacy violations or improper processing of data.

9.7.2 Ethical Frameworks for E-Health Systems

The ethical frameworks shape the proper use of technology in healthcare by acting as guiding principles. Stakeholders in the healthcare ecosystem can help create a more patient-centered and ethically sound digital health environment by putting patient autonomy, privacy, and well-being first, following professional ethical guidelines, maintaining research integrity, and taking social and cultural factors into account. Establishing ethical and moral guidelines that direct the correct creation, use, and use of these technologies is essential for innovation and advancement. The ethical issues surrounding WBANs and e-health systems are examined in this chapter, emphasizing the value of moral strategies that put patients' privacy, autonomy, and well-being first [23].

9.7.2.1 Patient-Centered Ethical Considerations

Informed consent, patient data confidentiality, and transparency are the three major ethical considerations that must be taken care of.

Informed Consent: Getting informed permission is crucial for WBANs and e-health systems. Patients should have complete knowledge of the goals, dangers, and advantages of using these technologies to make independent decisions about sharing and using their health information.

Confidentiality of Patient's Data: Respecting the privacy and confidentiality of patients is a basic ethical precept. To protect sensitive health data and guarantee patient privacy from unwanted access, e-health systems and WBANs need strong security mechanisms [24].

Transparency: Transparency in communication between healthcare providers, tech developers, and patients is emphasized by ethical frameworks. Patients need to understand how e-health systems and WBANs gather, use, and store personal data.

9.7.2.2 Professional and Research Ethics

Important ethical considerations need to be considered in the development, application, and research of WBAN and e-health technologies. The welfare of people engaged in research and utilizing these state-of-the-art health technologies, as well as honesty and transparency, are important considerations when examining the ethical guidelines that practitioners and researchers in e-health systems and WBANs adhere to.

Healthcare Professional Integrity: According to ethical principles, healthcare practitioners should always operate with integrity while interacting with e-health Systems and WBANs. This entails providing correct information, making objective decisions, and putting the patient's needs first.

Avoiding Conflicts of Interest: Ethical frameworks strongly emphasize avoiding conflicts of interest that could jeopardize the integrity of electronic health records or patient treatment. Prioritizing patients' best interests over self-interest or financial gain is a duty for healthcare providers.

Publication and Reporting Integrity: Research results about WBANs and e-health systems may be reported and published by ethical requirements. According to ethical principles, researchers should never manipulate or falsify data and instead give factual, objective information.

Human Subjects' Protection: It is crucial to follow ethical guidelines for the protection of human subjects when doing research with e-health systems and WBANs. Informed consent, risk minimization, and voluntary involvement are all requirements for researchers.

9.7.3 Legal Framework Considerations for E-Health Systems

Legal frameworks ensure ethical behavior, protect patient rights, and handle any concerns related to the collecting and handling health data by establishing the norms, regulations, and standards that regulate the use of digital health technologies. The significant considerations mentioned below must be taken into account to emphasize the importance of legal issues in e-health systems [25].

Regular Compliance Audits: Regular compliance auditing aids in locating and resolving any legal ambiguities or departures from accepted practices. WBANs and e-health systems gain from continuing evaluations to guarantee consistent conformance to changing legislative frameworks [26].

Legal Counsel: Engaging attorneys with technology and healthcare law expertise can help you navigate complex legal environments by offering vital insights. Legal counsel should be sought to guarantee that WBANs and e-health systems are established and implemented according to applicable rules.

Patient Advocacy and Education: Maintaining legal compliance is facilitated by patient rights advocacy and open lines of contact with users. Consumer trust in e-health systems and WBANs is increased by patient education regarding privacy rights, data usage regulations, and available legal protections.

9.8 CONCLUSION AND FUTURE INSIGHTS

Focusing on the critical aspects of security and privacy frameworks for WBANs and e-health systems, this chapter has emphasized the importance of data protection in the quickly evolving field of healthcare technologies. The chapter highlights the various frameworks for WBANs and e-health systems and stresses the importance of robust security and privacy rules. Examining existing frameworks and their applications can be helpful for researchers, practitioners, and policymakers. The chapter advocates for a proactive and adaptable approach to address emerging threats and ensure sensitive health data's integrity, availability, and confidentiality as technology advances. The discussion in this chapter will serve key foundation for the researchers and practitioners working in the security and privacy of e-health systems.

REFERENCES

[1] Butpheng, C., Yeh, K. H., & Xiong, H. (2020). Security and privacy in IoT-cloud-based e-health systems—A comprehensive review. *Symmetry*, 12(7), 1191.

[2] Singh, A. K., & Kumar, S. (2023). Performance analysis of ECC-based security solutions for Internet of Medical Things. In *International Conference on Data Analytics & Management* (pp. 337–347). Singapore: Springer Nature Singapore.

[3] Singh, A. K., Garg, A., & Nayyar, A. (2023). Blockchain for security and privacy in healthcare informatics. *Innovations in Healthcare Informatics: From Interoperability to Data Analysis*, 41, 157.

[4] Singh, A. K., & Kumar, S. (2024). An efficient and secure CLAKA protocol for blockchain-aided wireless body area networks. *Expert Systems with Applications*, 242, 122740.

[5] Sivan, R., & Zukarnain, Z. A. (2021). Security and privacy in a cloud-based e-health system. *Symmetry*, 13(5), 742.

[6] Garg, A., & Singh, A. K. (2022). Internet of Things (IoT): Security, cybercrimes, and digital forensics. In *Internet of Things and Cyber-Physical Systems* (pp. 23–50). CRC Press.

[7] Gupta, D. S., Mazumdar, N., Nag, A., & Singh, J. P. (2023). Secure data authentication and access control protocol for industrial healthcare systems. *Journal of Ambient Intelligence and Humanized Computing*, 1–12.

[8] Singh, A. K., Nayyar, A., & Garg, A. (2023). A secure elliptic curve-based anonymous authentication and key establishment mechanism for IoT and cloud. *Multimedia Tools and Applications*, 82(15), 22525–22576.

[9] Salehi Shahraki, A., Lauer, H., Grobler, M., Sakzad, A., & Rudolph, C. (2023). Access control, key management, and trust for emerging wireless body area networks. *Sensors*, 23(24), 9856.

[10] Khan, M. A., Din, I. U., Majali, T. E., & Kim, B. S. (2022). A survey of authentication in internet of things-enabled healthcare systems. *Sensors*, 22(23), 9089.

[11] Singh, A. K., & Patro, B. D. K. (2017). Performance comparison of signcryption schemes–a step towards designing lightweight cryptographic mechanism. *International Journal of Engineering and Technology (IJET)*, 9(2), 1163–1170.

[12] Odelu, V., Saha, S., Prasath, R., Sadineni, L., Conti, M., & Jo, M. (2019). Efficient privacy-preserving device authentication in WBANs for industrial e-health applications. *Computers & Security*, 83, 300–312.

[13] Sarosh, P., Parah, S. A., Bhat, G. M., & Muhammad, K. (2021). A security management framework for big data in smart healthcare. *Big Data Research*, 25, 100225.

[14] Singh, A. K., & Patro, D. B. (2019). A novel security protocol for wireless sensor networks based on elliptic curve Signcryption. *International Journal of Computer Networks & Communications (IJCNC)*, 11.

[15] Li, J., Liu, X., Liu, L., Sun, D., & Li, B. (2013). HiTrust: Building cross-organizational trust relationship based on a hybrid negotiation tree. *Telecommunication Systems*, 52, 1353–1365.

[16] De Haes, S., Van Grembergen, W., Joshi, A., & Huygh, T. (2020). COBIT as a framework for enterprise governance of IT. In *Enterprise Governance of Information Technology: Achieving Alignment and Value in Digital Organizations* (pp. 125–162). Cham: Springer.

[17] Granjal, J., & Pedroso, A. (2018). An intrusion detection and prevention framework for internet-integrated CoAP WSN. *Security and Communication Networks*, 2018.

[18] Masood, I., Wang, Y., Daud, A., Aljohani, N. R., & Dawood, H. (2018). Towards smart healthcare: Patient data privacy and security in sensor-cloud infrastructure. *Wireless Communications and Mobile Computing*, 2018, 1–23.

[19] Saleem, S., Ullah, S., & Kwak, K. S. (2011). A study of IEEE 802.15. 4 security framework for wireless body area networks. *Sensors*, 11(2), 1383–1395.

[20] Singh, A. K., & Patro, B. D. K. (2020). Signcryption-based security framework for low computing power devices. *Recent Advances in Computer Science and Communications (Formerly: Recent Patents on Computer Science)*, 13(5), 845–857.

[21] Yuan, B., & Li, J. (2019). The policy effect of the general data protection regulation (GDPR) on the digital public health sector in the European Union: An empirical investigation. *International Journal of Environmental Research and Public Health*, 16(6), 1070.

[22] Weiss, J. N. (2023). The health insurance portability and accountability act (HIPAA). In *Physician Crisis: Why Physicians Are Leaving Medicine, Why You Should Stay, and How to Be Happy* (pp. 79–81). Cham: Springer Nature Switzerland.

[23] Vandemeulebroucke, T., Denier, Y., Mertens, E., & Gastmans, C. (2022). Which framework to use? A systematic review of ethical frameworks for the screening or evaluation of health technology innovations. *Science and Engineering Ethics*, 28(3), 26.

[24] Garg, A., Singh, A. K., & Garg, M. (2023). Role of Internet of Things and artificial intelligence for healthcare informatics: An overview. *Innovations in Healthcare Informatics: From Interoperability to Data Analysis*, 41, 107.

[25] Compagnucci, M. C., Minssen, T., Fenwick, M., Aboy, M., & Liddell, K. (Eds.). (2024). *The Law and Ethics of Data Sharing in Health Sciences*. Springer Singapore: Springer Nature.

[26] Garg, A., & Singh, A. K. (2022). Internet of Things (IoT): Security, cybercrimes, and digital forensics. In *Internet of Things and Cyber Physical Systems* (pp. 23–50). Boca Raton: CRC Press.

10 A Comprehensive Security Framework for WBANs in the Healthcare Environment

Pabitha B., Shridhar Sanshi, and Karthik N.

10.1 INTRODUCTION TO WIRELESS BODY AREA NETWORKS IN E-HEALTHCARE

Wireless Body Area Networks (WBANs) have become a game-changing technology in the field of e-healthcare. Wireless communication technology has stepped into many fields like transport, agriculture, aquaculture, defense, and medicine, which is no exception to other industries. During and after COVID-19, e-medical services for healthcare were widely spread all over the world. The Internet of Things (IoT) has many applications, among which WBAN plays a trending role, which is enabled with sensor and actuator nodes according to the user's body condition. Based on the sensed sensor information, the medical physician decides whether the patient's condition is normal or abnormal. Then, the necessary treatment is done remotely, if required, based on the accessed dataset information from the sensor. In these situations, a wireless electronic device plays a vital role in electronic medical care, especially remotely for older adults, and is crucial in an emergency. The patient's status is securely shared with the family, patient database, doctor, and caretaker in a wireless network environment, as shown in Figure 10.1.

To provide Confidentiality, Integrity, and Availability (CIA) for WBAN data in e-healthcare applications, certain algorithms and frameworks have been discussed, which have been used in recent days efficiently with increased performance measures.

To keep everyone involved hopeful and acknowledged and to prevent direct bodily injury to any individual, which includes the possibility of mortality, communication over the constrained devices must be protected over the entire network being communicated [1]. For security hardness and efficiency, quite the opposite of certain cryptographic techniques used already, the hyper-elliptic curve cryptography of the proposed technique requires tiny parameters, key structure, and the like.

DOI: 10.1201/9781032635101-13

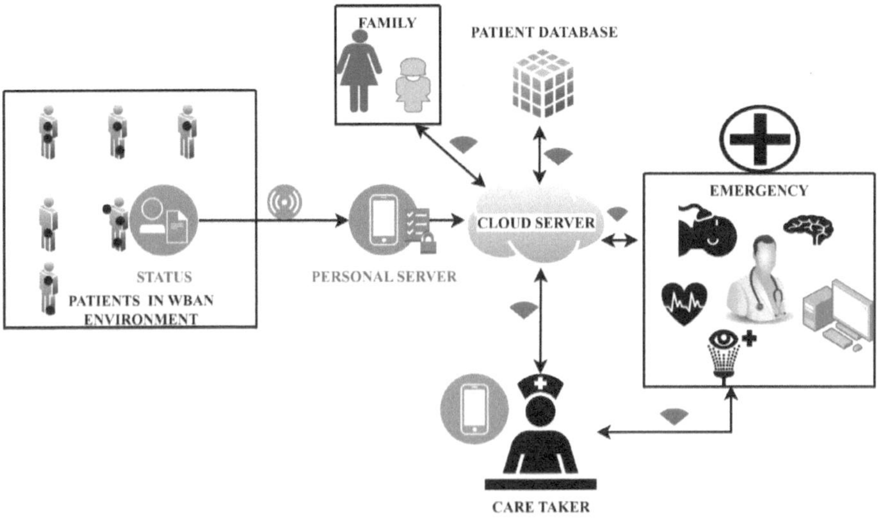

FIGURE 10.1 WBAN structure in e-healthcare.

The certificate-less sign encryption method was tested using a security testing and validation program called Automated Validation of Internet Security Protocols and Applications (AVISPA), which revealed that it was protected efficiently. However, as WBAN capabilities increase, concerns about protecting sensitive medical data logs also exist.

The objectives of the chapter include:

- Background, pros, and cons of WBAN in e-healthcare.
- Security requirements for WBAN in e-healthcare.
- Overview of possible threats to e-healthcare data.
- Different security frameworks with different layouts and protocols have been used in recent research.
- Intrusion detection and prevention in the WBAN.
- Secure communication protocols used against the threats.

- **Background**

Traditional healthcare systems have frequently relied on recurrent trips to the doctor for examinations and consultations. However, this strategy may be constrained, particularly for patients with chronic illnesses or those needing ongoing care. WBAN hinders regular doctor trips and gives patients a new way to incorporate WBAN effectively in mobile health applications [2]. Among the many different applications of WBANs, the majority are in the medical sector. WBANs help to overcome certain constraints by enabling the continuous and unobtrusive gathering of critical physiological data from patients in their natural settings, including temperature, blood pressure, and heart rate.

- **The Promise and Concerns**

WBANs have many benefits for both patients and healthcare professionals. Patients are given more authority over their health and the information they need to make wise choices. Healthcare professionals need to monitor patients 24×7 in any real-time applications, which allow for timely treatments and reduce the workload for hospitals. Notwithstanding these benefits, there are reasonable worries about the widespread adoption of WBANs, the most important of which are patient privacy and data security.

- Improved medical care
- Increased portability
- Cost-effectiveness
- Personalized consideration

- **Challenges in WBAN Security**

The decentralized and networked architecture of WBANs raises several security concerns. Because medical data is so sensitive, hackers would find it easy to target. Malicious actors could use network vulnerabilities to get unauthorized access to patient data, which could lead to privacy violations and even deadly consequences. Furthermore, implementing robust security measures may be challenging because wearable devices have limited computational power.

- Protection and security concerns: The sensitive nature of the health data acquired by WBANs raises significant privacy and security concerns.
- Interference: WBANs are frequently used in areas with high interference from other wireless devices.
- Battery life: The battery should be re-energized often for continuous patient monitoring for chronic diseases.
- Limited range usage: WBANs operate in a restricted range with limited resources and a shared medium.

10.2 SECURITY REQUIREMENTS FOR WBAN'S IN E-HEALTHCARE

WBAN plays an essential role in e-healthcare in tracking the status of the patient's health for chronic diseases, particularly for the elderly, through the digital (internet) medium by minimizing physical visits to doctors frequently. The body sensors of WBAN are used to collect the data and pass it through the gateway for storage and analysis purposes. Once the data is analyzed, the respective steps will be taken accordingly. So providing security for the patient's data from attackers is needed (Figure 10.2).

The specific requirements for WBAN in the healthcare environment are as follows:

- Low power consumption
- Interoperability
- Self-healing

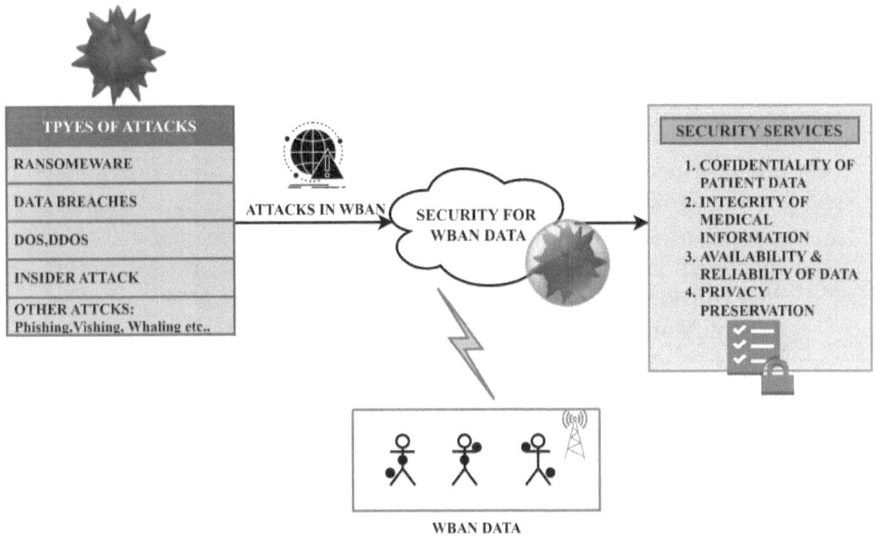

FIGURE 10.2 Attacks and services required for e-healthcare in WBAN.

- Security
  - Confidentiality of patient data
  - Integrity of medical Information
  - Availability and reliability of WBANs
  - Privacy preservation
- Low latency
- Popular threats or vulnerabilities in the e-healthcare system in the WBAN environment are as follows:
  a) Ransomware: Ransomware attack makes a file extremely challenging for the invader to access or manage on their own until compensation is done. Considering the healthcare industry, whenever a ransomware assault occurs, several critical processes are either completely stopped or rendered unusable. Training on ransomware safeguarding was conducted across the nation by the Multi-State Information Sharing and Analysis Centre (MSISAC) in collaboration with partners at the National Health Information Sharing and Analysis Centre (NH-ISAC) and Financial Services Information Sharing and Analysis Centre (FS-ISAC) [3].
  b) Breaches Over Data: Thieves can utilize personal health information for various illicit purposes, but more data fissures occur in the healthcare industry than in any other sector like defense, education, and so on. By turning that information against them, they make a profit.
  c) Distributed Denial-of-Service (DDoS) Attacks: This type of attack involves attackers taking down a network for a set amount of time, which

causes issues for those working in the healthcare industry who need to access the internet or a network to do a variety of tasks, like transferring emails about patients abnormal conditions, prescribing medication, keeping track of patient status, giving patients the care they need, and so forth.

d) Insider Threat: Companies frequently wrestle with safeguarding their company's integrity and view external threats to networks as extremely serious and significant hazards. An insider attack is the intruders with a thorough knowledge of the company from toe to head in all aspects. They can do anything from anywhere to any extreme without facing any issue in future.

Other attacks include phishing, vishing, smishing, whaling, and spear phishing.

10.2.1 Confidentiality of Patient Data

Data confidentiality protects information from unauthorized disclosure and is easily achieved by encryption. Recent research has successfully enabled secure transmission using cryptographic algorithms for wireless network protocols. As a result, lightweight cryptographic techniques are utilized to encrypt critical health-related information securely.

10.2.2 Integrity of Medical Information

Authentication is the most fundamental prerequisite for achieving network integrity and security [4]. Machine learning-based authentication techniques use some form of artificial intelligence (AI) to allow the machine to educate, train, and progress robotically from practice without the need for explicit programming like Artificial Neural Network (ANN), Support Vector Machine (SVM), Naive Bayes, and so on [5].

10.2.3 Availability and Reliability of WBANs

We require adequate security techniques, particularly in WBANs, to ensure continuous service availability; otherwise, their service performance would be effortlessly in jeopardy, resulting in network partitioning. Due to these threats, patients may reach a critical situation.

10.2.4 Privacy Preservation

Encryption should be used to protect the privacy of users' sensitive personal information and the illness models used by the healthcare center. In [6], a remote healthcare system, an efficient and privacy-preserving priority classification technique (PPC) is presented for classifying patients' encrypted information at the WBAN gateway.

10.3 A COMPREHENSIVE SECURITY FRAMEWORK FOR WBANS

10.3.1 End-to-End Data Encryption

- **Symmetric Versus Asymmetric Encryption**

- A single key is used for encryption and decryption in symmetric encryption. This makes it faster than asymmetric encryption, but it also necessitates the sender and receiver sharing the secret, which may be difficult in a WBAN. Many symmetric encryptions occur where Caesar's cipher plays a great role among all others. It is a simple, lightweight solution that requires fewer instances for the security process and is the ideal method for maintaining WBAN and Wireless Sensor Network (WSN) security. A few of the best-known symmetric key-based algorithms embrace data and advanced encryption standards, such as Blowfish and the like.
- The data transferred between the sensors and the gateway can be encrypted using symmetric encryption. This information could include things like heart rate, blood pressure, temperature, glucose level, and other medical parameters.
- Asymmetric encryption involves the use of two keys: one public and one private. The public key can be distributed to everyone, but the secret key must remain private. This increases its security over symmetric encryption, but it also slows it down. Examples of asymmetric cryptography algorithms are Rivest–Shamir–Adelman (RSA), elliptic curve cryptography (ECC), ElGamal, and so forth.
- Symmetric encryption may be the best option if data must be encrypted swiftly and effectively. Asymmetric encryption may be beneficial if the data is extremely safe.

- **Hybrid Cryptosystems**

- In WBANs, a hybrid approach is frequently adopted. The majority of the data is encrypted using symmetric encryption, while the symmetric key is encrypted using asymmetric encryption. This can offer the best of both worlds: rapid encryption and robust security.
- For validating a message widely, hashing generates a one-of-a-kind, rigid-sized signature to enhance the integrity of the data successfully.
- Hybrid cryptography provides the CIA with fast and secure communication.
- Hybrid encryption is provided by combining Advanced Encryption Standard (AES), ECC, and RSA methods. Authentication and integrity are additionally provided by the Secure Hash Algorithm (SHA-256). In terms of processing time, the new combo is proposed to give enriched results to outperform individual cryptographic techniques [7].
- The processing time for encoding and decoding for individuals as well as hybrid combinations of cryptographic algorithms is shown in Table 10.1, and the percentage of time is shown in the pie chart representation in Figure 10.3.

TABLE 10.1
Comparison of Individual and Hybrid Information Process Time [7]

Message Length (Bits)	Encoding–Decoding Time (seconds)				
	Individual Cryptographic			Hybrid Cryptographic	
	AES	ECC	RSA	AES-RSA	RSA-ECC
48	19.51	11.55	0.2306	10.2493	8.3266
64	28.46	15.48	0.2350	14.7360	14.8445
80	37.37	23.03	0.3184	21.8297	17.1274
96	42.97	35.19	0.3518	34.0297	21.6842

It assures that ECC provides higher security compared with other algorithm settings and key. ECC is a composite way of building with other cryptographic schemes to create a crossbreed scheme that is thought to be more resistant and safe than individual cryptographic systems (Figure 10.4).

- **Key Management and Distribution**

- Any end-to-end data encryption solution must include key management and distribution. They safeguard the security of sensitive information about patients throughout its lifecycle, from creation through storage and transfer. These keys are generated using cryptographically secure random number

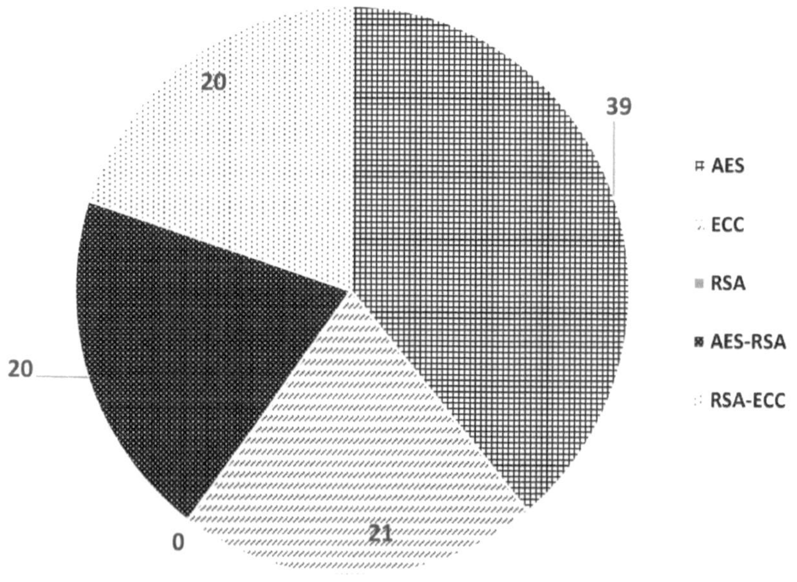

FIGURE 10.3 Pie chart of encode to decode time.

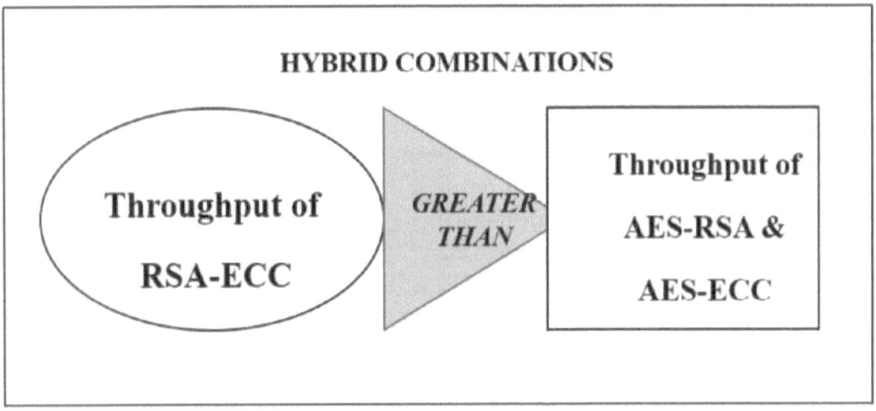

FIGURE 10.4 Throughput of hybrid cryptosystem.

generators. To avoid unauthorized access, encryption keys must be securely stored. Key exchange protocols, secure channels, and public key infrastructure (PKI) are all examples of secure key distribution techniques.

- In [8], the architecture for a three-tier network structure is constructed for heart disease diagnosis by adapting the IEEE 802.15.6 MAC address to determine the precedence level of the patient information. In the first tier, AES is used to encrypt data. In the second layer, bird swarm optimization is used to select the optimal channel for processing. At the diagnosis server, the sensed data is classified in the third layer to diagnose congenital heart diseases (CHD). For categorization, the neuro-fuzzy algorithm is integrated into the diagnosis server. CHD is identified on the diagnosis server using screening pulse oximetry as well as heart rate and breathing rate. A comparison is made between the three-tier network architecture with the Continuous Real-Time Wireless Body Area Network (CR-WBAN), classification, and four-tier architecture.

10.3.2 Secure Authentication Mechanisms and Authorization

- There should be a system that authenticates the message, and communication networks utilize the Message Authentication Code (MAC) for this purpose. Authentication is a technique that verifies the user's submitted credentials with the user's previously stored credentials, and if the result (pattern) is true (matched), then the resources can be accessed by the person who is authenticated and authorized or else they are not allowed for access. The authentication procedure is critical for maintaining data integrity and secrecy [3]. In the realm of healthcare networking, a three-way authentication mechanism is used:

1. Single-way process
2. Double-way process
3. Triple-way process

- **Biometric Authentication**

Non-biometric and biometric-based authentication techniques are the two broad types of authentication schemes. Each person has unique identifiers such as fingerprints, retina patterns, iris, voice, DNA, and so on. To uniquely identify a person, biometric identification techniques leverage all aspects of the vital signs of the body.

Biometric features are classified into two types: physiological (the structure of physical components of a human body like shape, size, texture of hair, etc.) and activities (human speech, walk, poses, etc.) [9]. Because there is no need to memorize passwords/credentials for authentication, this kind of authentication is simple to adopt and more secure. Biometric tools are used to verify the patient's identity so that the data can be passed through the WBAN.

- **Multi-Factor Authentication**

Multi-factor authentication (MFA), also known as two-factor authentication (2FA) or two-step verification (2SV), is a security mechanism used to increase the security of online accounts and systems. It uses two or more separate elements to achieve authentication. In addition to the usual username and password combination, it demands that users submit several kinds of authentication before giving access to their accounts. MFA helps reduce the dangers of stolen credentials or unauthorized access.

In general, three factors can be employed for authentication:

- **Something you know**: Typically, this is a password or a personal identification number (PIN). It is the most prevalent type of authentication, although it is subject to a variety of attacks when used alone (e.g., password/PIN).
- **Something You Have**: A physical object that only the account owner possesses, such as a smartphone, security token, or smart card, is used. These devices generate time-sensitive codes or respond to challenges, providing an additional layer of security (e.g., cryptographic identification device or token).
- **Something You Are:** This includes biometric features such as fingerprints, retina scans, voice recognition, and facial recognition. These are unique to each person and difficult to imitate (e.g., biometric).

MFA approaches that are commonly used include:

a) **Text messages or SMS:** A verification code is texted to your registered phone number.
b) **Authentication Apps:** Smartphone programs such as Google Authenticator, Authy, and Microsoft Authenticator produce time-sensitive codes that you enter upon login.
c) **Email Verification:** A link or code for verification is emailed to your registered email address.
d) **Push Notifications:** On your device, you will receive a notification to allow or deny access attempts.

e) Hardware tokens are physical devices that create time-sensitive codes and are commonly utilized in business settings.
f) **Biometric Verification:** Fingerprint, facial recognition, or other biometric scans are examples of biometric verification.

MFA has become increasingly important as online threats have evolved. While it improves security, it is critical to strike a balance between security and user comfort, as too-complex authentication methods can cause user irritation.

- **Token-Based Authentication**

Token-based authentication systems allow users to submit their credentials only once and receive an encrypted string of random characters in return. Here the generation of the token is very important, which is unique for each user and is generated digitally by the computer. Each token allows the user to access the device easily and quickly without any confusion among the users. Token-based authentication uses the RESTful application programming interfaces along with decentralized cryptography. The token has many advantages like easy to access, no need to remember the credentials for each login, highly secured, application-independent, and so on, all because of its digitization. Each user identity is verified, and then the token is generated based on the individual user details uniquely. So any website or network can be accessed by using this token reliably and eventually. There are two types of tokens: physical tokens and web tokens. Tokens can be implemented with certain drivers like user, client server, authorization server, and resource server. Token-based authentication works with the following steps:

- Request
- Verification
- Token validation
- Storage

In [10], in order to secure users' private data in WBAN, a 2FA system is developed that reduces the false acceptance rate (FAR) to nearly zero while maintaining a high false rejection rate (FRR). The system involves customized iris fusion as well as electrocardiogram (ECG) signal characteristics. By combining two features, patients are enhanced with the security features (i.e., the WBAN).

- The first level is the authentication of the iris.
- The first level is followed by the second level, the authentication of the ECG.

A general experiment has been conducted to authenticate the performance of the two factors using public databases. Finally, the results have been discussed for the FAR and the FRR.

For WBAN security applications, a method for creating iris coding templates is suggested. This approach incorporates concepts related to mapping, reducing the ode of iris, and logical operations, which drastically lowers the amount of resources used in WBAN.

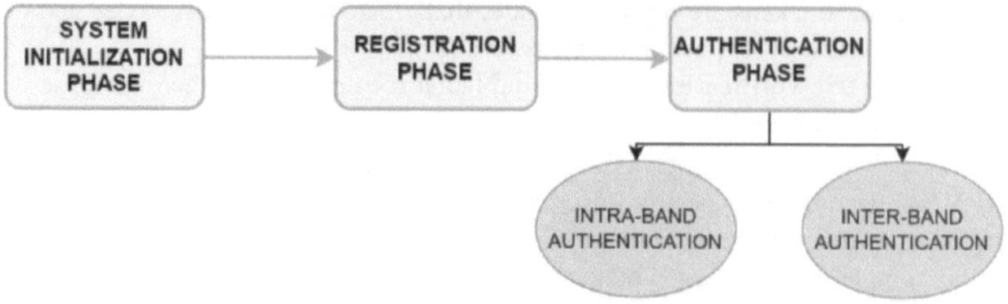

FIGURE 10.5 Flow of MSR technique.

Utilizing three effective inter-pulse interval (IPI) programming techniques, an enhanced ECG IPI) encoding methodology is used. Through testing, it demonstrated that the suggested two-factor authentication method employs reduced interval while maintaining acceptable unpredictability when compared to six other widely used encoding techniques. A [11] new security feature provides session unlink ability as well as forward secrecy to achieve great robustness against security evaluation tools and techniques while ensuring mutual authentication with BAN-logic.

Yang et al. [12] provided a more lightweight authentication method for WBANs that support mobile edge computing. The modular square roots (MSR) technique is used in two authentication phases: intra-BAN authentication between the sensor node and the edge node (EN) and inter-BAN authentication between the EN and the application provider. One of the key advantages of MSR over RSA and ECC is that it is significantly simpler and more efficient. The suggested approach provides robust security by doing a thorough security analysis of the eavesdropping attack, replay attack, impersonation attack, man-in-the-middle attack, and denial-of-service (DoS) attack. The MSR technique is executed in three phases, which are shown in Figure 10.5.

10.3.3 Access Control

In general, there are three types:

- Access control based on role (RBAC)
- Access control based on attributes (ABAC)
- Access control, that is, fine-grained

The CIA property is attained by the following:

C- data encryption circulates the secret key (E).
I- Data hashing which is represented by (H).
A- by generating MAC from the input data.

Finally, data access control is achieved using attribute-based encryption (ABE) to encrypt the data. Three-layer models are used to create a secure WBAN environment for e-healthcare.

- **First Layer: Data Collection:** Occurs within sensors and the base station (BS) and serves as the second layer's data source, with sensor interaction restricted to the patient's body.
- **Second Layer: Communication:** The BS and third layer are connected through communication to deliver data from the sensors to a repository over the internet.
- **Third Layer: Application:** To retrieve and analyze sensed data, tasks are carried out. Both responsibilities are carried out by accredited parties.

A simple cryptographic technique can ensure protection in a WBAN environment [13]:

- **Lightweight Block Encryption:** For encryption and decryption, a synchronized key is utilized to offer confidentiality. To process a block of 64/128 bit, a block cipher goes through multiple rounds.
- **Hash Function:** A cryptographic hash function H, commonly known to be checksum function, often ensures integrity. This approach generates a bit string that can be any size. The function makes use of all data bits to create an exclusive, unrepeatable code.
- **MAC:** The message's authenticity is verified using MAC values. The two arguments required by the MAC creation are k (key) and M (message). The sender and receiver are the only parties who are aware of the cryptographic hash function used to calculate MAC values, H to M combined with k.
- **ABE:** It is an efficient technique to grant access to encrypted data to a set of customers promptly. This type of encryption substitutes a collection of qualities that confirm an entity's certification and access rights for the use of an identity.

Lightweight cryptography comes into play to provide security services in constrained environments by trading off pre-existing algorithms like MAC, hashing, and so on. The performance and longevity of a WBAN decrease as the security level rises. The different security levels of lightweight ciphers are discussed in Table 10.2 based on

TABLE 10.2
Security Level of Lightweight Cryptographic Algorithms [13]

Year	Types of Services				Level of Security
	Confidentiality	Authenticity	Integrity	Access Control	
2012	Custom	Custom	×	v	24×5
2014	AES	×	×	ABE	27
2017	PRESENT	×	×	×	24×5
2017	LEA	×	×	×	27
2018	KBS/KAISC	MAC	×	×	—

TABLE 10.2 ***(Continued)***
Security Level of Lightweight Cryptographic Algorithms [13]

Year	Types of Services				Level of Security
	Confidentiality	Authenticity	Integrity	Access Control	
2019	×	EC2	Digital signature	ABE	24×5
	×	EC2	×	×	—
	2-fish	×	×	ABE	—
	×	PKC	SHA2	×	—
	×	EC2	ABE	ABE	24×5
2020	×	×	×	×	—
	CAESAR	EC2	SHA-512	×	27
	×	×	×	×	27
	×	×	×	ABE	24×5
	×	Custom	×	×	—
	Signcryption	Signcryption	Signcryption	×	27
2021	×	Custom	×	×	—
	×	IBE	×	×	24×5
	×	Custom	SHA	×	27
	×	IBE-PKC	×	×	—
	LEA	MAC	Spongent	ABE	≥27

attribute-based encryption. The gadgets' energy and memory requirements also rise as the security level does. When access control uses the ABE, more complex systems hurt overall performance, memory, energy usage, and the performance of the underlying computing hardware. A security framework for the WBAN with encryption, authentication, authorization, and access control is depicted in Figure 10.6. Security primarily depends on key distribution and management over the network structure in the healthcare environment.

- Attribute-Based Access Control (ABAC)

- The database has control over access to the data by using ciphertext policy-ABE (CP-ABE). Let's say, a hacker has access to the encrypted data as well as the encryption key that can be used to unlock it. CP-ABE ensures access control for the user who has the private key SKu linked to an authoritative set of characteristics and can retrieve the materials.
- Based on a collection of characteristics gathered from it, the BS establishes the encryption as a link policy in Layer 2. Only legitimate features fulfill the access policy in Layer 2 for encryption and decryption if the trusted server is sincere and safely deployed.

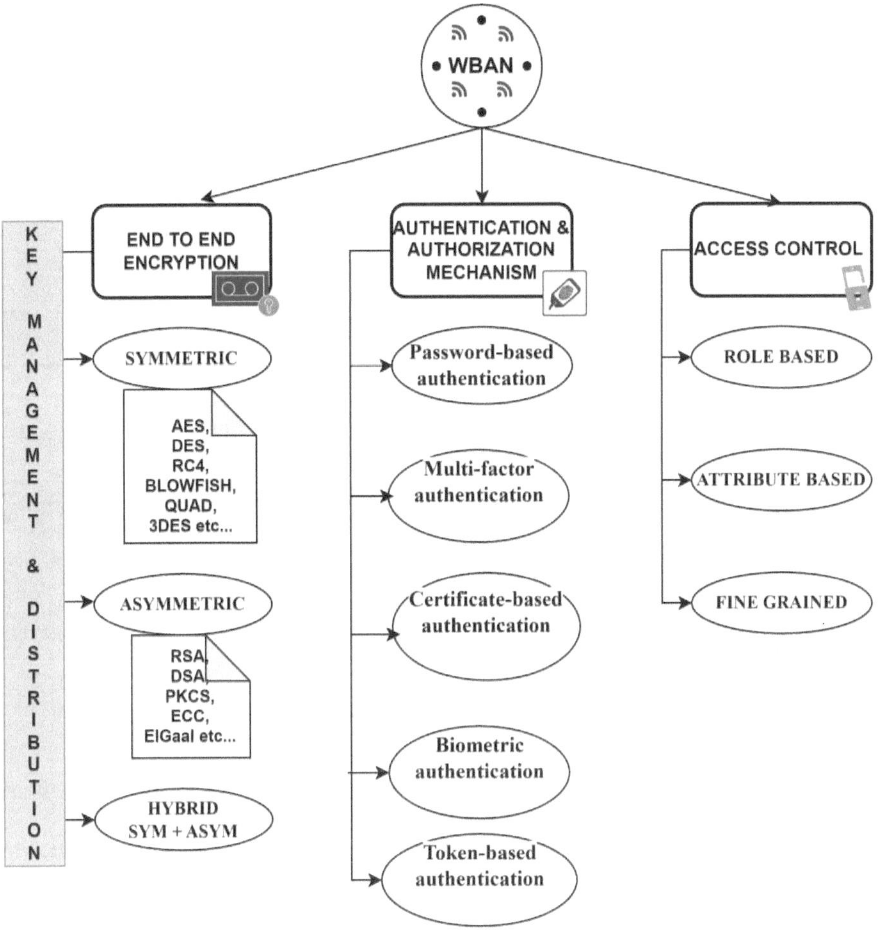

FIGURE 10.6 Security framework WBAN.

10.4 INTRUSION DETECTION/PREVENTION SYSTEMS

The means of detecting or signaling the possibility that privacy (sensitive or personal data) has been compromised somewhere within a communication network, that is, insider assailants or intruders have prevented or hindered device communication or system functionality, is termed intrusion detection. The work [14] will provide existing security mechanisms or tools for intrusion detection and prevention in the healthcare environment.

A network needs to be designed and built to notify the abnormal behavior of unauthorized user access. A system collects and keeps track of the malicious behavior from normal behavior, when it reaches a threshold level it alarms about the critical situation as information to the preventive system. Types of intrusions are shown in Figure 10.7.

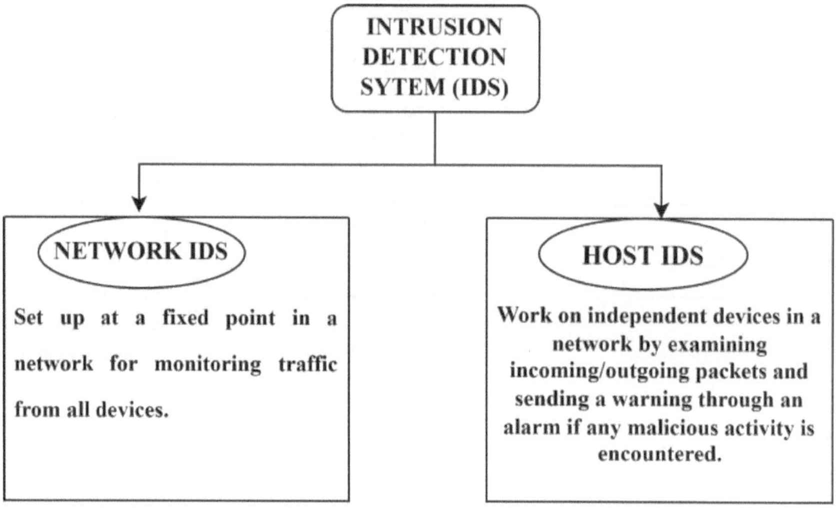

FIGURE 10.7 Types of intrusion detection system.

There are two intrusion detection systems (IDS):

1. Anomaly-based IDS
2. Signature-dependent IDS

The above methods can be used together, separately depending on the application and their constraints.

Some of the tools and techniques employed for intrusion detection are as follows:

- An event manager for providing secrecy, Solar Winds is used effectively.
- Snort is a Cisco Systems tool, which is benign to use and leads to IDS based on the network.
- OSSEC is a great host intrusion detection solution that is free to use.
- Suricata is a system for network-based intrusion detection that improves application layer visibility.
- Bro is a system that uses networks for intrusion prevention and network control.
- Sagan is a log analysis program that can integrate documents created from snort data, acting as both a Host-Based Intrusion Detection System (HIDS) and a Network-Based Intrusion Detection System (NIDS).
- Security Onion: Software for network security and monitoring that incorporates parts of applications
- AIDE is a host-based IDS for multiple operating systems in the Advanced Intrusion Detection Environment.
- Open WIPS-NG: A system for preventing intrusions and IDS based on wireless networks from the company that created Aircrack-NG. A simple host-based intrusion detection tool for different OS is called Samhain.
- Fail2Ban: For different operating systems, this host-based IDS is lightweight.

There are other latest techniques that are in trending, that is, machine learning algorithms used for intrusion detection. However, these algorithms have cryptography as a base for their performance metrics. IEEE 802.15.6 is a protocol specially designed for WBAN security. Many lightweight protocols, hybrid versions of cryptography techniques, and multiple-factor integrity checks for authentication have evolved in the current era. Generally, any type of network is in need of minimum security and privacy around all aspects. Blockchain helps a lot in achieving security enhancement for WBAN data transmission for sensitive biological data.

- Real-Time Monitoring and Response

WBAN provides a wide range of healthcare system evaluations. Different sensors can be placed under the skin and attached to the body or clothing. Real-time parameter calculations are made from a patient's body using the sensors [15]. Based on significant results from a survey on BAN in the healthcare system, an Android-based heart monitoring and reporting system was developed. The recommended solution depends on body temperature and heart rate metrics and uses essential WBAN characteristics.

10.5 SECURE COMMUNICATION PROTOCOLS

There are different secure communication protocols employed in WBAN [16]. Some of them are listed in Table 10.3.

TABLE 10.3
Protocol and Its Functions for Secured Communication

S. No.	Security Protocol	Functions of the Protocol
1	Zigbee Security	Zigbee is a low-power wireless communication protocol widely utilized in WBANs. It includes security functions such as encryption, message integrity, and device authentication.
2	Bluetooth Low Energy (BLE) Security	BLE is another standard protocol used in WBANs. It includes encryption, privacy protections to prevent tracking, and pairing techniques to establish secure connections.
3	Health Level 7 (HL7) Security	HL7 is a set of standards for electronic health information transmission, integration, sharing, and retrieval. It contains security standards to safeguard patient data during communication.
4	Transport Layer Security (TLS)	TLS provides communication security between a WBAN and third-party systems such as hospital servers or cloud platforms. It protects the confidentiality and integrity of data during transmission.

It should be noted that the protocol and security measures selected should be based on the unique use case, the amount of sensitive information being transferred, and the regulatory requirements of the medical area.

In addition to a physiological signal, an electroencephalogram (EEG) results in the presentation of a twofold chaotic encryption method with natural randomness [17]. To guarantee the intrinsic unpredictability of the beginning values, the three-dimensional eigenvalues—which consist of deep multi-domain features—are extracted as the first elements of the chaos-based pseudo-random number generator. Next, two chaotic maps are proposed to create a hybrid pseudo-random number generator with high randomness and low hardware complexity: a piecewise linear map (PLCM) and a coupled logistic-tent map (LTM). Finally, a double chaotic encryption technique for WBANs based on the EEG eigenvalue is shown. The encryption method can repel a variety of attacks, such as noisy and draining ones, and is robust enough to safeguard the security of the system.

A Secure and Anonymous Mutual Authentication and Key Agreement (SAMAKA) system which is safe and mutually authenticated using key agreement based is employed in [18] to protect all WBAN security elements against different types of security intrusions. The SAMAKA scheme is displayed in Figure 10.8.

- The SA accomplishes the setup and enrollment processes using a private and secure communication channel. During the agreement (MAKA) phase of mutual authentication, the sensor and chief nodes safely construct a session key across a public, insecure communication channel through the mobile node. Finally, a fresh sensor is dynamically inserted into the current system during the dynamic node addition phase.
- SAMAKA provides a safe and privacy-preserving authentication system that uses a mid-node (MN) to mutually authenticate the sensor node as well as the chief node and secure session key agreement by using basic hash and XOR operations.
- The security evaluation tools and techniques are used to validate the security of SAMAKA against a variety of adversarial attacks, achieve shared validation, and assure safe agreement. In addition, the unofficial safety scrutiny is demonstrated to show that SAMAKA achieves the required confidence characteristics and guards against adversarial attacks.
- Finally, SAMAKA's capabilities are evaluated against comparable systems regarding communication, computing, and storage overhead.

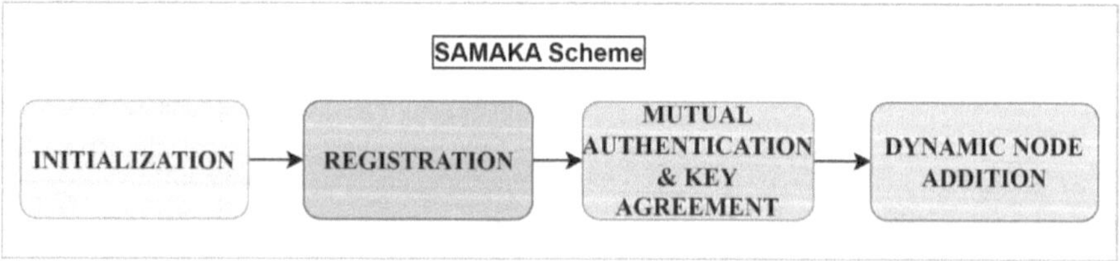

FIGURE 10.8 Four phases of SAMAKA.

The physical layer attacks are created due to more vulnerable wireless interfaces. Indeed, attempts to jam WBAN infrastructures have grown to be a severe issue. To combat jamming attacks in medical settings, Cherifi et al. [19] propose CFD-BD: Chaotic FH/DS Based Dissimulation, a new security protocol that combines frequency-hopping (FHSS) and direct sequence (DSS) spread spectrum. The approach depends on dissimulation and the confused signal. Combining these two approaches guarantees a strong defense against jamming attacks while consuming minimal amounts of communication burden, complexity, and storage.

The CFD-BD scheme involves four stages:

1. Initialization: The Diffie-Hellman method is used at this phase to create an encryption key (K) between the sensors to safeguard the messy significance.
2. To find chaotic value: Calculate the messy significance $X_{i.}$
3. Dissimulation: Which includes X_i's messy significance to the transited frame.
4. FHSS/DS requires encrypting the message (M) to be conveyed using X_i and selecting in advance the level of frequency bouncing to be utilized between the two points of transmission.

- **Secure Routing Protocols for WBANs**

The security of patient data is proposed, along with an effective extended and upgraded IoT-based radio-frequency identification (RFID) authentication technique for WBANs that might fend off various threats and guarantee reciprocal validation between tag and medical server [20]. The elliptic curve digital signature with message recovery (ECDSMR) and ECC encryption mechanisms have been used for this. The elliptic curve cryptosystem's primary building element is thought to be point multiplication (PM). Using hybrid ECC, public key encryption, and hashing function, this protocol aims to offer common validation between the tag, the reader, and the medical server. Furthermore, this protocol uses session recovery and a digital signature to encrypt data exchanged between the device reading it and the server.

To ensure that data is exchanged across secure channels within a WBAN, protocols like Authenticated Routing (ARAN) and Distance Vector (SEAD) for ad hoc networks to be secured efficiently might be utilized.

While patients are receiving therapy, a wireless body sensor network (WBSN) collaborates, collects data, organizes the data, and constantly monitors them for abnormalities from normal conditions. To overcome the security issue with WBAN routing, it relies on a cryptographic network policy for proactive trust ware routing, which generates encrypted communication in WBSN [21]. The creation of the Fork-Hook method of encryption is based on the Secured Data Centric Routing Gateway Protocol (SDCRGP) with data-centric routing. There are three steps to be accomplished by using blockchain principles, Fork-Hook encryption policy (FHEP), and so on. It starts with hashing the packets on the transmission media using blockchain principles. Then, navigate through establishing a Successive Hop Index (SHI) using the FHEP. Lastly, confirmation about the secret key associated with each transmission node is active. This is how the sleeper node threading protocol encrypts data packets: it removes the path's non-authenticated slumber cell nodes. Each node receives the encrypted packet from the associated overpowering node after it has been confirmed by the dominant node grouping by the Hop Count Rollback Node Aggregator (HCRNA). After broadcasting the cluster head-on, the controller that manages node aggregation ultimately determines the hop calculation adjustment. The authenticator then offers the crucial key to unlock the data. In terms of secure transit and identification, the suggested method outperforms protected adaptive heads of cluster aggregation.

10.6 SUMMARY OF THE SECURITY FRAMEWORK

To safeguard the privacy, confidentiality, reliability, and integrity of sensitive medical data acquired by wearable and implantable devices, WBAN security employs a combination of encryption, authentication, access control, data integrity, and security communication protocols. Individuals would be able to benefit from continuous health monitoring while retaining their privacy and data protection. Machine learning algorithms can be involved in efficient anomaly detection in health monitoring. Many tools available for IDS are discussed. Symmetric, asymmetric, and certificate-less encryption techniques and types were reviewed in this chapter. Apart from many traditional techniques, many lightweight cipher techniques have been proposed for constrained devices with hashing, MAC, ABE, and so forth in WBAN environments for e-healthcare applications. The AVISPA tool is most prominently employed to check security performance. The results of different recent lightweight ciphers, along with the hybrid combination, were discussed. In the future, new variants of ciphers can be created and employed for security purposes in WBAN e-healthcare. Blockchain can also be implemented for high security in medical fields in a WBAN environment.

REFERENCES

[1] Ullah, I., Zeadally, S., Amin, N. U., Khan, M. A., & Khattak, H. (2021). Lightweight and provable secure cross-domain access control scheme for the Internet of Things (IoT) based wireless body area networks (WBAN). *Microprocessors and Microsystems, 81*, 103477.

[2] Bouazizi, A., Zaibi, G., Samet, M., & Kachouri, A. (2017, February). Wireless body area network for e-health applications: Overview. In *2017 International Conference on Smart, Monitored and Controlled Cities (SM2C)* (pp. 64–68). IEEE.

[3] Kumar, P., & Sharma, A. (2017). Survey on authentication process in the body area network. *The International Journal of Electronics Engineering Research, 9*(6), 913–921.

[4] Hussain, M., Mehmood, A., Khan, S., Khan, M. A., & Iqbal, Z. (2019). Authentication techniques and methodologies used in wireless body area networks. *Journal of Systems Architecture, 101*, 101655.

[5] Andročec, D., & Vrček, N. (2018, July). Machine learning for the Internet of things security: A systematic. In *13th International Conference on Software Technologies* (Vol. 4120, p. 97060). SCITEPRESS.

[6] Wang, G., Lu, R., & Guan, Y. L. (2019). Achieve privacy-preserving priority classification on patient health data in remote eHealthcare system. *IEEE Access, 7*, 33565–33576.

[7] Subedar, Z., & Araballi, A. (2020). Hybrid cryptography: Performance analysis of various cryptographic combinations for secure communication. *International Journal of Mathematical Sciences and Computing (IJMSC)*, *6*(4), 35–41.
[8] Reddy, S. R. N., & Kumar, D. (2022). Early congenital heart defect diagnosis in neonates using novel WBAN-based three-tier network architecture. *Journal of King Saud University-Computer and Information Sciences*, *34*(6), 3661–3672.
[9] Al-Shayea, Q., & Al-Ani, M. (2018). Biometric face recognition based on enhanced histogram approach. *International Journal of Communication Networks and Information Security*, *10*(1), 148–154.
[10] Jiang, D., Zhang, G., Samuel, O. W., Liu, F., & Xiao, H. (2022). Dual-factor WBAN enhanced authentication system based on Iris and ECG descriptors. *IEEE Sensors Journal*, *22*(19), 19000–19009.
[11] Attir, A., Naït-Abdesselam, F., & Faraoun, K. M. (2023). Lightweight anonymous and mutual authentication scheme for wireless body area networks. *Computer Networks*, *224*, 109625.
[12] Yang, X., Yi, X., Khalil, I., Luo, J., Bertino, E., Nepal, S., & Huang, X. (2021). Secure and lightweight authentication for mobile-edge computing-enabled WBANs. *IEEE Internet of Things Journal*, *9*(14), 12563–12572.
[13] Morales-Sandoval, M., De-La-Parra-Aguirre, R., Galeana-Zapién, H., & Galaviz-Mosqueda, A. (2021). A three-tier approach for Lightweight data security of body area networks in E-health applications. *IEEE Access*, *9*, 146350–146365.
[14] Mishra, S., Agarwal, K., & Tyagi, A. K. (2020). Intrusion detection in Internet of Things based e-healthcare system-a systematic review. *ICT for Competitive Strategies*, 651–667.
[15] Sathya, G., & Evanjaline, D. J. (2018). Body area network in health care system with future analysis. *Advances in Computational Sciences and Technology*, *11*(1), 77–83.
[16] Garg, A., Singh, A. K., & Garg, M. (2023). Role of Internet of Things and artificial intelligence for healthcare informatics: An overview. *Innovations in Healthcare Informatics: From Interoperability to Data Analysis*, *41*, 107.
[17] Wang, J., Huang, D., Fan, S., Han, K., Jeon, G., & Rodrigues, J. J. (2023). PSDCE: Physiological signal-based double chaotic encryption for instantaneous E-healthcare services. *Future Generation Computer Systems*, *141*, 116–128.
[18] Narwal, B., & Mohapatra, A. K. (2021). SAMAKA: Secure and anonymous mutual authentication and key agreement scheme for wireless body area networks. *Arabian Journal for Science and Engineering*, *46*(9), 9197–9219.
[19] Cherifi, F., Omar, M., Chenache, T., & Radji, S. (2022). Efficient and lightweight protocol for anti-jamming communications in wireless body area networks. *Computers & Electrical Engineering*, *98*, 107698.
[20] Izza, S., Benssalah, M., & Drouiche, K. (2021). An enhanced scalable and secure RFID authentication protocol for WBAN within an IoT environment. *Journal of Information Security and Applications*, *58*, 102705.
[21] Ramalingam, V., Saminathan, R., & Baalamurugan, K. M. (2023). Fork-Hook encryption policy-based secured Data Centric Routing Gateway for proactive trust ware data transmission in WBSN. *Measurement: Sensors*, *27*, 100760.

11 Machine Learning-Based Security Approaches for Wireless Body Area Networks

Ankit Garg, Ajay Kumar, and Anuj Kumar Singh

11.1 INTRODUCTION

Wireless Body Area Networks (WBANs) represent a transformative technology in healthcare, enabling the seamless integration of wearable and implantable medical devices for continuous monitoring and management of patient health data [1]. WBANs consist of interconnected sensors, actuators, and communication modules that collect physiological information such as heart rate, body temperature, and blood glucose levels in real time. By wirelessly transmitting this data to healthcare providers, WBANs facilitate early detection of medical conditions, personalized treatment plans, and remote patient monitoring, thus improving the quality of care and patient outcomes. Moreover, WBANs empower patients to actively participate in their own healthcare by enabling self-monitoring and health tracking. With the potential to revolutionize various healthcare applications, including chronic disease management, eldercare, and fitness tracking, WBANs have garnered significant attention from researchers, clinicians, and industry stakeholders alike [2]. Their significance lies in their ability to enhance healthcare delivery, promote preventive medicine, and empower individuals to lead healthier lifestyles through continuous monitoring and personalized interventions. WBANs introduce several security challenges that must be addressed to ensure the confidentiality, integrity, and availability of sensitive medical data [3]. One primary concern is data privacy, as WBANs transmit personal health information wirelessly, raising the risk of unauthorized access and breaches. Without robust encryption and access control mechanisms, patient data may be vulnerable to interception and exploitation by malicious actors [4]. Authentication issues also pose a significant challenge, as WBAN devices must verify the identities of both users and other network components to prevent unauthorized access and tampering. Moreover, the dynamic and heterogeneous nature of WBAN environments can introduce potential vulnerabilities, including sensor tampering, data injection attacks, and denial-of-service (DoS) attacks [5]. Ensuring the security of WBANs requires a comprehensive approach that encompasses encryption, authentication protocols, Intrusion Detection

 DOI: 10.1201/9781032635101-14

System (IDS), and secure communication channels to mitigate these risks and safeguard patient privacy and data integrity.

11.1.1 Data Privacy Concerns [6]

- WBANs transmit highly sensitive medical data, including patient health records, physiological parameters, and treatment information.
- Without adequate encryption and access control mechanisms, this data is susceptible to interception, eavesdropping, and unauthorized access.
- Privacy regulations such as Health Insurance Portability and Accountability Act (HIPAA) impose strict requirements on the protection of patient health information, necessitating robust security measures within WBANs [7].

11.1.2 Authentication Issues [8]

- Ensuring the authenticity and integrity of data within WBANs requires effective authentication mechanisms.
- WBAN devices must verify the identities of users, sensors, and network components to prevent unauthorized access and data manipulation.
- Traditional authentication methods, such as passwords and cryptographic keys, may be inadequate for WBANs due to the limited resources and computational capabilities of wearable and implantable devices.

11.1.3 Vulnerabilities in WBAN Environments [9]

- The dynamic and heterogeneous nature of WBAN environments introduces various potential vulnerabilities.
- Sensor tampering: Malicious actors may tamper with sensors or implantable devices to manipulate data or disrupt functionality.
- Unauthorized parties may inject false or malicious data into the WBAN through data injection attacks, leading to inaccurate medical diagnoses or treatment decisions.
- Attackers may attempt to overload WBAN communication channels or disrupt network operations through DoS attacks, causing service interruptions and compromising patient care.

11.1.4 Resource Constraints [10]

- Wearable and implantable devices within WBANs are typically resource-constrained in terms of processing power, memory, and battery life.
- Implementing robust security measures while minimizing resource consumption poses a significant challenge in WBAN environments.
- Security protocols and cryptographic algorithms must be optimized to operate efficiently within the constraints of WBAN devices, balancing security requirements with resource limitations.

11.1.5 Interoperability and Standards [11]

- Interoperability issues between different WBAN devices and systems can create security vulnerabilities and compatibility challenges.
- A lack of standardized security protocols and communication interfaces may hinder secure data exchange and interoperability between WBAN components.
- Establishing industry-wide standards and protocols for secure communication, data encryption, and authentication is crucial to addressing interoperability challenges and ensuring the security of WBAN deployments.

Addressing aforementioned security challenges requires a holistic approach that integrates encryption, authentication, intrusion detection, and secure communication protocols tailored to the unique requirements and constraints of WBAN environments. By implementing comprehensive security measures, healthcare providers can mitigate risks, safeguard patient privacy, and ensure the integrity of medical data transmitted over WBANs.

11.2 APPLICABILITY OF MACHINE LEARNING IN WBANS

Machine learning (ML) has emerged as a powerful tool for addressing security threats in WBANs, offering adaptive and intelligent solutions to mitigate risks and safeguard sensitive medical data. At its core, ML involves the development of algorithms and models that enable computers to learn from data, identify patterns, and make predictions or decisions without explicit programming [12]. In the context of WBAN security, ML techniques leverage data-driven insights to detect anomalies, predict potential threats, and enhance overall network resilience. The applicability of ML in WBAN security is multifaceted and encompasses various below-mentioned aspects of threat detection, anomaly identification, and behavioral analysis. These aspects have been shown in Figure 11.1.

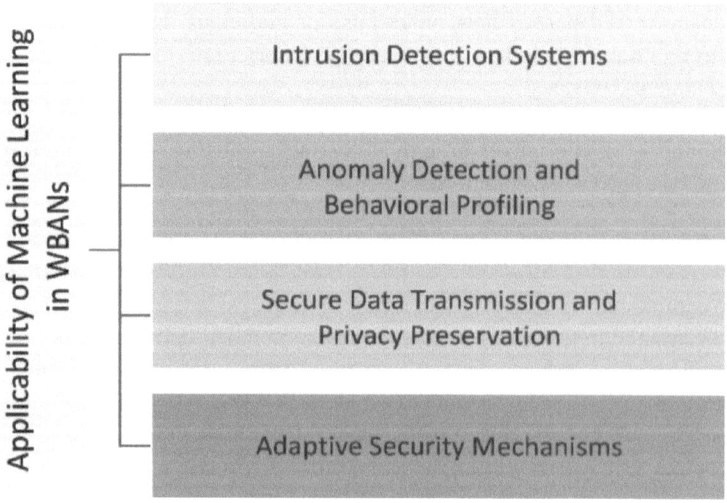

FIGURE 11.1 Applicability of ML in resolving issues of WBANs.

11.2.1 Intrusion Detection Systems

ML algorithms can analyze network traffic patterns and device behavior to detect anomalous activities indicative of security breaches. Supervised learning techniques, such as Support Vector Machines (SVM) and Random Forests, learn from labeled datasets to classify normal and malicious network behavior. Unsupervised learning methods, such as clustering and anomaly detection algorithms, identify deviations from normal patterns without requiring explicit training data, making them well-suited for detecting novel threats in WBAN environments [13].

11.2.2 Anomaly Detection and Behavioral Profiling

ML enables the development of models that learn the normal behavioral profiles of individual WBAN devices and users. By analyzing sensor data and device interactions, ML algorithms can detect deviations from established patterns and identify potentially malicious activities, such as sensor tampering or data injection attacks. Techniques such as clustering, Principal Component Analysis (PCA), and Hidden Markov Models (HMMs) enable the identification of anomalous behavior and facilitate proactive threat mitigation in WBANs [14].

11.2.3 Secure Data Transmission and Privacy Preservation

ML-driven encryption schemes and cryptographic protocols enhance the security of data transmission over WBANs while preserving patient privacy. ML algorithms optimize encryption parameters and key management strategies to mitigate the risk of data interception and manipulation. In addition, twwechniques such as Differential Privacy and Homomorphic Encryption leverage ML to anonymize and mask sensitive information, ensuring the confidentiality and integrity of medical data in transit [15,16].

11.2.4 Adaptive Security Mechanisms

ML empowers WBANs to adaptively respond to emerging security threats in real time. By leveraging historical attack data, threat intelligence feeds, and reinforcement learning frameworks, ML algorithms can anticipate and preemptively defend against known vulnerabilities and attack vectors [17]. Adaptive security mechanisms enable autonomous decision-making and policy enforcement, enhancing the resilience of WBANs against adversarial activities and evolving threat landscapes.

In summary, ML offers a diverse array of tools and techniques for addressing security threats in WBANs. By harnessing the power of data-driven insights and adaptive decision-making, ML-driven security approaches enhance the integrity, confidentiality, and availability of medical data transmitted over WBANs, thereby enabling safer and more secure healthcare delivery.

11.3 SECURITY CHALLENGES IN WBANs

WBANs present unique security challenges that stem from their distinctive characteristics, including limited resources, dynamic topology, and the requirement for real-time monitoring. Understanding below-listed challenges is essential for designing effective security mechanisms tailored to the specific constraints of WBAN environments [18].

11.3.1 Limited Resources

- WBAN devices, including wearable sensors and implantable medical devices, are typically constrained in terms of processing power, memory, and energy supply.
- Implementing robust security mechanisms within these resource-constrained devices presents challenges in terms of computational complexity, memory overhead, and energy consumption.
- Security protocols and cryptographic algorithms must be optimized to operate efficiently within the limitations of WBAN devices while ensuring adequate protection against security threats [19].

11.3.2 Dynamic Topology [20]

- WBANs exhibit dynamic and unpredictable network topologies, with devices frequently joining or leaving the network as users move or engage in activities.
- The dynamic nature of WBAN topology complicates the establishment and maintenance of secure communication channels between devices.
- Security mechanisms must be adaptive and resilient to changes in network topology, enabling seamless authentication, encryption, and access control across evolving WBAN configurations.

11.3.3 Real-Time Monitoring Requirements [21]

- WBANs are often deployed for real-time monitoring of physiological parameters and health-related data.
- Security mechanisms must not introduce significant latency or overhead that could compromise the timeliness and accuracy of data transmission and analysis.
- Balancing the need for real-time monitoring with stringent security requirements poses a challenge in designing efficient and scalable security solutions for WBANs.

11.3.4 Interference and Reliability [22]

- WBANs operate in crowded and dynamic wireless environments, where interference from other devices and environmental factors can degrade signal quality and reliability.

- Security mechanisms must be resilient to signal interference and transmission errors, ensuring the integrity and confidentiality of data even in adverse conditions.
- Techniques such as error correction codes and signal processing algorithms can enhance the reliability of data transmission and reception in WBANs.

11.3.5 Heterogeneity of Devices and Protocols [23]

- WBANs comprise diverse devices and sensors with varying capabilities, communication protocols, and data formats.
- Ensuring interoperability and compatibility between heterogeneous devices while maintaining security and privacy standards is a significant challenge in WBAN deployments.
- Standardization efforts and interoperability frameworks play a crucial role in addressing the heterogeneity of WBAN devices and promoting seamless integration and communication within the network.

Addressing these unique security challenges requires a multidisciplinary approach that combines expertise in wireless communications, embedded systems, cryptography, and ML. By understanding the constraints and requirements of WBAN environments, security researchers and practitioners can develop innovative solutions that mitigate risks, safeguard sensitive medical data, and ensure the integrity and reliability of WBANs in healthcare settings.

11.4 TRADITIONAL SECURITY MECHANISMS AND LIMITATIONS IN THE CONTEXT OF WBANS

Traditional security mechanisms, while effective in many contexts, face several limitations when applied to WBANs. The limitations in the security of WBANs have been highlighted in Figure 11.2. These limitations arise due to the unique characteristics and constraints of WBAN environments. Below are some of the traditional security mechanisms and their challenges in the context of WBANs [24].

11.4.1 Encryption [25]

- *Challenge:* Encryption is a fundamental security mechanism used to protect data confidentiality in transit. However, the computational overhead of strong encryption algorithms can be prohibitive for resource-constrained WBAN devices.
- *Limited Resources:* WBAN devices, such as wearable sensors and implantable medical devices, have limited processing power and memory, making it challenging to perform complex encryption operations in real-time.
- *Energy Consumption:* Strong encryption algorithms require significant computational resources and energy consumption, leading to faster battery depletion in WBAN devices, which is undesirable for long-term monitoring applications.

11.4.2 Authentication [26]

- *Challenge:* Authentication mechanisms are crucial for verifying the identities of users and devices within WBANs. Traditional authentication methods, such as passwords and cryptographic keys, may not be well-suited for WBAN environments.
- *User Interaction:* User authentication mechanisms that rely on passwords or PINs require user interaction, which may not be practical for wearable or implantable devices that lack user interfaces.
- *Key Management:* Managing cryptographic keys securely in WBAN environments is challenging due to the dynamic nature of network topologies and the frequent addition or removal of devices.

11.4.3 Access Control [27]

- *Challenge:* Access control mechanisms regulate the permissions and privileges granted to users and devices within a network. In WBANs, access control mechanisms must be adaptive and granular to accommodate dynamic network configurations and evolving user roles.
- *Fine-Grained Control:* Traditional access control models may lack the granularity needed to enforce fine-grained access policies in WBANs, where different users may have varying levels of access to sensitive medical data.
- *Dynamic Authorization:* Implementing dynamic authorization mechanisms that adapt to changes in network topology and user context is challenging in WBAN environments.

11.4.4 Integrity Checking [28]

- *Challenge:* Integrity checking mechanisms verify the integrity of data transmitted over the network to detect tampering or modification attempts. In WBANs, ensuring data integrity is critical for maintaining the accuracy and reliability of medical information.
- *Overhead:* Traditional integrity checking mechanisms, such as cryptographic hashes and checksums, introduce additional computational overhead and bandwidth consumption, which may be prohibitive for resource-constrained WBAN devices.
- *Real-Time Monitoring:* Performing real-time integrity checks on streaming data in WBANs requires efficient algorithms and protocols that minimize latency and processing overhead.

In summary, while traditional security mechanisms provide foundational principles for securing network communications, their direct application to WBANs faces significant challenges due to resource constraints, dynamic network topologies, and the need for real-time monitoring. Addressing these limitations requires the development of innovative security solutions tailored to the specific requirements and constraints of WBAN environments, including lightweight encryption algorithms, adaptive authentication

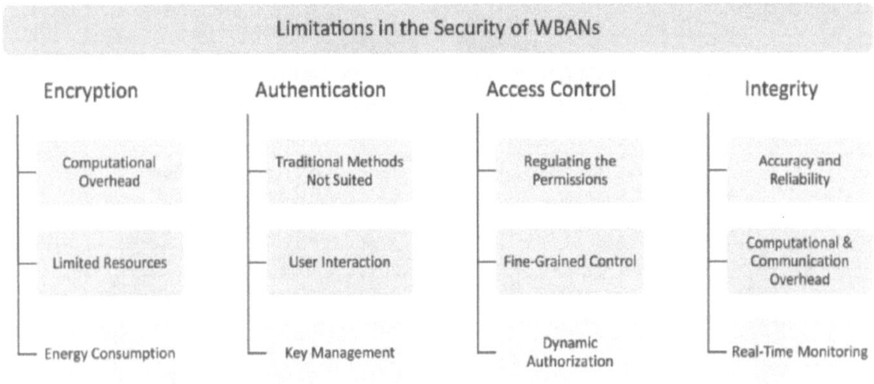

FIGURE 11.2 Limitations in the security of WBANs.

mechanisms, and efficient access control policies. In addition, interdisciplinary collaboration between security researchers, wireless engineers, and healthcare professionals is essential for designing effective security architectures that protect sensitive medical data while ensuring the reliability and integrity of WBANs in healthcare settings.

11.5 ML FOR INTRUSION DETECTION IN WBANS

ML holds significant promise for enhancing IDS in WBANs. By leveraging ML algorithms, WBANs can effectively detect and mitigate security threats in real time, thereby safeguarding sensitive medical data and ensuring the integrity of healthcare services [29]. ML-based IDS analyze network traffic patterns and device behavior to identify anomalous activities indicative of security breaches. Supervised learning techniques such as SVM, Random Forests, and Neural Networks can learn from labeled datasets to classify normal and malicious network behavior. Unsupervised learning methods, including clustering and anomaly detection algorithms, enable the detection of novel threats and deviations from established patterns without explicit training data. ML-driven IDS offer adaptive and intelligent threat detection capabilities, enabling WBANs to proactively defend against emerging security threats and maintain the confidentiality and availability of medical data. An examination of ML techniques for intrusion detection in WBANs is given below [30,31]. These ML approaches for intrusion detection in WBANs have been listed in Figure 11.3.

11.5.1 Supervised Learning Techniques

- Supervised learning algorithms such as SVM, Random Forests, and Neural Networks are commonly employed for intrusion detection in WBANs.
- These algorithms learn from labeled datasets comprising normal and malicious network behavior, enabling them to classify and identify security threats with high accuracy.
- Supervised learning techniques are well-suited for detecting known attack patterns and can adapt to changes in WBAN environments over time.

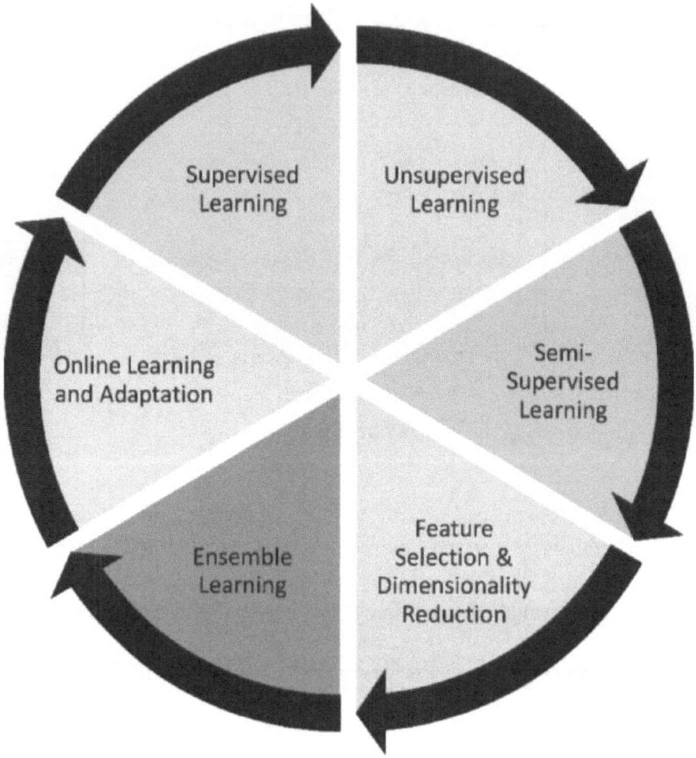

FIGURE 11.3 Machine learning approaches for intrusion detection in WBANs.

11.5.2 Unsupervised Learning Techniques

- Unsupervised learning methods, including clustering and anomaly detection algorithms, are utilized for intrusion detection in WBANs.
- Unlike supervised learning, unsupervised techniques do not require labeled training data and can identify anomalous behavior without prior knowledge of attack signatures.
- Clustering algorithms group similar instances of network traffic or device behavior, enabling the detection of outliers and deviations from normal patterns indicative of security breaches.

11.5.3 Semi-Supervised Learning Approaches

- Semi-supervised learning techniques combine elements of supervised and unsupervised learning to enhance intrusion detection in WBANs.
- These approaches leverage a small amount of labeled data combined with a larger pool of unlabeled data to improve the accuracy and robustness of IDS.
- Semi-supervised learning techniques are particularly useful in scenarios where obtaining labeled training data is challenging or expensive.

11.5.4 Feature Selection and Dimensionality Reduction [32]

- ML-based IDS in WBANs often employ feature selection and dimensionality reduction techniques to improve efficiency and performance.
- Feature selection methods identify the most relevant attributes or features from raw sensor data, reducing computational overhead and improving detection accuracy.
- Dimensionality reduction techniques, such as PCA or t-distributed Stochastic Neighbor Embedding (t-SNE), reduce the complexity of the dataset while preserving critical information for intrusion detection.

11.5.5 Ensemble Learning

- Ensemble learning techniques combine multiple ML models to enhance the robustness and reliability of IDS in WBANs.
- Methods such as bagging, boosting, and stacking integrate diverse classifiers to improve detection accuracy and mitigate the risk of false positives and false negatives.
- Ensemble learning approaches leverage the collective wisdom of multiple models, offering superior performance compared to individual classifiers.

11.5.6 Online Learning and Adaptation

- ML-based IDS in WBANs often employ online learning and adaptation mechanisms to accommodate changes in network behavior and evolving security threats.
- Online learning algorithms continuously update the model parameters based on incoming data streams, enabling real-time adaptation to dynamic WBAN environments.
- Adaptive IDS leverage historical data and feedback mechanisms to refine detection algorithms and improve performance over time.

In summary, ML techniques offer versatile and effective approaches for intrusion detection in WBANs, enabling proactive threat detection and mitigation while ensuring the integrity and confidentiality of medical data. By leveraging supervised, unsupervised, semi-supervised, and ensemble learning methods, WBANs can enhance their security posture and safeguard against evolving security threats in healthcare environments.

11.6 ML ALGORITHMS FOR ANOMALY DETECTION AND THREAT MITIGATION

An overview of supervised, unsupervised, and semi-supervised learning algorithms suitable for anomaly detection and threat mitigation in various contexts, including WBANs, is as follows.

11.6.1 Supervised Learning Algorithms [33]

- Supervised learning algorithms learn from labeled training data, where each instance is associated with a class label. These algorithms aim to predict the class labels of unseen instances based on their features.
- In the context of anomaly detection and threat mitigation, supervised learning algorithms can be trained on historical data containing both normal and anomalous instances.
- Examples of supervised learning algorithms for anomaly detection include:
  - SVM
  - Decision Trees
  - Random Forests
  - Neural Networks
  - k-Nearest Neighbors (k-NN)
- Supervised learning algorithms are effective for detecting known anomalies but may struggle with detecting novel or previously unseen threats.

11.6.2 Unsupervised Learning Algorithms [34]

- Unsupervised learning algorithms do not require labeled training data. Instead, they aim to identify patterns, clusters, or anomalies in the data without prior knowledge of class labels.
- Unsupervised learning is particularly useful for identifying anomalies in situations where labeled data is scarce or unavailable.
- Examples of unsupervised learning algorithms for anomaly detection include:
  - K-means Clustering
  - DBSCAN (Density-Based Spatial Clustering of Applications with Noise)
  - Isolation Forest
  - Autoencoders
  - One-Class Support Vector Machines (OCSVM)
- Unsupervised learning algorithms excel at detecting novel or unknown anomalies but may produce more false positives compared to supervised approaches.

11.6.3 Semi-Supervised Learning Algorithms [35]

- Semi-supervised learning algorithms leverage a combination of labeled and unlabeled data for training. They aim to exploit the abundance of unlabeled data to improve the performance of models trained on limited labeled data.
- In the context of anomaly detection and threat mitigation, semi-supervised learning algorithms can leverage a small amount of labeled data for known anomalies while identifying novel anomalies from the unlabeled data.
- Examples of semi-supervised learning algorithms include:
  - Self-Training
  - Co-Training
  - Label Propagation
  - Generative Adversarial Networks (GANs)

- Semi-supervised learning algorithms strike a balance between the accuracy of supervised approaches and the flexibility of unsupervised approaches, making them suitable for scenarios with limited labeled data.

In summary, supervised, unsupervised, and semi-supervised learning algorithms offer complementary approaches for anomaly detection and threat mitigation in various domains, including WBANs. By leveraging the strengths of each approach, organizations can develop robust and adaptive security systems capable of detecting both known and unknown threats while minimizing false positives and false negatives.

11.7 CASE STUDIES TO SHOWCASE EFFECTIVENESS OF ML-BASED IDS

Case studies and examples showcasing the effectiveness of ML-based IDS in WBANs environments provide valuable insights into real-world applications and their impact on healthcare security. Below are some illustrative examples.

11.7.1 Smart Healthcare Wearable Devices [36]

- Consider a scenario where a hospital deploys smart wearable devices equipped with physiological sensors to monitor patients' vital signs in real time via WBANs.
- ML-based IDS integrated into the WBAN infrastructure continuously analyze sensor data streams to detect anomalies indicative of security breaches, such as unauthorized access or data tampering.
- By leveraging supervised learning algorithms trained on labeled datasets of normal and malicious activities, the IDS can accurately classify and mitigate security threats without human intervention.
- Case studies demonstrating the effectiveness of ML-based IDS in detecting sensor tampering, data injection attacks, and unauthorized access can highlight the importance of proactive security measures in safeguarding patient health data.

11.7.2 Implantable Medical Devices [36,37]

- In another scenario, consider the use of implantable medical devices, such as pacemakers or insulin pumps, connected to WBANs for remote monitoring and control of patients' health conditions.
- ML-driven IDS embedded within the WBAN infrastructure continuously monitor device communications and physiological signals to detect anomalies indicative of cyber-attacks or device malfunctions.
- Unsupervised learning algorithms, such as anomaly detection models based on clustering or autoencoders, can identify deviations from normal device behavior and trigger timely alerts or interventions.

- Case studies showcasing the effectiveness of ML-based IDS in detecting anomalous device activities and preventing potential security threats can underscore the critical role of security in ensuring the safety and reliability of implantable medical devices.

11.7.3 Real-Time Threat Detection and Response [38]

- In a dynamic healthcare environment, ML-based IDS play a crucial role in enabling real-time threat detection and response in WBANs.
- By leveraging reinforcement learning algorithms, IDS can adaptively adjust security policies and response strategies based on evolving network conditions and emerging threat patterns.
- Through continuous monitoring and analysis of network traffic and device behavior, ML-driven IDS can proactively identify and mitigate security incidents, minimizing the risk of data breaches and service disruptions.

Overall, case studies and examples demonstrating the effectiveness of ML-based IDS in WBAN environments provide valuable evidence of their ability to enhance security, protect patient health data, and ensure the integrity and reliability of connected healthcare systems. These examples serve as compelling testimonials to the transformative potential of ML-driven security solutions in addressing the unique challenges of WBANs and advancing patient-centric healthcare delivery.

11.8 PRIVACY PRESERVATION AND DATA CONFIDENTIALITY

Privacy preservation and data confidentiality are paramount concerns in WBANs, where sensitive medical data is transmitted wirelessly between wearable and implantable devices. Robust privacy-preserving techniques are essential to safeguarding patient privacy and ensuring the confidentiality of personal health information. Encryption algorithms, such as homomorphic encryption and differential privacy mechanisms, play a crucial role in protecting data integrity and preventing unauthorized access during data transmission and storage [39]. In addition, anonymization techniques and access control mechanisms are employed to limit access to sensitive information, ensuring that only authorized individuals or entities can access patient data. By implementing comprehensive privacy preservation measures, WBANs can maintain patient trust, comply with regulatory requirements, and uphold the confidentiality of medical data, thereby promoting the adoption and acceptance of connected healthcare technologies. Importance of preserving patient privacy and ensuring data confidentiality in WBANs is listed below.

11.8.1 Legal and Ethical Considerations [40]

- Preserving patient privacy and ensuring data confidentiality in WBANs are not only legal requirements but also ethical imperatives.
- Healthcare regulations such as the HIPAA in the United States mandate the protection of patient health information, including data transmitted via WBANs.

- Upholding patient privacy rights fosters trust between healthcare providers, patients, and other stakeholders, enhancing the integrity of healthcare delivery.

11.8.2 Sensitive Nature of Medical Data

- Medical data transmitted through WBANs often includes sensitive information such as patient diagnoses, treatment plans, and physiological parameters.
- Unauthorized access or disclosure of this data can lead to privacy breaches, identity theft, and even potential harm to patients.
- Safeguarding the confidentiality of medical data in WBANs is essential to protect patient privacy and prevent unauthorized use or exploitation of sensitive information.

11.8.3 Trust and Patient Engagement [41]

- reserving patient privacy and ensuring data confidentiality in WBANs are critical for fostering patient trust and engagement in healthcare.
- Patients are more likely to participate in WBAN-based monitoring and treatment programs if they trust that their personal health information will be kept confidential and secure.
- Maintaining patient privacy encourages open communication between patients and healthcare providers, facilitating informed decision-making and improving health outcomes.

11.8.4 Regulatory Compliance

- Healthcare organizations and providers must adhere to stringent regulatory requirements regarding patient privacy and data confidentiality, including those outlined in HIPAA and the General Data Protection Regulation (GDPR).
- Noncompliance with these regulations can result in severe penalties, legal consequences, and reputational damage for healthcare entities.
- Implementing robust privacy preservation measures in WBANs ensures compliance with regulatory standards and mitigates the risk of legal and financial liabilities.

11.8.5 Mitigating Risks of Data Breaches

- WBANs are susceptible to various security threats, including data breaches, unauthorized access, and interception of sensitive information.
- Preserving patient privacy and ensuring data confidentiality involve implementing encryption, access control, and authentication mechanisms to mitigate the risk of data breaches and unauthorized disclosures.
- By proactively addressing security vulnerabilities and safeguarding patient data, healthcare organizations can mitigate the potential impact of data breaches on patient trust and organizational reputation.

In summary, preserving patient privacy and ensuring data confidentiality in WBANs are essential components of ethical healthcare delivery. By prioritizing privacy preservation measures and implementing robust security mechanisms, healthcare organizations can protect patient confidentiality, foster trust, and uphold regulatory compliance, thereby promoting the widespread adoption and acceptance of WBAN technologies in modern healthcare ecosystems.

11.9 ML IN SECURE TRANSMISSION, ENCRYPTION, AND PRIVACY-ENHANCING TECHNOLOGIES

Exploration of ML-driven techniques for secure data transmission, encryption, and privacy-enhancing technologies involves leveraging ML algorithms to enhance the confidentiality, integrity, and privacy of data in various contexts, including WBANs. The breakdown of these techniques is given below.

11.9.1 Secure Data Transmission [42]

- ML-driven techniques can improve the security of data transmission in WBANs by identifying and mitigating potential vulnerabilities in communication protocols and channels.
- ML algorithms can analyze network traffic patterns, detect anomalies, and identify suspicious activities indicative of security threats, such as eavesdropping or man-in-the-middle attacks.
- By continuously monitoring and analyzing data transmissions, ML-based IDS can proactively detect and respond to security incidents, ensuring the integrity and confidentiality of data during transit.

11.9.2 Encryption and Cryptography [43]

- ML algorithms can enhance encryption and cryptographic protocols used to protect data confidentiality in WBANs.
- ML-driven approaches enable the optimization of encryption parameters, key generation, and management processes to improve security while minimizing computational overhead and resource consumption.
- ML algorithms can also identify and mitigate vulnerabilities in cryptographic algorithms, such as identifying weaknesses in encryption schemes or detecting patterns indicative of cryptographic attacks.

11.9.3 Privacy-Enhancing Technologies [44]

- ML-driven privacy-enhancing technologies aim to protect sensitive information and preserve user privacy in WBANs.
- Differential privacy mechanisms, for example, leverage ML algorithms to anonymize and aggregate data in a way that protects individual privacy while still allowing meaningful analysis at the aggregate level.

- Privacy-preserving data mining techniques, such as federated learning and secure multiparty computation, enable collaborative analysis of distributed datasets without sharing sensitive information.
- ML algorithms can also be used to develop predictive models that maintain privacy by design, ensuring that sensitive information is not exposed or exploited during data analysis and inference processes.

11.9.4 Anomaly Detection and Threat Mitigation

- ML-based anomaly detection techniques can identify abnormal behavior and potential security threats in WBANs, enhancing the overall security posture of the network.
- ML algorithms trained on historical data can learn patterns of normal behavior and detect deviations indicative of security breaches or malicious activities.
- By continuously adapting to evolving threat landscapes, ML-driven anomaly detection systems can improve their effectiveness over time and mitigate emerging security risks in WBAN environments.

In summary, ML-driven techniques offer versatile and effective solutions for enhancing the security, privacy, and integrity of data in WBANs. By leveraging ML algorithms to optimize encryption, detect anomalies, and preserve privacy, WBANs can ensure the confidentiality of sensitive medical information, protect against security threats, and foster trust among users and stakeholders in healthcare environments.

11.10 ADVANCED CRYPTOGRAPHIC METHODS FOR WBAN SECURITY

Examination of advanced cryptographic methods, including differential privacy, homomorphic encryption, and other techniques, sheds light on their applicability to enhancing security in WBANs. Advanced cryptographic techniques for enhanced security in WBANs are listed below and have been highlighted in Figure 11.4.

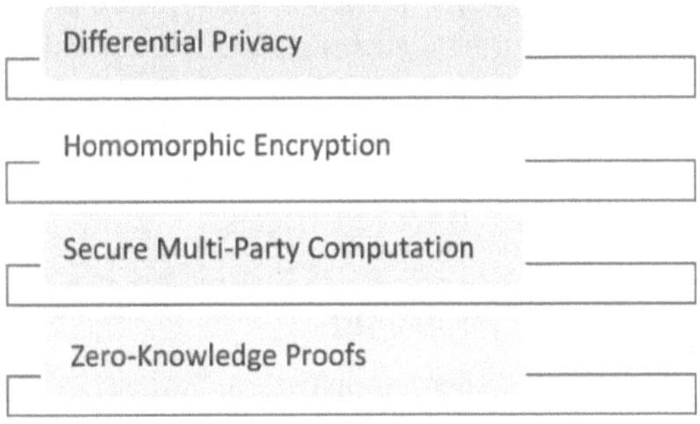

FIGURE 11.4 Advanced cryptographic approaches for WBAN security.

11.10.1 Differential Privacy [45]

- Differential privacy is a privacy-preserving technique that aims to enable the analysis of sensitive data while protecting the privacy of individual records.
- In the context of WBANs, where personal health information is transmitted and stored, differential privacy can mitigate the risk of privacy breaches and unauthorized access to sensitive medical data.
- Differential privacy achieves privacy protection by adding noise to query responses or perturbing data in a manner that preserves statistical accuracy while ensuring individual privacy.
- WBANs can leverage differential privacy mechanisms to aggregate and analyze patient data across multiple devices without revealing sensitive information, thereby enabling collaborative research and analysis while preserving privacy.

11.10.2 Homomorphic Encryption [46]

- Homomorphic encryption is a form of encryption that allows computations to be performed on encrypted data without decrypting it first.
- In WBAN environments, homomorphic encryption enables secure data processing and analysis while maintaining data confidentiality.
- By performing computations on encrypted data, homomorphic encryption protects sensitive medical information from exposure to unauthorized parties, including network operators and service providers.
- Homomorphic encryption enables secure data aggregation, analysis, and ML tasks in WBANs, facilitating remote monitoring, diagnosis, and treatment without compromising patient privacy.

11.10.3 Secure Multi-Party Computation [47]

- Secure multi-party computation (SMPC) enables multiple parties to jointly compute a function over their private inputs without revealing individual inputs to each other.
- In WBANs, SMPC can facilitate collaborative data analysis and processing while preserving the privacy of individual patient data.
- SMPC protocols allow WBAN devices and healthcare providers to collectively analyze aggregated data, such as population health trends or treatment outcomes, without disclosing individual patient records.
- By enabling secure and privacy-preserving computations, SMPC enhances data confidentiality and fosters trust among patients, healthcare providers, and other stakeholders in WBAN ecosystems.

11.10.4 Zero-Knowledge Proofs [48]

- Zero-knowledge proofs enable one party (the prover) to prove to another party (the verifier) that they possess certain knowledge or information without revealing the knowledge itself.

- In WBANs, zero-knowledge proofs can be used to authenticate users, verify transactions, and establish trust between devices without exchanging sensitive information.
- Zero-knowledge proofs ensure data integrity and authentication while preserving privacy and confidentiality in WBAN communications.
- By leveraging zero-knowledge proofs, WBANs can implement secure and privacy-preserving authentication mechanisms, ensuring that only authorized users and devices can access sensitive medical data.

In summary, advanced cryptographic methods such as differential privacy, homomorphic encryption, SMPC, and zero-knowledge proofs offer powerful tools for enhancing security and privacy in WBANs. By leveraging these techniques, WBANs can protect sensitive medical data, facilitate secure data analysis and communication, and uphold patient privacy rights in healthcare environments.

11.11 ANOMALY DETECTION AND BEHAVIORAL ANALYSIS: CLUSTERING, PCA, AND HMMS

Clustering, PCA, and HMMs are powerful techniques used for anomaly detection and behavioral analysis in various domains, including WBANs. A discussion on how these methods contribute to enhancing security and identifying abnormal behavior within WBAN environments is given below.

11.11.1 Clustering [49]

- Clustering algorithms group similar data points together based on their characteristics or features.
- In the context of WBANs, clustering can be used to identify clusters of normal behavior by grouping together instances with similar patterns of device interactions, data transmissions, and user activities.
- Clustering techniques such as k-means clustering or density-based clustering enable the detection of outliers or anomalies that fall outside of established clusters, indicating potential security threats or abnormal behavior.
- By analyzing cluster memberships and distances from cluster centroids, clustering algorithms can identify deviations from normal behavior and trigger alerts for further investigation.

11.11.2 PCA [50]

- PCA is a dimensionality reduction technique used to identify the underlying structure and patterns in high-dimensional datasets.
- In WBANs, PCA can be applied to reduce the dimensionality of sensor data and extract principal components that capture the most significant variations in device behavior and user interactions.

- By projecting high-dimensional sensor data onto a lower-dimensional subspace, PCA facilitates visualization and interpretation of complex behavioral patterns within WBAN environments.
- Anomalies detected through PCA analysis may manifest as deviations from the principal components or as outliers in the reduced-dimensional space, indicating potential security threats or abnormal activities that warrant further investigation.

11.11.3 HMMs [50]

- HMMs are probabilistic models used to capture temporal dependencies and sequential patterns in time-series data.
- In WBANs, HMMs can model the dynamic nature of device behavior and user interactions over time, enabling the detection of anomalous sequences or transitions in network activity.
- By learning the underlying state transitions and emission probabilities from observed data, HMMs can identify deviations from expected behavioral sequences and detect abnormalities in real time.
- HMMs are particularly well-suited for modeling complex, sequential behaviors within WBANs, such as patient activity patterns, device usage sequences, and network traffic flows.

In summary, clustering, PCA, and HMMs offer complementary approaches for anomaly detection and behavioral analysis in WBANs. By leveraging these techniques, WBANs can identify deviations from normal behavior, detect security threats and anomalies in real time, and enhance the overall security posture of healthcare environments. However, the selection and integration of these methods should be carefully tailored to the specific requirements and characteristics of WBAN deployments, taking into account factors such as data volume, dimensionality, and temporal dynamics.

11.12 ML FOR SENSOR TAMPERING AND DATA INJECTION ATTACK DETECTION: CASE STUDIES

Case studies showcasing the application of ML in detecting sensor tampering and data injection attacks highlight the effectiveness of ML-driven techniques in enhancing the security of WBANs. The two illustrative examples are given below.

11.12.1 Detection of Sensor Tampering [51]

- *Case Study:* A hospital deploys WBANs to monitor patients' vital signs using wearable sensors. The integrity of sensor data is crucial for accurate diagnosis and timely medical interventions. However, adversaries may attempt to tamper with sensor readings to disrupt patient care or deceive healthcare providers.

- *ML Approach:* The hospital implements an ML-based IDS tailored to detect sensor tampering attacks in WBANs. The IDS analyzes sensor data streams in real time and employs supervised learning algorithms, such as SVM or Random Forests, trained on labeled datasets of normal and tampered sensor readings.
- *Results:* Through continuous monitoring and analysis of sensor data, the ML-driven IDS successfully detects anomalies indicative of sensor tampering. By comparing incoming sensor readings against learned patterns of normal behavior, the IDS identifies deviations and triggers alerts for further investigation by healthcare professionals.

11.12.2 Detection of Data Injection Attacks [52]

- *Case Study:* A research institution develops a WBAN platform for remote patient monitoring and telemedicine applications. The platform relies on data transmitted from wearable sensors to make critical medical decisions. However, malicious actors may attempt to inject false data into the system, leading to incorrect diagnoses or treatment recommendations.
- *ML Approach:* The research institution integrates ML-based anomaly detection techniques into the WBAN platform to identify data injection attacks. The anomaly detection system employs unsupervised learning algorithms, such as clustering or autoencoders, to model normal patterns of sensor data and detect deviations indicative of injected anomalies.
- *Results:* By analyzing incoming sensor data streams, the ML-driven anomaly detection system identifies unexpected patterns or outliers that deviate significantly from learned norms. Through continuous learning and adaptation, the system enhances its ability to differentiate between genuine sensor readings and injected anomalies, thus improving the reliability and security of the WBAN platform.

In both case studies, ML plays a pivotal role in detecting and mitigating security threats in WBAN environments. By leveraging supervised and unsupervised learning algorithms, healthcare organizations and research institutions can enhance the integrity, confidentiality, and availability of medical data transmitted through WBANs, ultimately improving patient care and ensuring the trustworthiness of connected healthcare systems.

11.13 ADAPTIVE SECURITY MECHANISMS AND THREAT INTELLIGENCE

Adaptive security mechanisms and threat intelligence are crucial components of modern cybersecurity strategies, especially in dynamic and evolving environments such as WBANs. Adaptive security mechanisms continuously assess the threat landscape and adjust security controls and policies in response to emerging threats and changing network conditions [53]. By leveraging threat intelligence feeds, ML algorithms, and real-time monitoring techniques, adaptive security mechanisms enable proactive threat detection, rapid incident response, and effective mitigation of security risks within WBAN environments. By staying ahead of evolving threats and vulnerabilities, adaptive security mechanisms enhance the resilience and robustness of WBANs, ensuring the confidentiality, integrity, and availability of sensitive medical data transmitted within the network. Overview of adaptive security mechanisms empowered by ML in WBANs is given below.

11.13.1 Continuous Monitoring and Analysis [54]

- Adaptive security mechanisms in WBANs leverage ML algorithms to continuously monitor and analyze network traffic, device behavior, and user interactions.
- ML models learn from historical data to establish baseline behavior and detect deviations indicative of security threats or anomalies.

11.13.2 Dynamic Threat Assessment [55]

- ML algorithms enable dynamic threat assessment by analyzing real-time data streams and identifying patterns indicative of emerging threats or suspicious activities.
- Adaptive security mechanisms adjust security controls and policies based on the evolving threat landscape, ensuring proactive threat detection and response.

11.13.3 Behavioral Profiling and Anomaly Detection

- ML techniques such as clustering, PCA, and HMMs enable behavioral profiling and anomaly detection in WBAN environments.
- By learning from historical attack data and user behavior, adaptive security mechanisms identify abnormal activities and deviations from expected norms, triggering alerts and mitigating security risks.

11.13.4 Adaptive Access Control and Authentication [56]

- ML algorithms enhance access control and authentication mechanisms in WBANs by dynamically adjusting user privileges and authentication requirements based on risk profiles and contextual information.
- Adaptive security mechanisms leverage reinforcement learning to adapt authentication policies and access controls in response to changing user behaviors and environmental conditions.

11.14 PROACTIVE THREAT DETECTION

Adaptive security mechanisms use reinforcement learning algorithms to learn optimal security policies and response strategies through trial and error, maximizing the efficacy of threat detection and response mechanisms over time [57]. The foundational elements that can contribute to proactive threat detection strategies are given below.

11.14.1 Threat Intelligence Feeds

- Threat intelligence feeds provide real-time information on emerging threats, vulnerabilities, and attack techniques.
- Adaptive security mechanisms in WBANs integrate threat intelligence feeds to enrich contextual information and prioritize security alerts based on the severity and relevance of threats.

11.14.2 Historical Attack Data

- Historical attack data serves as a valuable resource for training ML models and identifying patterns of malicious behavior.
- ML algorithms learn from historical attack data to detect known attack signatures, predict future attack trends, and adapt security controls to mitigate similar threats in the future.

11.14.3 Reinforcement Learning

- Reinforcement learning enables autonomous decision-making and policy enforcement in WBAN security.

11.15 CHALLENGES AND FUTURE DIRECTIONS

Challenges and future directions in the realm of securing WBANs are multifaceted. Addressing the limited computational resources and energy constraints of WBAN devices while implementing robust security measures remains a significant challenge. In addition, ensuring interoperability and compatibility among diverse WBAN platforms and devices is crucial for establishing standardized security protocols [58]. As WBANs continue to evolve and integrate with emerging technologies such as the Internet of Things (IoT) and 5G networks, novel security threats and attack vectors may arise, necessitating continuous research and innovation in adaptive security mechanisms, threat intelligence, and ML-driven approaches. Future directions include the development of lightweight encryption algorithms, decentralized authentication mechanisms, and privacy-preserving techniques tailored to the unique requirements of WBAN environments. Moreover, fostering collaboration between academia, industry, and regulatory bodies is essential for shaping policies, standards, and best practices that promote the security and resilience of WBAN ecosystems in the face of evolving cyber threats. The challenges and future directions for ML in the security of WBANs have been illustrated in Figure 11.5. The identified challenges and limitations associated with integrating ML into WBAN security frameworks are listed below.

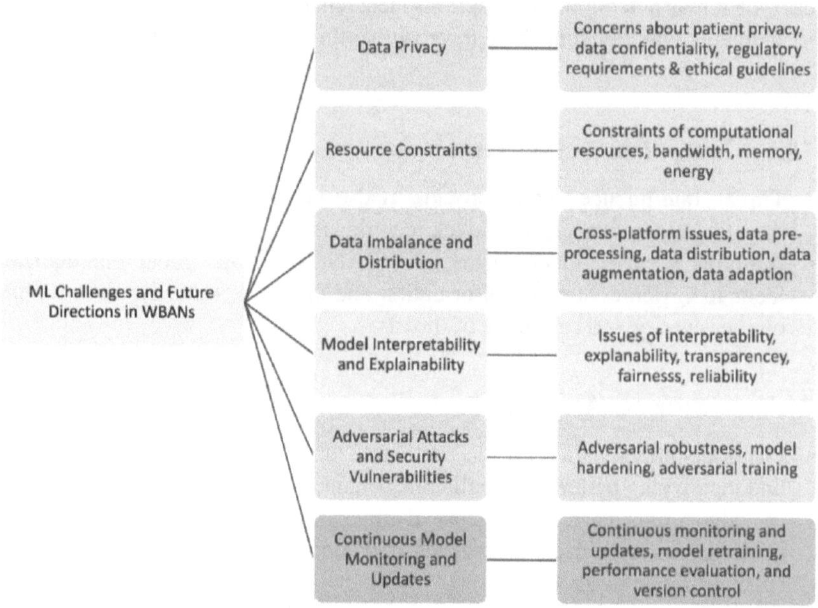

FIGURE 11.5 Challenges and future directions for ML in the security of WBANs.

11.15.1 Data Privacy Concerns

- ML algorithms require access to sensitive medical data for training and model development, raising concerns about patient privacy and data confidentiality.
- Integrating ML into WBAN security frameworks requires careful handling of personal health information to ensure compliance with regulatory requirements and ethical guidelines.

11.15.2 Resource Constraints [59]

- WBAN devices typically have limited computational resources, memory, and battery life, posing challenges for implementing computationally intensive ML algorithms.
- Developing lightweight ML models optimized for resource-constrained environments is essential to minimize computational overhead and energy consumption without compromising security.

11.15.3 Data Imbalance and Distribution

- WBAN datasets may exhibit imbalanced class distributions and heterogeneous data sources, leading to biased or skewed models that may fail to generalize across diverse patient populations or device types.

- Addressing data imbalance and distribution challenges requires robust data preprocessing techniques, data augmentation strategies, and domain adaptation methods to ensure model fairness and generalizability.

11.15.4 Model Interpretability and Explainability

- ML-driven security frameworks in WBANs may produce complex and opaque models that lack interpretability and explainability, making it challenging to understand and trust model predictions.
- Enhancing the interpretability of ML models through techniques such as feature importance analysis, model explanation methods, and transparent algorithm selection is crucial for fostering trust and acceptance among healthcare professionals and end-users.

11.15.5 Adversarial Attacks and Security Vulnerabilities [60]

- ML models deployed in WBAN security frameworks are susceptible to adversarial attacks and exploitation of security vulnerabilities, potentially compromising the integrity and reliability of security mechanisms.
- Adversarial robustness testing, model hardening techniques, and adversarial training are essential for mitigating the risks posed by adversarial threats and safeguarding ML-driven security solutions against exploitation.

11.15.6 Continuous Model Monitoring and Updates

- ML models deployed in WBAN security frameworks require continuous monitoring and updates to adapt to evolving threats, changing data distributions, and emerging attack vectors.
- Establishing mechanisms for model retraining, performance evaluation, and version control is essential to ensure the effectiveness and reliability of ML-driven security solutions over time.

Addressing these challenges and limitations is essential for harnessing the full potential of ML in enhancing the security and resilience of WBANs. By addressing data privacy concerns, optimizing resource utilization, ensuring model fairness and interpretability, mitigating adversarial threats, and implementing robust model monitoring and update mechanisms, WBAN security frameworks can leverage ML to effectively detect, prevent, and mitigate security risks in healthcare environments.

11.16 WBAN FUTURE RESEARCH: FEDERATED LEARNING, EDGE COMPUTING, AND COLLABORATION

Exploration of future research directions in WBANs encompasses innovative approaches such as federated learning, edge computing, and interdisciplinary collaborations. The overview of each approach is given below.

11.16.1 Federated Learning [61]

- Federated learning enables model training across distributed WBAN devices without centralized data aggregation.
- Future research in federated learning for WBANs involves developing privacy-preserving algorithms that allow WBAN devices to collaboratively train ML models while keeping sensitive data localized.
- By leveraging federated learning, WBANs can enhance data privacy, reduce communication overhead, and enable collaborative model training across diverse healthcare environments.

11.16.2 Edge Computing

- Edge computing brings computational capabilities closer to WBAN devices, enabling real-time data processing and analysis at the network edge.
- Future research in edge computing for WBANs focuses on optimizing ML algorithms and model inference techniques for resource-constrained edge devices.
- By offloading computation-intensive tasks to edge servers and gateways, WBANs can reduce latency, conserve bandwidth, and improve the responsiveness of healthcare applications.

11.16.3 Interdisciplinary Collaborations [62]

- Future research in WBANs emphasizes interdisciplinary collaborations between researchers, healthcare practitioners, policymakers, and industry stakeholders.
- Interdisciplinary collaborations enable the integration of diverse expertise and perspectives to address complex challenges in WBAN deployment, security, and usability.
- By fostering collaboration across disciplines such as healthcare, engineering, computer science, and ethics, WBAN research can develop holistic solutions that meet the needs of patients, healthcare providers, and regulatory frameworks.

11.16.4 Security and Privacy Enhancements [63]

- Future research in WBANs focuses on enhancing security and privacy mechanisms to protect sensitive medical data and mitigate emerging threats.
- Research directions include the development of lightweight encryption algorithms, privacy-preserving techniques, and secure communication protocols tailored to the unique requirements of WBAN environments.
- By integrating advanced security measures and privacy-enhancing technologies, WBANs can ensure the confidentiality, integrity, and availability of patient data while facilitating secure and trustworthy healthcare delivery.

11.16.5 Standardization and Regulatory Compliance [64,65]

- Future research efforts in WBANs prioritize standardization and regulatory compliance to promote interoperability, compatibility, and adherence to healthcare regulations.
- Research directions include the development of industry standards, interoperability frameworks, and certification processes for WBAN devices and applications.
- By establishing clear guidelines and best practices, WBAN research can facilitate seamless integration into existing healthcare infrastructures and promote widespread adoption of connected healthcare technologies.

In summary, future research directions in WBANs encompass a broad range of interdisciplinary collaborations, innovative technologies, and security enhancements aimed at advancing the capabilities, usability, and security of WBAN-enabled healthcare systems. By embracing federated learning, edge computing, and interdisciplinary collaborations, WBAN research can drive transformative advancements in patient care, disease management, and healthcare delivery while addressing the unique challenges and opportunities presented by connected healthcare ecosystems.

11.17 CONCLUSION

WBANs have emerged as a transformative technology in healthcare, enabling seamless integration of wearable and implantable medical devices for real-time monitoring and management of patient health data. However, the adoption of WBANs also introduces significant security challenges that must be addressed to ensure the confidentiality, integrity, and availability of sensitive medical information. This chapter proposal explores the multifaceted landscape of security challenges in WBANs, including data privacy concerns, authentication issues, vulnerabilities in WBAN environments, and resource constraints. Traditional security mechanisms face limitations in mitigating these risks effectively. The chapter delves into the applicability of ML approaches to enhance security in WBANs. ML techniques offer promising solutions for intrusion detection, anomaly detection, and privacy preservation in dynamic and heterogeneous WBAN environments. By leveraging ML algorithms, WBANs can proactively identify and mitigate security threats, ensuring the integrity of patient data and the reliability of healthcare services. In the chapter, the presented case studies highlight the effectiveness of ML-based intrusion detection systems in real-world scenarios, including smart healthcare wearable devices and implantable medical devices. Furthermore, the exploration of advanced cryptographic methods and privacy-enhancing technologies underscores the importance of secure data transmission and encryption in protecting sensitive medical information. The chapter also addresses key challenges and future directions in WBAN security, emphasizing the need for continuous research and innovation to address evolving security threats, regulatory compliance requirements, and emerging technologies such as federated learning and edge computing. In conclusion, the integration of ML-based security approaches represents a critical step toward safeguarding the integrity and privacy of patient data in WBANs while unlocking the full potential of these innovative health-care technologies.

REFERENCES

[1] Hajar, M. S., Al-Kadri, M. O., & Kalutarage, H. K. (2021). A survey on wireless body area networks: Architecture, security challenges and research opportunities. *Computers & Security, 104*, 102211.

[2] Majeed, J. H., & Aish, Q. (2021). A remote patient monitoring based on WBAN implementation with internet of thing and cloud server. *Bulletin of Electrical Engineering and Informatics, 10*(3), 1640–1647.

[3] Yaghoubi, M., Ahmed, K., & Miao, Y. (2022). Wireless body area network (WBAN): A survey on architecture, technologies, energy consumption, and security challenges. *Journal of Sensor and Actuator Networks, 11*(4), 67.

[4] Vyas, A., & Pal, S. (2020). Preventing security and privacy attacks in WBANs. In *Handbook of Computer Networks and Cyber Security: Principles and Paradigms* (pp. 201–225). Springer.

[5] Roy, M., Chowdhury, C., & Aslam, N. (2020). Security and privacy issues in wireless sensor and body area networks. In *Handbook of Computer Networks and Cyber Security: Principles and Paradigms* (pp. 173–200). Springer.

[6] Soni, M., & Singh, D. K. (2023). New directions for security attacks, privacy, and malware detection in WBAN. *Evolutionary Intelligence, 16*(6), 1917–1934.

[7] Paul, P. C., Loane, J., McCaffery, F., & Regan, G. (2021). Towards design and development of a data security and privacy risk management framework for WBAN based healthcare applications. *Applied System Innovation, 4*(4), 76.

[8] Joshi, A., & Mohapatra, A. K. (2019). Authentication protocols for wireless body area network with key management approach. *Journal of Discrete Mathematical Sciences and Cryptography, 22*(2), 219–240.

[9] Mucchi, L., Jayousi, S., Martinelli, A., Caputo, S., & Marcocci, P. (2019, May). An overview of security threats, solutions and challenges in WBANs for healthcare. In *2019 13th International Symposium on Medical Information and Communication Technology (ISMICT)* (pp. 1–6). IEEE.

[10] Zahoor, S., & Mir, R. N. (2020). Design and modeling of resource-constrained IoT based body area networks. *Scalable Computing: Practice and Experience, 21*(3), 497–514.

[11] Hasan, K., Biswas, K., Ahmed, K., Nafi, N. S., & Islam, M. S. (2019). A comprehensive review of wireless body area network. *Journal of Network and Computer Applications, 143*, 178–198.

[12] Negra, R. (2022, February). Leveraging machine learning for WBANs. In *International Workshop on Distributed Computing for Emerging Smart Networks* (pp. 38–59). Springer International Publishing.

[13] Saif, S., Karmakar, K., Biswas, S., & Neogy, S. (2022). MLIDS: Machine learning enabled intrusion detection system for health monitoring framework using BA-WSN. *International Journal of Wireless Information Networks, 29*(4), 491–502.

[14] Awad, M., Sallabi, F., Shuaib, K., & Naeem, F. (2022). Artificial intelligence-based fault prediction framework for WBAN. *Journal of King Saud University-Computer and Information Sciences, 34*(9), 7126–7137.

[15] Punj, R., & Kumar, R. (2022). Data dissemination approach using machine learning techniques for WBANs. *Concurrency and Computation: Practice and Experience, 34*(5), e6688.

[16] Khatun, M. A., Memon, S. F., Eising, C., & Dhirani, L. L. (2023). Machine learning for healthcare-IoT security: A review and risk mitigation. *IEEE Access, 11*, 145869–145896.

[17] Tunc, M. A., Gures, E., & Shayea, I. (2021). A survey on IoT smart healthcare: Emerging technologies, applications, challenges, and future trends. *arXiv preprint arXiv:2109.02042.*

[18] Yaghoubi, M., Ahmed, K., & Miao, Y. (2022). Wireless body area network (WBAN): A survey on architecture, technologies, energy consumption, and security challenges. *Journal of Sensor and Actuator Networks, 11*(4), 67.

[19] Salayma, M., Al-Dubai, A., Romdhani, I., & Nasser, Y. (2018). Reliability and energy efficiency enhancement for emergency-aware wireless body area networks (WBANs). *IEEE Transactions on Green Communications and Networking, 2*(3), 804–816.

[20] Samanta, A., & Misra, S. (2018). Dynamic connectivity establishment and cooperative scheduling for QoS-aware wireless body area networks. *IEEE Transactions on Mobile Computing, 17*(12), 2775–2788.

[21] Al Barazanchi, I., Hashim, W., Alkahtani, A. A., Abbas, H. H., & Abdulshaheed, H. R. (2021, October). Overview of WBAN from literature survey to application implementation. In *2021 8th International Conference on Electrical Engineering, Computer Science and Informatics (EECSI)* (pp. 16–21). IEEE.

[22] Singh, A. K., & Garg, A. (2021). Applications of signal processing. In *Machine Learning in Signal Processing* (pp. 73–95). Chapman and Hall/CRC.

[23] Nandikanti, A., & Sahu, K. N. (2022). WBAN technology: Challenges and security attacks. In *ICDSMLA 2020: Proceedings of the 2nd International Conference on Data Science, Machine Learning and Applications* (pp. 1519–1525). Springer.

[24] Hussain, S., Ullah, S. S., Uddin, M., Iqbal, J., & Chen, C. L. (2022). A comprehensive survey on signcryption security mechanisms in wireless body area networks. *Sensors, 22*(3), 1072.

[25] Al Barazanchi, I., Hashim, W., Alkahtani, A. A., Abbas, H. H., & Abdulshaheed, H. R. (2021, October). Overview of WBAN from literature survey to application implementation. In *2021 8th International Conference on Electrical Engineering, Computer Science and Informatics (EECSI)* (pp. 16–21). IEEE.

[26] Sindhu, S., Vashiath, S., & Chakarvarti, S. K. (2016). A review on wireless body area network (WBAN) for health monitoring system, implementation protocol. *Communication on Applied Electronics, Foundation of Computer Science, New York, USA, 4*(7).

[27] Singh, S., & Prasad, D. (2022). Wireless body area network (WBAN): A review of schemes and protocols. *Materials Today: Proceedings, 49*, 3488–3496.

[28] Teshome, A. K., Kibret, B., & Lai, D. T. (2018). A review of implant communication technology in WBAN: Progress and challenges. *IEEE Reviews in Biomedical Engineering, 12*, 88–99.

[29] Medina, S. E., & Kholidy, H. A. (2020). *An Analysis of a Signature-based Approach for an Intrusion Detection System in a Wireless Body Area Network (WBAN) Using Data Mining Techniques.* SUNY Polytechnic Institute.

[30] Alsubaie, F., Al-Akhras, M., & Alzahrani, H. A. (2020, November). Using machine learning for intrusion detection system in wireless body area network. In *2020 First International Conference of Smart Systems and Emerging Technologies (SMARTTECH)* (pp. 100–104). IEEE.

[31] Verner, A., & Butvinik, D. (2017, December). A machine learning approach to detecting sensor data modification intrusions in WBANs. In *2017 16th IEEE International Conference on Machine Learning and Applications (ICMLA)* (pp. 161–169). IEEE.

[32] Anitha, P., & Priya, R. (2022). A technical aspect of WBAN security protocols and their challenges: Brief survey. *NeuroQuantology, 20*(8), 9029.

[33] Poornima, I. G. A., & Paramasivan, B. (2020). Anomaly detection in wireless sensor network using machine learning algorithm. *Computer Communications, 151*, 331–337.
[34] Ifzarne, S., Tabbaa, H., Hafidi, I., & Lamghari, N. (2021). Anomaly detection using machine learning techniques in wireless sensor networks. In *Journal of Physics: Conference Series* (Vol. 1743, No. 1, p. 012021). IOP Publishing.
[35] Baraneetharan, D. E. (2020). Role of machine learning algorithms intrusion detection in WSNs: A survey. *Journal of Information Technology and Digital World, 2*(3), 161–173.
[36] Tabasum, A., Safi, Z., AlKhater, W., & Shikfa, A. (2018, August). Cybersecurity issues in implanted medical devices. In *2018 International Conference on Computer and Applications (ICCA)* (pp. 1–9). IEEE.
[37] Baranchuk, A., Refaat, M. M., Patton, K. K., Chung, M. K., Krishnan, K., Kutyifa, V., . . . & American College of Cardiology's Electrophysiology Section Leadership. (2018). Cybersecurity for cardiac implantable electronic devices: What should you know? *Journal of the American College of Cardiology, 71*(11), 1284–1288.
[38] Garg, A., & Singh, A. K. (2021). Applications of Internet of Things (IoT) in green computing. In *Intelligence of Things: AI-IoT Based Critical-Applications and Innovations* (pp. 1–34). Springer.
[39] Chahal, A., Kumar, A., & Rani, A. (2014). Secure key management in Ad-hoc Network: A review. *International Journal of Advances in Engineering & Technology, 7*(3), 1009.
[40] Maram, B., Majji, R., Gopisetty, G. K. D., Garg, A., Daniya, T., & Kumar, B. S. (2023, December). Lightweight cryptography based deep learning techniques for securing IoT based e-healthcare system. In *2023 2nd International Conference on Automation, Computing and Renewable Systems (ICACRS)* (pp. 1334–1341). IEEE.
[41] Saini, M. L., Tripathi, B., Kaushal, J., & Garg, A. (2023, November). A hybrid model for diagnosis cardiovascular disease using clinical features, ECG and MRI. In *2023 International Conference on Ambient Intelligence, Knowledge Informatics and Industrial Electronics (AIKIIE)* (pp. 1–6). IEEE.
[42] Singh, A. K., Nayyar, A., & Garg, A. (2023). A secure elliptic curve based anonymous authentication and key establishment mechanism for IoT and cloud. *Multimedia Tools and Applications, 82*(15), 22525–22576.
[43] Dutta, M., Singh, A. K., & Kumar, A. (2013, February). An efficient signcryption scheme based on ECC with forward secrecy and encrypted message authentication. In *2013 3rd IEEE International Advance Computing Conference (IACC)* (pp. 399–403). IEEE.
[44] Singh, A. K., Solanki, A., Nayyar, A., & Qureshi, B. (2020). Elliptic curve signcryption-based mutual authentication protocol for smart cards. *Applied Sciences, 10*(22), 8291.
[45] Singh, A. K., & Vaisla, K. S. (2014). A lightweight signcryption scheme based on elliptic curve cryptography. In *Proceedings of the 1st International Conference on Communication and Computer Engineering* (Vol. 1, pp. 7–10). BTKIT.
[46] Singh, A. K., & Patro, B. D. K. (2020). Elliptic curve signcryption based security protocol for RFID. *KSII Transactions on Internet and Information Systems (TIIS), 14*(1), 344–365.
[47] Zhao, C., Zhao, S., Zhao, M., Chen, Z., Gao, C. Z., Li, H., & Tan, Y. A. (2019). Secure multi-party computation: Theory, practice and applications. *Information Sciences, 476*, 357–372.
[48] Ma, L., Ge, Y., & Zhu, Y. (2014). TinyZKP: A lightweight authentication scheme based on zero-knowledge proof for wireless body area networks. *Wireless Personal Communications, 77*, 1077–1090.
[49] Punj, R., & Kumar, R. (2022). Data dissemination approach using machine learning techniques for WBANs. *Concurrency and Computation: Practice and Experience, 34*(5), e6688.

[50] Fernandes, G., Rodrigues, J. J., Carvalho, L. F., Al-Muhtadi, J. F., & Proença, M. L. (2019). A comprehensive survey on network anomaly detection. *Telecommunication Systems, 70*, 447–489.

[51] Illiano, V. P., & Lupu, E. C. (2015). Detecting malicious data injections in event detection wireless sensor networks. *IEEE Transactions on Network and Service Management, 12*(3), 496–510.

[52] Elnour, M., Noorizadeh, M., Shakerpour, M., Meskin, N., Khan, K., & Jain, R. (2023). A machine learning based framework for real-time detection and mitigation of sensor false data injection cyber-physical attacks in industrial control systems. *IEEEAccess, 11*, 86977–86998.

[53] Sharma, D., Singhal, T. K., & Singh, D. (2022). Threat intelligence model to secure IoT based body area network and prosthetic sensors. *ECS Transactions, 107*(1), 15417.

[54] Wang, D., Zhou, J., Masdari, M., Qasem, S. N., & Sayed, B. T. (2023). Security in wireless body area networks via anonymous authentication: Comprehensive literature review, scheme classification, and future challenges. *Ad Hoc Networks*, 103332.

[55] Muthuvel, S., Rajagopal, S., & Subramaniam, S. K. (2022). Analysis of security issues in wireless body area networks in heterogeneous networks. *Sensors, 22*(19), 7588.

[56] Gupta, M., & Kumar, B. S. (2023). Lightweight secure session key protection, mutual authentication and access control (LSSMAC) for WBAN assisted IoT NetWork. *IEEE Sensors Journal, 23*(17), 20283–20293.

[57] Kumar, A., Singh, A. K., & Garg, A (2023). Evaluation of machine learning techniques for heart disease prediction using multi-criteria decision making. *Journal of Intelligent & Fuzzy Systems* (Preprint), 1–15.

[58] Preethichandra, D. M. G., Piyathilaka, L., Izhar, U., Samarasinghe, R., & De Silva, L. C. (2023). Wireless body area networks and their applications–A review. *IEEEAccess, 11*, 9202–9220.

[59] Liu, Z., Liu, B., Chen, C., & Chen, C. W. (2015, December). Energy-efficient resource allocation with QoS support in wireless body area networks. In *2015 IEEE Global Communications Conference (GLOBECOM)* (pp. 1–6). IEEE.

[60] Chakraborty, A., Alam, M., Dey, V., Chattopadhyay, A., & Mukhopadhyay, D. (2018). Adversarial attacks and defences: A survey. *arXiv preprint arXiv:1810.00069.*

[61] Hafi, H., Brik, B., Frangoudis, P. A., Ksentini, A., & Bagaa, M. (2024). Split federated learning for 6G enabled-networks: Requirements, challenges and future directions. *IEEE Access, 12*, 9890–9930.

[62] Shibghatullah, A. S., & Al Barazanchi, I. (2014). An analysis of requirements for efficient protocols in WBAN. *Journal of Telecommunication, Electronic and Computer Engineering (JTEC), 6*(2), 19–22.

[63] Dhanvijay, M. M., & Patil, S. C. (2021). Optimized mobility management protocol for the IoT based WBAN with an enhanced security. *Wireless Networks, 27*, 537–555.

[64] Kohno, R., Hamaguchi, K., Li, H. B., & Takizawa, K. (2008, September). R&D and standardization of body area network (BAN) for medical healthcare. In *2008 IEEE International Conference on Ultra-Wideband* (Vol. 3, pp. 5–8). IEEE.

[65] Gupta, S. H., & Devarajan, N. (2023). Performance exploration of on-body WBAN using CM3A-IEEE 802.15. 6 channel model. *Journal of Ambient Intelligence and Humanized Computing, 14*(5), 5761–5772.

12 Security Threat Mitigation in WBAN Using Tiny Machine Learning

Vinod K. S. and ED Kanmani Ruby

12.1 INTRODUCTION

This section consists of two parts: the first part is about WBAN or Wireless Body Area Network, and the second part discusses TinyML, which is Tiny Machine Learning.

12.1.1 Wireless Body Area Network

WBAN refers to a network of interconnected devices worn on or implanted into the human body, typically for medical, health monitoring, or personal communication purposes. These devices can include sensors, monitors, actuators, and communication modules that gather information from the body and transmit it wirelessly to a central hub or other devices within the network.

The primary goal of WBANs is to enable real-time monitoring of various physiological parameters, such as heart rate, temperature, blood pressure, glucose levels, and more. This technology has significant applications in healthcare, as it allows for continuous and remote monitoring of patient's health conditions, enabling early detection of anomalies and providing valuable data to healthcare professionals for diagnosis and treatment.

WBANs are designed to be low-power and lightweight, ensuring they can be comfortably worn without causing discomfort to the wearer. The data collected by WBAN devices can be transmitted through wireless communication technologies such as Bluetooth, Zigbee, or other custom protocols optimized for low-power and short-range communication.

Overall, WBANs can revolutionize healthcare by providing continuous and personalized health monitoring, improving patient care, and enabling preventive measures based on real-time health data.

Figure 12.1 shows the block diagram of a typical WBAN. Different sensors like electrocardiogram (ECG) sensors, blood pressure sensors, pulse oximeters,

DOI: 10.1201/9781032635101-15

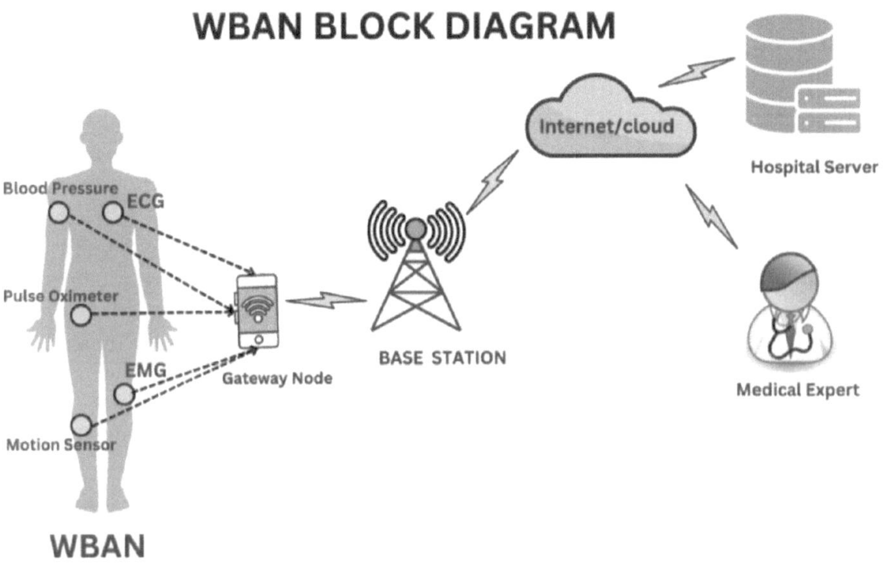

FIGURE 12.1 WBAN block diagram.

electromyography (EMG) sensors, motion sensors, and the like are placed on the human body to read the different vitals of the person. These measured vitals are then transmitted to the cloud via gateway and base station. The hospital server accesses these data for processing these vital information to find any emergency situation in the patient.

Security is a critical consideration in WBANs, especially given the sensitive nature of the health data being transmitted. Here are some common security techniques used in WBAN communication:

Encryption: Encryption is used to protect the confidentiality of data during transmission. Data is encoded using cryptographic algorithms, and only authorized parties with the appropriate decryption keys can access the original data. This prevents unauthorized access and eavesdropping.

Authentication: Authentication ensures that only authorized devices can participate in the WBAN. This can involve mechanisms like username–password combinations, digital certificates, or biometric authentication to verify the identity of devices or users.

Access Control: Access control mechanisms restrict the interactions and data exchanges between devices within the WBAN. This prevents unauthorized devices from joining the network or accessing certain resources.

Data Integrity: Data integrity mechanisms ensure that the data remains unchanged during transmission. Hash functions and checksums are often used to verify that data has not been tampered with.

Key Management: Secure key management is crucial for encryption and decryption processes. Techniques such as Diffie-Hellman key exchange, key distribution protocols, and secure storage of keys are employed to protect the confidentiality of keys.

Secure Pairing: When devices are added to the WBAN, secure pairing protocols ensure that the devices are properly authenticated before establishing a connection. This prevents unauthorized devices from masquerading as legitimate ones.

Physical Layer Security: Techniques can be applied at the physical layer to enhance security. For example, frequency hopping can make it more difficult for eavesdroppers to intercept data, and directional antennas can limit the range over which transmissions can be received.

Intrusion Detection and Prevention: Intrusion detection systems monitor network traffic for signs of malicious activities or unauthorized access attempts. If detected, these systems can trigger alerts or take preventive actions.

Privacy Preservation: WBANs often deal with sensitive health data. Techniques like data anonymization, aggregation, and differential privacy can be used to protect the privacy of individuals while still allowing useful analysis of the data.

Firewalls and Network Segmentation: Employing firewalls and segmenting the network into different zones with varying levels of trust can help isolate critical health data from less secure components of the network.

Secure Routing: Ensuring that data is transmitted through secure routes is important. Routing protocols that consider security metrics, such as energy-efficient secure routing, can be used to prevent data interception and tampering.

Firmware and Software Security: Ensuring the security of device firmware and software is essential. Regular updates, secure boot mechanisms, and code signing can prevent unauthorized access and exploitation of vulnerabilities.

The choice of security techniques will depend on the specific use case, the level of sensitivity of the data being transmitted, and the potential threats to the WBAN. It's crucial to implement a comprehensive security strategy that addresses various aspects of security to protect the integrity, confidentiality, and availability of data within a WBAN.

12.1.2 Tiny Machine Learning

TinyML is a field of study that focuses on deploying ML models on extremely resource-constrained devices, such as microcontrollers and other embedded systems. These devices often have limited computational power, memory, and energy resources, making running traditional machine-learning models designed for more powerful hardware challenging.

Figure 12.2 shows the block diagram of TinyML. TinyML involves designing and optimizing ML algorithms and models to be efficient enough to run on constrained hardware while providing useful insights or performing specific tasks. This opens up new possibilities for bringing ML capabilities to a wide range of applications and devices that were previously not feasible due to limitations in size, power, and cost.

- Key aspects of TinyML include:
  - **Model Optimization:** Algorithms and architectures are designed to be lightweight and efficient, often involving trade-offs between model complexity and accuracy. Techniques like quantization (reducing the

precision of numbers used in computations), model pruning (removing unnecessary model parameters), and efficient activation functions are employed.

- **Low-Power Hardware:** TinyML focuses on running models on hardware platforms with minimal energy consumption. This might involve optimizing the hardware architecture or co-designing hardware and software to ensure optimal performance.
- **Data Collection and Processing:** In many cases, data is collected and processed directly on the device, reducing the need for constant communication with cloud servers. This can help save bandwidth and enhance privacy.
- **Real-Time Inference:** TinyML models are often used for real-time inference tasks, where quick decisions need to be made based on incoming data. This can be particularly useful in applications like sensor data analysis, anomaly detection, and immediate feedback systems.
- **Applications:** TinyML has applications in a wide range of domains, including wearable devices, Internet of Things (IoT) sensors, edge computing, health monitoring, environmental monitoring, industrial automation, agriculture, and more.

The development of TinyML involves a combination of ML expertise, software development skills, and knowledge of embedded systems. Tools and libraries specifically designed for TinyML, such as TensorFlowLite for Microcontrollers, Edge Impulse, and Apache TVM, have emerged to assist developers in creating and deploying models on resource-constrained devices.

The goal of TinyML is to democratize ML by making it accessible to a broader range of devices and applications, enabling them to make intelligent decisions without relying on constant connectivity to powerful servers.

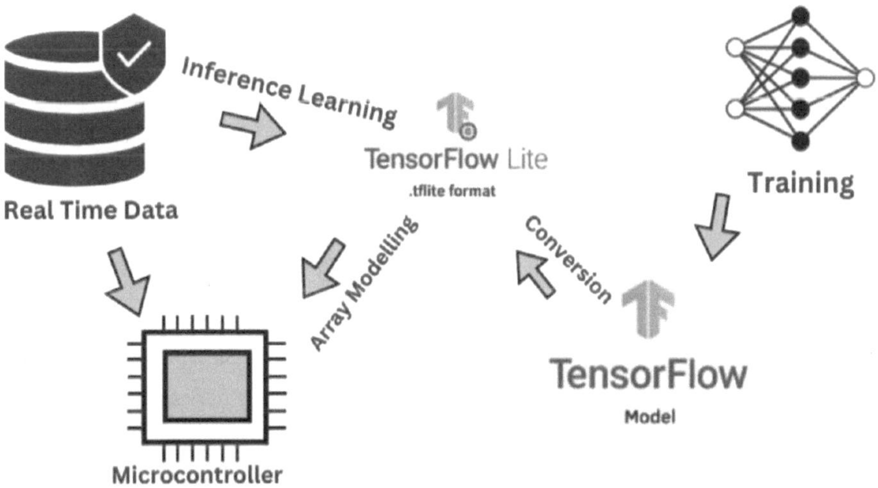

FIGURE 12.2 TinyML block diagram.

12.2 BACKGROUND

WBANs have gained significant attention in recent years due to their potential to revolutionize healthcare and monitoring systems. WBANs consist of interconnected wearable and implantable devices that collect and transmit physiological data from the human body to external devices for analysis and decision-making. This section provides an overview of key research trends, challenges, and advancements in the field of WBANs.

12.2.1 Evolution and Applications of WBANs

The concept of WBANs originated from the broader field of wireless sensor networks and personal area networks. Research by Akyildiz and colleagues [1] introduced the concept of using miniature sensors for healthcare monitoring, leading to the birth of WBANs as a specialized subset of wireless networks dedicated to healthcare applications. Since then, WBANs have found applications in various domains, including medical diagnosis, patient monitoring, sports performance analysis, and ambient assisted living [1,2].

- **Technical Challenges and Solutions in WBAN**
- While WBANs offer promising capabilities, several technical challenges have been identified:
  - **Energy Efficiency:** Due to limited power resources in wearable devices, energy-efficient communication protocols and power management strategies are crucial. The paper [3] proposed an energy-aware routing protocol for WBANs that dynamically adjusts transmission power to minimize energy consumption while maintaining communication reliability.
  - **Security and Privacy:** Given the sensitive nature of health data, ensuring secure communication and patient privacy is vital. The work [4] introduced a secure data aggregation scheme that employs cryptographic techniques to protect data integrity and confidentiality in WBANs.
  - **Interference and Coexistence:** WBANs operate in a crowded radio frequency spectrum. Mitigating interference and ensuring coexistence with other wireless technologies is essential. The study [5] investigated the interference management strategies for the coexistence of WBANs with Wi-Fi networks.
  - **Reliability and Quality of Service:** WBANs must ensure reliable data transmission and maintain a certain quality of service (QoS) for medical applications. The work [6] proposed a QoS-aware routing algorithm for WBANs that considers energy efficiency and reliability metrics.
- **Advancements in Signal Processing and ML**
- Signal processing and ML techniques have been instrumental in improving WBAN performance and data analysis:
  - **Feature Extraction:** Feature extraction from physiological signals is crucial for health monitoring. The research [7] developed a novel feature extraction method based on wavelet transform for accurate ECG classification in WBANs.

- **Anomaly Detection:** Detecting anomalies in physiological data is essential for early disease detection. The paper [8] presented an anomaly detection framework using a recurrent neural network (RNN) for real-time monitoring in WBANs.
- **Energy-Efficient ML Models:** Developing energy-efficient ML models is vital for resource-constrained devices. The work [9] proposed a lightweight convolutional neural network (CNN) architecture for gesture recognition in WBANs with reduced computational complexity.

- **Security in WBANs**
- WBANs have emerged as a transformative technology for healthcare monitoring and communication. However, the sensitive nature of health data and the unique characteristics of WBANs necessitate robust security measures to ensure data privacy, integrity, and confidentiality. This section reviews key research trends, challenges, and advancements in the security aspects of WBANs.
- **Security Challenges in WBANs**
    - **Data Privacy and Confidentiality:** The personal and sensitive health data transmitted by WBAN devices requires strong privacy protection. The work [23] emphasized the need for data encryption and patient consent management to safeguard against unauthorized data access.
    - **Authentication and Access Control:** Ensuring that only authorized devices and users can access the network is crucial. A study by Sowjanya and collegues [24] proposed a lightweight authentication scheme using elliptic curve cryptography to achieve secure device pairing in WBANs.
    - **Energy-Efficient Security:** WBAN devices are resource-constrained, necessitating energy-efficient security mechanisms. An energy-efficient trust-based security framework was proposed by Bairwa and Joshi [25] that balances security and energy consumption in WBANs.
- **Secure Communication Protocols**
    - **Key Management:** Effective key management is fundamental to secure communication. The paper [26] introduced a distributed key management protocol for WBANs that efficiently manages encryption keys among wearable devices.
    - **Secure Routing:** Secure routing protocols are essential to prevent unauthorized access and data tampering during transmission. The researchers [27] presented a safe and energy-efficient routing protocol that uses a trust model to select optimal routes in WBANs.
    - **Secure Data Aggregation:** Aggregating data from multiple sensors is common in WBANs. A privacy-preserving data aggregation scheme was proposed by the researchers [28] that masks data values to ensure confidentiality while preserving aggregated information.
- **Intrusion Detection and Prevention**
    - **Anomaly Detection:** Anomaly detection techniques help identify unusual behavior in the network. Yakub Kayode Saheed and colleagues [29] developed an anomaly detection framework based on ML to detect abnormal activities in WBANs, improving intrusion detection capabilities.

- **Dynamic Intrusion Response:** Rapid response to detected intrusions is vital. The researchers [30] introduced a dynamic intrusion response mechanism that adapts security measures based on the severity of detected intrusions in WBANs.
- **Emerging Technologies and Future Directions**
  - **Blockchain for WBAN Security:** The use of blockchain technology for enhancing security and privacy in WBANs has gained attention [21] and proposed a blockchain-based patient-centric access control mechanism to ensure data privacy and patient consent management.
  - **Post-Quantum Cryptography:** With the advent of quantum computing, post-quantum cryptographic methods are being explored for securing WBAN communication. The researchers [31] investigated the suitability of post-quantum cryptographic algorithms for WBANs' resource-constrained environment.

12.2.2 TinyML and Its Applications

TinyML, integrating ML into resource-constrained devices, has emerged as a transformative technology with applications spanning various domains. This section provides an overview of key research trends, challenges, and advancements in TinyML and its diverse applications.

- **Evolution of TinyML**
- TinyML emerged as a response to the growing demand for running ML models on devices with limited computational resources. Research by the authors [10] demonstrated the feasibility of on-device ML for activity recognition using sensor data. This marked the beginning of TinyML as a specialized field that combines ML and embedded systems.
- **Applications of TinyML**
  - **Wearable Health Monitoring:** Wearable devices equipped with TinyML models have found applications in health monitoring. The work [11] developed a TinyML-based heart rate monitoring system that provides real-time insights into wearable devices.
  - **Edge Intelligence:** TinyML enables edge devices to perform intelligent tasks without relying on cloud servers. EdgeML by the study [12] showcases an edge inference engine for resource-constrained devices, enhancing real-time decision-making.
  - **Agriculture and IoT:** In agriculture, TinyML aids in crop monitoring and pest detection. The paper [13] introduced a crop disease classification system using TinyML on low-power devices.
  - **Energy Management:** TinyML models assist in optimizing energy consumption in buildings. The research [14] demonstrated an energy-efficient heating, ventilation, and air-conditioning (HVAC) control system powered by TinyML.
- **Technical Challenges and Solutions**

    - **Model Optimization:** Optimizing ML models for resource-constrained devices is crucial. The paper [15] presented a neural architecture search method to create efficient neural networks for TinyML.
    - **Quantization and Compression:** Shrinking model sizes through quantization and compression is essential. The work [16] proposed techniques to reduce the memory footprint of neural networks while maintaining performance.
    - **Energy-Efficient Inference:** Energy-efficient inference strategies are required to prolong battery life. A low-energy neural network design was introduced by the researchers [17] for TinyML applications.
- **Real-World Deployments**
    - **Speech Recognition:** TinyML models enable recognition on devices like hearables and voice assistants. The paper [18] presented a TinyML-based voice command recognition system for wearable devices.
    - **Gesture Recognition:** Gesture recognition on wearables benefits from TinyML's efficiency. The research [19] developed a gesture recognition system using TinyML for human–computer interaction.
    - **Vibration Analysis:** TinyML aids in analyzing vibrations for fault detection. The paper [20] showcased a vibration-based anomaly detection system on small devices.

Using TinyML for security purposes in WBANs involves implementing lightweight and efficient ML models to address specific security challenges. Here's how TinyML can be applied to enhance security in WBANs:

- **Intrusion Detection:** TinyML can be utilized to develop efficient anomaly detection models that run directly on the WBAN devices. These models can continuously monitor the data generated by various sensors in the WBAN and identify unusual patterns that may indicate an intrusion or unauthorized access. By processing data locally, these models can provide real-time alerts without constant communication with external systems, improving the responsiveness of intrusion detection.
- **Device Authentication:** TinyML can assist in implementing lightweight authentication mechanisms on WBAN devices. ML models can be trained to recognize unique patterns or behaviors associated with authorized devices or users. When a new device attempts to join the network, the model can assess its legitimacy based on these patterns, preventing unauthorized devices from gaining access.
- **Behavioral Analysis:** ML models can be trained to analyze the behavioral patterns of users or devices in the WBAN. By establishing baselines for normal behavior, the models can detect deviations from these patterns, which may indicate a security breach or unusual activity. This approach can help identify compromised devices or abnormal user behavior.
- **Resource-Efficient Encryption:** TinyML can be used to optimize encryption and decryption processes in WBANs. By developing lightweight cryptographic algorithms or optimizing existing ones, the computational burden

of encryption can be reduced, making it more feasible to implement strong encryption on resource-constrained devices.
- **Secure Communication:** TinyML models can assist in verifying the integrity of communication within the WBAN. By analyzing data payloads for signs of tampering or data corruption, these models can help ensure that the information exchanged between devices remains intact and secure.
- **Privacy-Preserving Data Analysis:** TinyML can analyze sensitive health data locally on WBAN devices while preserving user privacy. This approach can reduce the need to transmit raw data to external servers for analysis, mitigating privacy risks associated with data transfer.
- **Real-Time Threat Mitigation:** TinyML models can be designed to trigger specific actions or countermeasures in response to detected security threats. For example, if an intrusion is detected, the WBAN device can take immediate steps to isolate the compromised device, enhance encryption, or trigger alerts to the user.

It's essential to tailor the TinyML-based security approach to the specific security requirements and constraints of the WBAN environment. In addition, ongoing model monitoring and updates should be considered to adapt to evolving security threats and maintain the effectiveness of the security measures.

Security is an ongoing process, and staying up to date with the latest security research and best practices is essential to ensure the continued effectiveness of your TinyML-based security solution in the WBAN environment.

Security Challenges in WBANs: Security challenges in WBANs arise due to the sensitive nature of the health data being transmitted and the unique characteristics of the network environment. Addressing these challenges is crucial to ensure data privacy, confidentiality, integrity, and availability. Here are some key security challenges in WBANs:

Data Privacy and Confidentiality: WBANs collect sensitive health data, including personal medical records and physiological measurements. Ensuring the privacy and confidentiality of this data is paramount. Unauthorized access to health data can lead to identity theft, exposure to personal medical conditions, and potential misuse of the information.

Solutions: Encryption techniques such as symmetric and asymmetric encryption can secure data in transit and at rest. Privacy-preserving data aggregation methods can reduce the granularity of data while still enabling useful analysis.

Authentication and Access Control: WBANs often consist of multiple wearable devices, requiring robust authentication mechanisms to ensure that only authorized devices can join the network. Malicious devices might infiltrate the network without proper authentication, leading to unauthorized data access or denial of service attacks.

Solutions: Two-factor authentication, digital certificates, and biometric authentication can enhance device and user authentication. Access control policies should be enforced to restrict device interactions.

Resource Constraints: WBAN devices are resource-constrained in processing power, memory, and energy. Implementing traditional security measures can consume excessive resources, leading to performance degradation and shorter battery life.

Solutions: Design lightweight encryption algorithms and authentication protocols tailored for resource-constrained devices. Optimization techniques like data aggregation and compression can reduce communication overhead.

Wireless Communication Vulnerabilities: WBANs communicate wirelessly, making them vulnerable to eavesdropping, interception, and jamming attacks. The short-range communication nature of WBANs limits the coverage area, but attackers within proximity can still exploit vulnerabilities.

Solutions: Implement secure communication protocols like IEEE 802.15.6, including encryption and authentication provisions. Frequency hopping and spread spectrum techniques can enhance resistance to interference.

Data Integrity: Ensuring data integrity is crucial, as tampering with health data can lead to incorrect diagnoses and treatment. Unauthorized modifications can also compromise the trustworthiness of medical records and research outcomes.

Solutions: Hash functions, digital signatures, and message authentication codes can be used to detect data tampering and ensure data integrity.

Interoperability and Standardization: WBANs might interact with other networks, such as hospital information systems or cloud servers, leading to potential security gaps at the interfaces. A lack of standardized security protocols and practices can result in vulnerabilities when data crosses different network boundaries.

Solutions: Adopt standardized security protocols and communication frameworks that ensure secure data exchange between different components of the WBAN ecosystem.

User Privacy and Consent: User privacy must be respected, and individuals' explicit consent should be obtained for data collection, storage, and sharing. Improper handling of personal health data can violate privacy regulations and lead to legal consequences.

Solutions: Implement mechanisms for obtaining informed consent, anonymizing data when necessary, and giving users control over their data-sharing preferences.

Dynamic Network Topology: WBANs can have active and frequently changing network topologies due to the movement of wearable devices. This dynamic nature can make maintaining secure and stable communication links challenging.

Solutions: Develop adaptive security mechanisms that can accommodate changes in network topology. Use trust-based routing protocols to enhance communication reliability.

Addressing these security challenges requires a multifaceted approach involving cryptographic techniques, secure protocols, intrusion detection systems, and awareness of the unique constraints of WBAN devices. As WBAN technology advances, ongoing research and collaboration are essential to stay ahead of emerging security threats.

12.3 SECURITY MECHANISMS AND PROTOCOLS IN WBAN

Security mechanisms and protocols in WBANs play a crucial role in ensuring sensitive health data's privacy, integrity, and confidentiality. Here are some key security mechanisms and protocols commonly used in WBANs:

- **Authentication and Key Establishment:**
  - **Device Authentication:** Devices in a WBAN need to authenticate themselves to ensure that only authorized devices can join the network. Lightweight authentication methods like symmetric key authentication or digital signatures can be used.
  - **User Authentication:** Users accessing WBAN data may need to authenticate themselves. Biometric authentication, PINs, or passwords can be used for user identity verification.
  - **Key Establishment:** Secure key exchange protocols like Diffie-Hellman can establish session keys between communicating devices to enable encrypted communication.
- **Encryption:**
  - **Data Encryption:** Encrypting data during transmission and storage is crucial to prevent unauthorized access and eavesdropping. Symmetric and asymmetric encryption algorithms can be used to ensure confidentiality.
  - **End-to-end Encryption:** Applying end-to-end encryption ensures that data remains encrypted from the sender to the recipient, preventing intermediate devices from accessing the data.
- **Secure Communication Protocols:**
  - **IEEE 802.15.6:** This standard defines a secure communication protocol for WBANs, incorporating features such as channel hopping, encryption, and authentication to enhance communication security.
  - **Lightweight Protocols:** Due to resource constraints, lightweight communication protocols like TinyDTLS (Datagram Transport Layer Security) can be used for secure communication on WBAN devices.
- **Privacy-Preserving Techniques:**
  - **Data Aggregation:** Aggregating data from multiple sensors before transmission can help preserve individual privacy while enabling useful analysis.
  - **Data Masking:** Sensitive data can be masked or perturbed before transmission, reducing the risk of unauthorized data exposure.
- **Intrusion Detection and Prevention:**
  - **Behavioral Analysis:** ML models can be trained to recognize normal behavior patterns in the network. Any deviation from these patterns can trigger alerts for potential intrusion attempts.
  - **Anomaly Detection:** Real-time anomaly detection can identify abnormal activities or patterns that may indicate security breaches.
- **Secure Routing Protocols:**
  - **Trust-Based Routing:** Trust levels can be assigned to devices, and routing decisions can be made based on trust levels. This prevents compromised devices from participating in routing decisions.
  - **Energy-Aware Routing:** Secure routing protocols can also consider energy consumption to optimize routes and prolong device battery life.

- **Access Control:**
  - **Role-Based Access Control (RBAC):** Assigns specific roles and permissions to devices or users based on their responsibilities and authorization levels.
  - **Attribute-Based Access Control (ABAC):** Access control decisions are based on attributes of the requester, the resource, and the environment.
- **Secure Bootstrapping:**
  - **Initial Trust Establishment:** During the initial deployment, mechanisms for establishing trust between devices can be employed. Public key infrastructure (PKI) and certificate-based authentication can be used for secure bootstrapping.
- **Blockchain for Audit Trails:**
  - **Immutable Records:** Employing blockchain technology can create an immutable audit trail of events, ensuring transparency and accountability in data handling and access.
- **Physical Security Measures:**
  - **Tamper Detection:** Implement mechanisms to detect physical tampering of devices, such as sensors and wearables, which can compromise data integrity.

Addressing security in WBANs requires a combination of these mechanisms and protocols tailored to the specific requirements and constraints of the network. As WBAN technology evolves, continuous research and development are essential to stay ahead of emerging security challenges.

12.4 PROPOSED RESEARCH METHODOLOGY

In our proposed algorithm (see Figure 12.3), the data is processed in the edge device, a microcontroller (Arduino Nano 33 BLE Sense). The sensory data is collected from different sensors that are placed in the different parts of human body, and the data is processed by the TinyML model that is compiled and uploaded in the Arduino. The model detects any emergency situation and, in case of emergency, triggers an alert in the form of SMS, which is sent to the medical expert or doctor and the patient's close family members, who can then take care of the patient.

12.4.1 How TinyML Is Used for Security in WBAN

Using TinyML for security in WBANs involves implementing lightweight ML models on resource-constrained devices to address security challenges. Here's a step-by-step guide on how to use TinyML for security in WBANs:

Identify Security Challenges: Identify the specific security challenges you want to address in your WBAN. These could include intrusion detection, device authentication, anomaly detection, secure communication, and more.

Select a Security Approach: Determine the appropriate security approach for your WBAN based on the identified challenges. For example, if you want to detect anomalies in physiological data, you could choose anomaly detection as your approach.

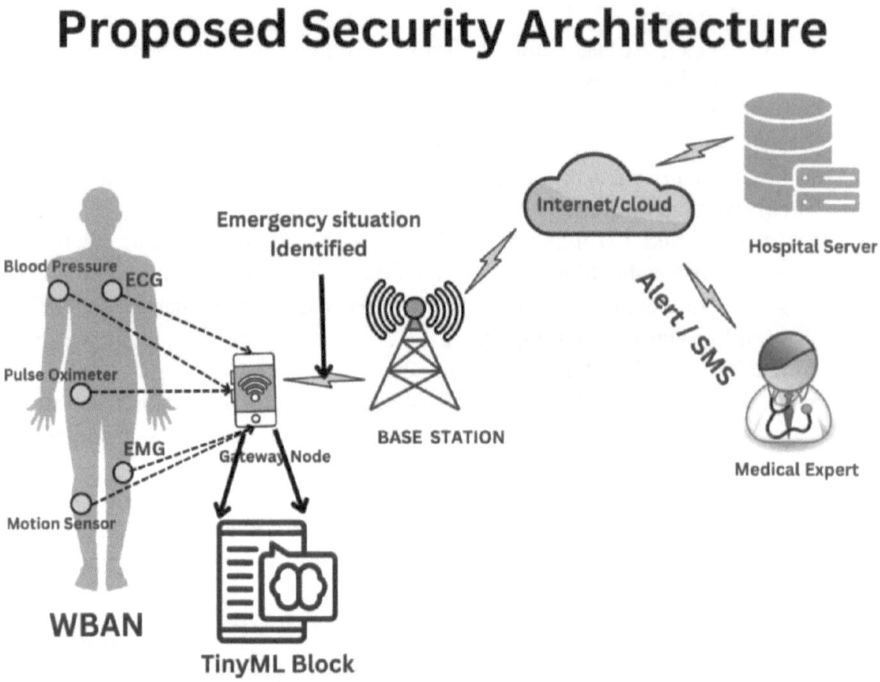

FIGURE 12.3 TinyML-based security mitigation.

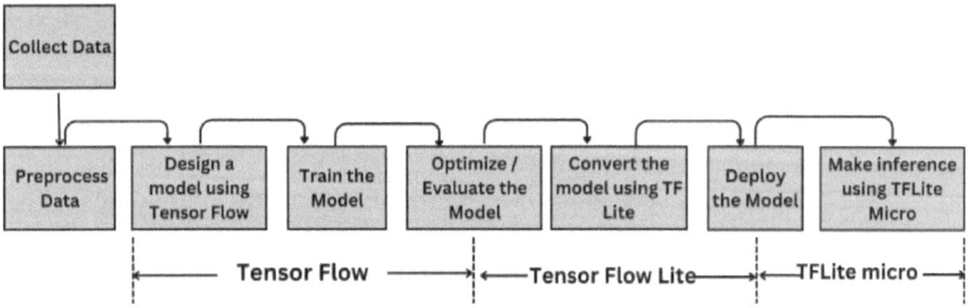

FIGURE 12.4 TinyML flow diagram.

Figure 12.4 shows the TinyML flow diagram. The first step is collecting data, and the next is to preprocess the data. Then design a model using TensorFlow. Then split the data into training, validation, and test set, and then train the model. After training, the model is evaluated and optimized. The model is then converted using TFLite for deployment in the microcontroller. This model is used for inference using inference engine like TFLite micro.

Collect and Prepare Data: Gather the necessary data for training and testing your TinyML security model. This could include normal and abnormal data samples related to the security challenge you're addressing. Ensure the data is representative of the WBAN environment and the potential security threats.

Choose a TinyML Framework: Select a TinyML framework that is suitable for your WBAN devices. Frameworks like TensorFlow Lite for Microcontrollers and Edge Impulse offer tools and libraries for developing and deploying TinyML models on resource-constrained devices.

Feature Extraction and Model Design: Design your TinyML model architecture. Depending on the chosen security approach, you may need to extract relevant features from the data and design an ML model that can run efficiently on the WBAN device. Consider using lightweight model architectures such as small neural networks or decision trees.

Training the Model: Split your data into training and testing sets. Train your TinyML model using the training data, considering the constraints of memory and processing power on the WBAN device. Regularize the model to prevent overfitting, and optimize hyperparameters accordingly.

Quantization and Compression: Since WBAN devices have limited resources, perform model quantization and compression to reduce the model's size and memory footprint while maintaining acceptable performance. This step is crucial to ensure the model can run efficiently on the device.

Model Deployment: Deploy the trained and optimized TinyML model onto the WBAN device. The deployment process may involve converting the model to the appropriate format supported by the selected TinyML framework.

Real-Time Inference: Integrate the deployed TinyML model into the WBAN device's software architecture to perform real-time inference on incoming data. The model will analyze the data and make security-related decisions based on the chosen security approach.

Testing and Validation: Test the deployed model using real-world data to ensure it accurately detects security threats or anomalies. Validate the model's performance against predefined metrics, such as accuracy, false positive rate, and false negative rate.

Continuous Monitoring and Updating: Security threats evolve over time. Monitor the performance of your TinyML security model, and update it as needed to adapt to new challenges and maintain effectiveness.

Alert Generation: If the model detects an anomaly, generate alerts or notifications. These alerts could be sent to a monitoring dashboard, a mobile app, or an email system, depending on the application's requirements.

Feedback Loop: Continuously gather data from the pulse rate sensor and periodically retrain the model to adapt to changing patterns and reduce false positives/negatives. This creates a feedback loop that helps improve the model's accuracy over time.

Circuit Diagram Explanation: In Figure 12.5 the pulse rate data is collected by the pulse rate sensor. The model training is done by using Python Heart rate variability dataset of Harvard that is used to train the model. TensorFlow library is used to train and evaluate the model. The model is optimized, and then it is converted to TinyML model using TFLite Converter. The converted model is very lightweight, and the code is verified and uploaded to the Arduino Nano 33 BLE Sense. Inference is made using the TFLite micro. When the anomaly in pulse rate is detected, the alert is sent to the patient's family and doctor for further action. The SMS is sent using a Global System for Mobile Communications (GSM) module, which is connected to the Arduino, as shown in Figure 12.5.

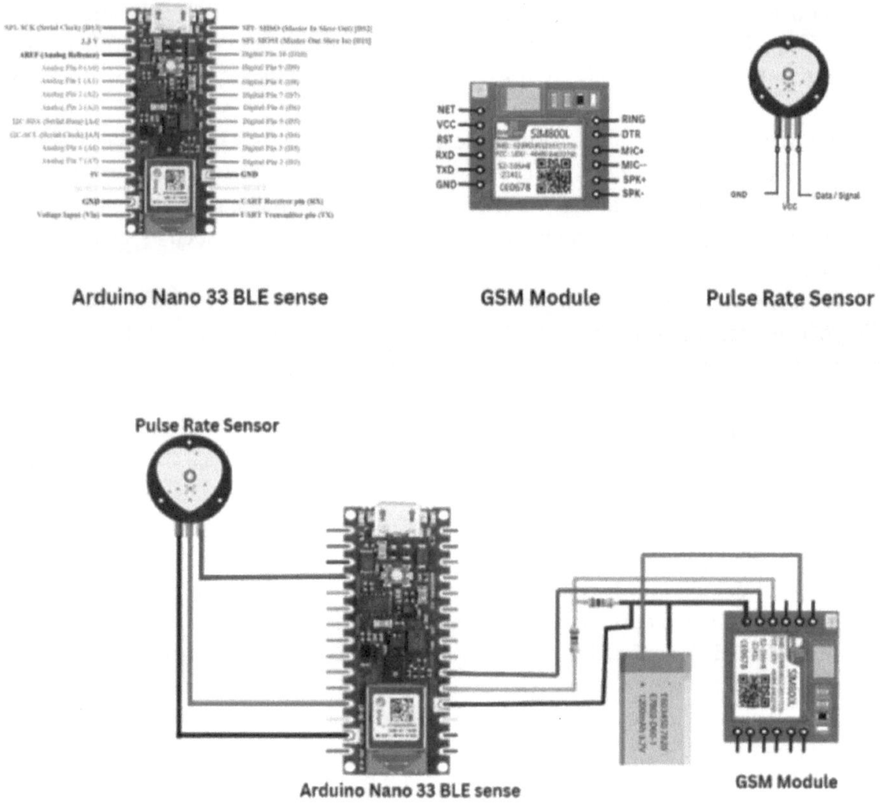

FIGURE 12.5 Circuit diagram for the proposed TinyML system.

12.4.2 Advantages of Using TinyML

Table 12.1 summarizes the different advantages of using TinyML in the WBAN system. We see that power and energy requirement is very less in the proposed WBAN system, since all the computation and processing are done in the edge devices, which is a microcontroller Arduino Nano 33 BLE Sense. Since power requirement is less (~mW), the battery powering the Arduino lasts longer, that is, battery life is more. Next valuable resource is the bandwidth; in traditional WBAN system, there is continuous flow of data to and from the cloud and hospital server and to the processing device, and this requires a lot of bandwidth in GBPS or hundreds of MBPS, but when we use TinyML, the bandwidth or data rate required is few kilobytes. This helps to make use of the spectrum for other purposes, thereby making use of the spectrum more efficiently. Latency of the entire system is also reduced because of the reduced communication overhead in the TinyML-based security architecture in the WBAN. One of the most important resources is the memory; the memory requirement is reduced to few kilobytes to megabytes from hundreds of gigabytes used in the traditional WBAN system. Finally, security threat is successfully mitigated in the TinyML-based WBAN security architecture as compared with the traditional WBAN system, which is prone

TABLE 12.1
Summary of Advantages of using TinyML versus Traditional WBAN Security

Parameters	Without Using TinyML	Using TinyML
Power requirements	Huge of order of watts or kilowatts	Comparatively less and order of milliwatts
Bandwidth	There is continuous need for large bandwidth, since sensor data needs to be sent to the cloud continuously	Less since there is no continuous communication, only in case of emergency the channel is used to send alert
Latency	More	Less
Memory	Large memory is needed	Comparatively, very little memory is needed since the model is compressed
Security	Prone to security issues, since communication is continuous and there is more chance of eavesdropping	Security threat is mitigated, and there is less chance of successful attacks on the system

to security threats and issues because of the continuous communication; as a result, there is more chance of eavesdropping and man-in-the-middle attacks.

12.5 CONCLUSION

WBANs hold immense promise for transforming healthcare and personalized monitoring. While significant progress has been made, challenges such as energy efficiency, security, and data analysis remain areas of active research. Advancements in signal processing and ML continue to enhance the capabilities of WBANs, enabling accurate health monitoring and real-time decision-making.

Security is paramount in WBANs due to the sensitive nature of health data and the unique challenges posed by resource-constrained devices. Researchers have made strides in addressing security challenges through encryption, authentication, secure communication protocols, and intrusion detection mechanisms. As WBANs continue to advance, incorporating emerging technologies like blockchain and post-quantum cryptography will be critical to ensuring a robust security framework.

TinyML's integration of ML into resource-constrained devices has opened new horizons for applications ranging from healthcare to IoT. While challenges persist, ongoing research into model optimization, energy efficiency, and deployment strategies continues to propel TinyML's impact across diverse domains

12.6 CHALLENGES AND FUTURE DIRECTIONS

Domain-Specific Models: Developing domain-specific models tailored to individual applications could enhance TinyML's efficiency and effectiveness [32].

Federated Learning: Applying federated learning techniques to TinyML models could enhance privacy and collaboration across distributed devices [22].

REFERENCES

[1] I. F. Akyildiz, W. Su, Y. Sankarasubramaniam and E. Cayirci (2002). Wireless Sensor Networks: A Survey. *Computer Networks*, 38(4), 393–422.

[2] E. Jovanov, A. Milenkovic, C. Otto, and P. C. de Groen (2005). A Wireless Body Area Network of Intelligent Motion Sensors for Computer Assisted Physical Rehabilitation. *Journal of NeuroEngineering and Rehabilitation*, 2(1), 6.

[3] M. Al Ameen, J. Liu and K. Kwak (2014). Security and Privacy Issues in Wireless Sensor Networks for Healthcare Applications. *Journal of Medical Systems*, 38(9), 1–8.

[4] J. Sen (2012, March). Secure and Privacy-Preserving Data Aggregation Protocols for Wireless Sensor Networks. *Cryptography and Security in Computing*, pp. 133–163. IntechOpen, Croatia.

[5] Thaier Hayajneh, Ghada Almashaqbeh, Sana Ullah and Athanasios V. Vasilakos (2014, November). A Survey of Wireless Technologies Coexistence in WBAN: Analysis and Open Research Issues. *Wireless Networks.* Springerlink.

[6] Naveed Ahmad, Mohammad Daud Awan, Malik Sikander Hayat Khiyal, Muhammad Imran Babar, Abdelzahir Abdelmaboud, Hamza Awad Ibrahim and Nadir O. Hamed, (2022, November). Improved QoS Aware Routing Protocol (Im-Qrp) For WBAN Based Healthcare Monitoring System, *IEEE Access*, 10.

[7] Shameer Faziludeen and P. V. Sabiq (2013). ECG Beat Classification Using Wavelets and SVM. *Proceedings of 2013 IEEE Conference on Information and Communication Technologies (ICT 2013).* IEEE.

[8] Jithin S. Sunny, C. Pawan K. Patro, Khushi Karnani, Sandeep C. Pingle, Feng Lin, Misa Anekoji, Lawrence D. Jones, Santosh Kesari and Shashaanka Ashili (2022). Anomaly Detection Framework for Wearables Data: A Perspective Review on Data Concepts. *Data Analysis Algorithms and Prospects*, Sensors 2022. MDPI.

[9] Soheil Zabihi, Elahe Rahimian, Amir Asif and Arash Mohammadi (2023, May). Light-Weight CNN-Attention Based Architecture for Hand Gesture Recognition Via Electromyography. *IEEE Explore*, 2023, 1–5.

[10] N. D. Lane, S. Bhattacharya, P. Georgiev, C. Forlivesi, F. Kawsar and F. Hamilton (2016). An Early Resource Characterization of Deep Learning on Wearables, Smartphones and Internet of Things Devices. *arXiv preprint arXiv:1610.00211.*

[11] Matteo Risso, Alessio Burrello, Daniele Jahier Pagliari and Simone Benatti (2021). Robust and Energy-Efficient PPG-Based Heart-Rate Monitoring. *IEEE International Symposium on Circuits and Systems (ISCAS)*, pp. 1–5. https://ieeexplore.ieee.org/abstract/document/9401282

[12] Soumyalatha Naveen, Manjunath R. Kounte and Mohammed Riyaz Ahmed (2021). Low Latency Deep Learning Inference Model for Distributed Intelligent IoT Edge Clusters. *IEEE Access*, 9, 160607–160621.

[13] Arunabha M. Roy and Jayabrata Bhaduri (2021). A Deep Learning Enabled Multi-Class Plant Disease Detection Model Based on Computer Vision, *MDPI Open Access*, 2(3), 413–428.

[14] Liang Yu, Yi Sun, Zhanbo Xu, Chao Shen, Dong Yue, Tao Jiang and Xiaohong Guan (2021, January). Multi-Agent Deep Reinforcement Learning for HVAC Control in Commercial Buildings. *IEEE Transactions on Smart Grid*, 12(1), 407–419.

[15] Anshumaan Chauhan, Siddhartha Bhattacharyya and S. Vadivel (2023). DQNAS: Neural Architecture Search Using Reinforcement Learning. *arXiv:2301.06687.*

[16] I. Hubara, M. Courbariaux, D. Soudry, R. El-Yaniv and Y. Bengio (2017). Quantized Neural Networks: Training Neural Networks with Low Precision Weights and Activations. *Journal of Machine Learning Research*, 18(1), 6869–6898.

[17] Daniele Atzeni, Davide Bacciu, Daniele Mazzei and Giuseppe Prencip (2022). *A Systematic Review of Wi-Fi and Machine Learning Integration with Topic Modeling Techniques*. MDPI.

[18] Mary Hewitt and Hamish Cunningham (2022), Taxonomic Classification of IoT Smart Home Voice Control. *arXiv:2210.15656v1.*

[19] Marcos Negreiros Rylo, Renan L. P. de Medeirosb and Vicente F. de Lucena Jr (2022). Gesture Recognition of Wrist Motion Based on Wearables Sensors. *Science Direct.* Elsevier.

[20] Nouman Shamim, Muhammad Asim, Thar Baker and Ali Ismail Awad (2023). Efficient Approach for Anomaly Detection in IoT Using System Calls. *MDPI Sensors.* MDPI.

[21] Wen-Xin Yuan, Bin Yan, Wen Li, Liu-Yao Hao and Hong-Mei Yang (2023). Block Chain-Based Medical Health Record Access Control Scheme with Efficient Protection Mechanism and Patient Control. *Multimedia Tools and Applications*, 82, 16279–16300.

[22] Taki Hasan Rafi, Faiza Anan Noor, Tahmid Hussain and Dong-Kyu Chae (2023). Fairness and Privacy-Preserving in Federated Learning: A Survey. *arXiv:2306.08402v2* [cs.CR] 14 July 2023.

[23] Muhammad Irfan Khalid, Mansoor Ahmed and Jungsuk Kim (2023). Enhancing Data Protection in Dynamic Consent Management Systems: Formalizing Privacy and Security Definitions with Differential Privacy. *Decentralization, and Zero-Knowledge Proofs, Sensors 2023*, 23(17), 7604. https://doi.org/10.3390/s23177604

[24] K. Sowjanya, Mou Dasgupta and Sangram Ray (2021, May). Elliptic Curve Cryptography Based Authentication Scheme for Internet of Medical Things. *Journal of Information Security and Applications*, 58, 102761

[25] Amit Kumar Bairwa and Sandeep Joshi (2023, February 6). ETB-SRF: An Efficient Trust-Based Secure Routing Framework in Mobile Ad-Hoc Networks. *IEEE Xplore.*

[26] Yong Ding, Hui Xu and Yujue Wang (2021, September). Group Authentication and Key Distribution for Sensors in Wireless Body Area Network. *International Journal of Distributed Sensor Networks*, 17(9).

[27] Ruby Dass, Manikandan Narayanan, Gayathri Ananthakrishnan, Tamilarasi Kathirvel Murugan, Musiri Kailasanathan Nallakaruppan, Siva Rama Krishnan Somayaji, Kannan Arputharaj, Surbhi Bhatia Khan and Ahlam Almusharraf (2023). A Cluster-Based Energy-Efficient Secure Optimal Path-Routing Protocol for Wireless Body-Area Sensor Networks. *Sensors*, 23(14), 6274.

[28] Xiaodong Shen, Liehuang Zhu, Chang Xu, Kashif Sharifa and Rongxing Lu (2020, April). A Privacy-Preserving Data Aggregation Scheme for Dynamic Groups in Fog Computing. *Science Direct, Information Sciences*, 514, 118–130.

[29] Yakub Kayode Saheed, Aremu Idris Abiodun, Sanjay Misra, Monica Kristiansen Holone and Ricardo Colomo-Palacios (2022, December). A Machine Learning-Based Intrusion Detection for Detecting Internet of Things Network Attacks. *Alexandria Engineering Journal, Science Direct*, 61(12), 9395–9409.

[30] Shaymaa Adnan Abdulrahman, Ehsan Qahtan Ahmed, Zahraa A. Jaaz and Alya'a R. Ali (2023). Intrusion Detection in Wireless Body Area Network using Attentive with Graphical Bidirectional Long-Short Term Memory. *International Journal of Online and Biomedical Engineering (iJOE)*, 19(6).

[31] Jose-Antonio Septien-Hernandez, Magali Arellano-Vazquez, Marco Antonio Contreras-Cruz and Juan-Pablo Ramirez-Paredes (2022). A Comparative Study of Post-Quantum Cryptosystems for Internet-of-Things Applications, *Sensors*, 22(2), 489.

[32] Albatul Albattah and Murad A. Rassam (1951). A Correlation-Based Anomaly Detection Model for Wireless Body Area Networks Using Convolutional Long Short-Term Memory Neural Network, MDPI. *Sensor*, 22(5).

13 Lightweight Solutions for Securing WBAN

P. Rajasekar, H. Mangalam, P. Chandrasekar, and P. Muthu Krishnammal

13.1 INTRODUCTION

Wireless Body Area Networks (WBANs) are tiny networks comprising intelligent, low-power sensor nodes that monitor the patient's health. These intelligent low-powered sensor nodes may be implanted on the body to link the numerous sensors and the network [1]. Based on the application, IEEE standard 802.15.16 divides WBANs into two categories, namely, medical application and nonmedical application. A medical application enables the continuous monitoring of a patient's medical condition in critical situations by promptly detecting any alterations through the use of attached sensors [2]. These details are to be promptly diagnosed and the treatment is to be identified to preserve the patient's life. Indeed, the process's data is crucial since it must be reliable, trustworthy, and secure. WBANs offer cost saving and flexibilities for both patients and healthcare professionals. Regardless of a person's location or level of physical activity, these networks are made up of sensors or other devices that are used to track and transmit physiological signals of various body functions, such as electrocardiogram (ECG), heart rate, temperature, blood pressure (BP), respiration, and other variables [3]. A WBAN offers two significant advantages in the current patient monitoring system, namely, a location-independent monitoring system model and autonomous communication system to transmit the data to remote server system [4]. In addition, the WBANs differ significantly from other Wireless Sensor Networks (WSNs) pertaining to the mobility of the device and energy consumption. WBAN users can move with sensor network with same patterns, whereas WSNs need stationary pattern [5]. WBANs were created as a result of pervasive wireless communication and the shrinking of actuators, sensors, and central processing units (CPUs). WBAN technology's mobility has also improved due to smartphone technology. WBAN may now be linked to cellular, Wi-Fi, Bluetooth, ZigBee, and internet networks. Wi-Fi is the most extensively utilized technology for wearable sensor nodes due to its low cost and rapid data rate (see Figure 13.1) [3,6].

In addition, the data is transmitted across an insecure route; therefore, data protection and authentication must be ensured. Different security methods may be employed in data transmission to ensure the privacy. However, crucial factors, including energy usage in WBANs, play a vital role. Hence, there are numerous research activities in lightweight security algorithms.

DOI: 10.1201/9781032635101-16

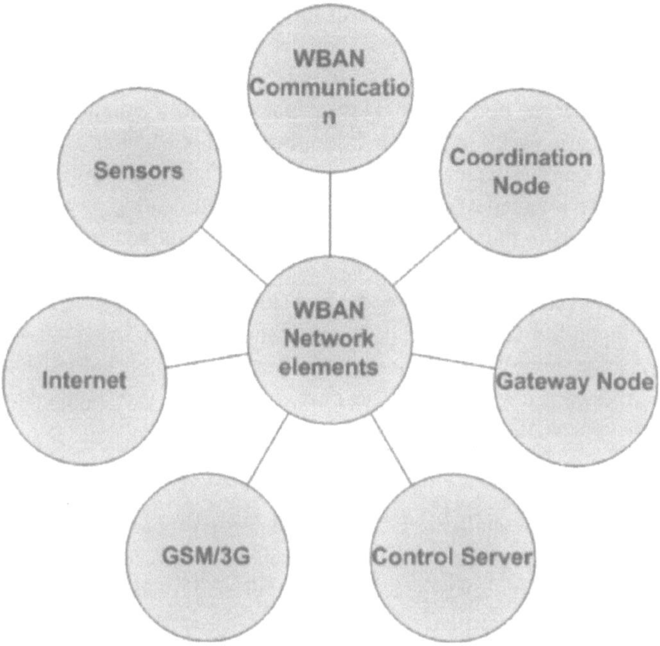

FIGURE 13.1 Components in WBAN.

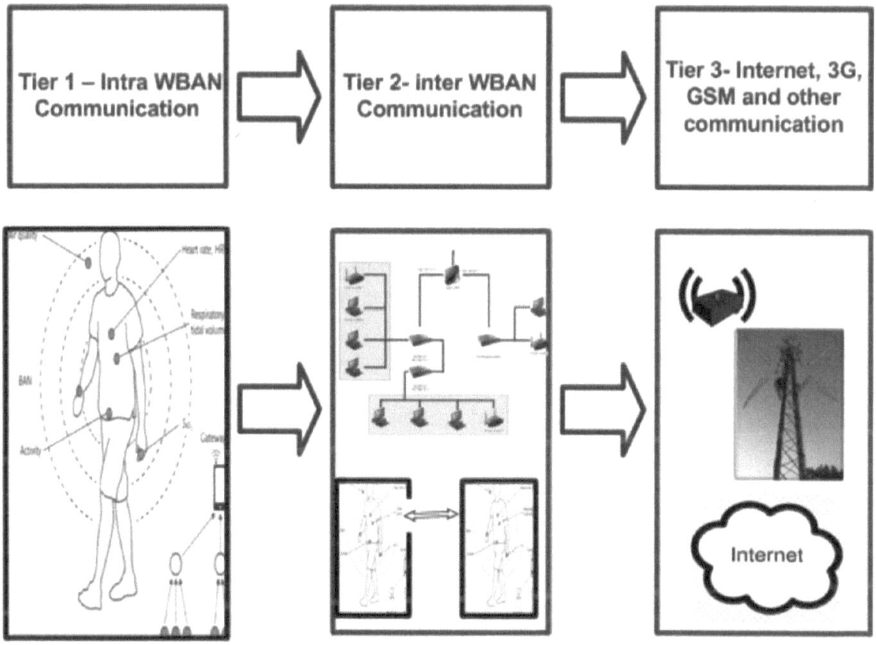

FIGURE 13.2 Types of tier communication.

There are three communication tiers in the WBAN: tier 1, tier 2, and tier 3 (see Figure 13.2). Sensor data from the human body is sent to the next communication layer via tier 1 communication, sometimes referred to as intra-BAN communication [7]. The sensor devices gather the body's signal and transform it into a digital signal. The next level of communication is known as tier 2 communication, which is used to process the inward data from tier 1. This level may be sent to the data in the higher level if necessary. Tier 2 is often referred to as the inter-BAN communication paradigm, and tier 3 communication refers to the usage of decision-making in the received data [8].

Regardless of whether the WBAN is two-tiered or three-tiered, WBAN entities can be divided into two categories: sensor nodes and sink or gateway nodes. The sensor node collects information from the human body, while the sink node sends it to other servers and communication links. These communication networks can include various networks such as mobile network, a wireless local area network (WLAN), a hospital, among others.

13.2 IMPORTANCE OF WBANS AND SECURITY

WBANs face numerous challenges, including energy efficiency, quality of service (QoS), and security issues, such as authentication, data loss, access control, and privacy concerns [9]. Several investigators have developed robust security mechanisms for WBAN to protect life-critical data.

In WBAN, internal communication requires security in data authenticity, confidentiality, integrity, and availability (see Figure 13.3). An attacker might insert false data into the WBAN. Public key cryptography techniques are used to ensure the validity of data. Data is indeed revealed to unauthorized parties as a result of data confidentiality. The usage of Message Authentication Code (MAC) or hashed MAC ensures data integrity. The reverse technique for generating authentication codes maintains integrity. The most common security need for WBAN is data availability. Because of the criticality of the physical system, WBAN assures data availability. For the best services, all of the aforementioned characteristics must be met in the WBAN [10].

In external communication, threats to WBAN include message alteration and unauthorized access. Consider a solid security approach for protecting transmission between external users and WBAN, where each user must confirm their validity and then access the data based on their privileges. The physical, MAC, transport, and network layers are all vulnerable to communication protocol attacks (see Figure 13.4) [11,12].

One of the most notable attacks that occurred at the physical layer (PHY) was the jammer attack (see Table 13.1). An attacker generates a random frequency, which interferes with sensor signals. A collision attack raises the back of a specific packet, resulting in a defective packet in transmission service, which leads to denial-of-service

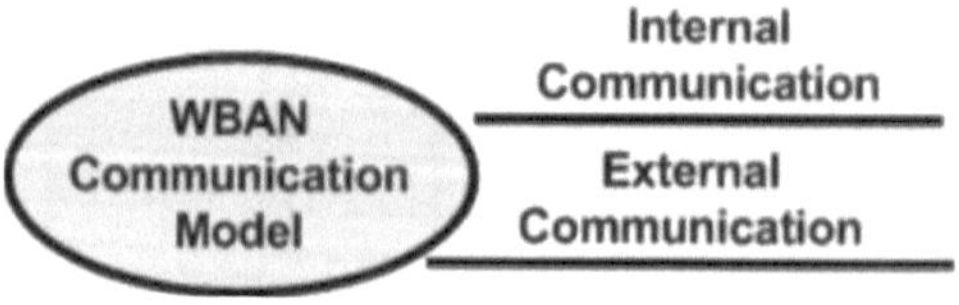

FIGURE 13.3 Types of communication.

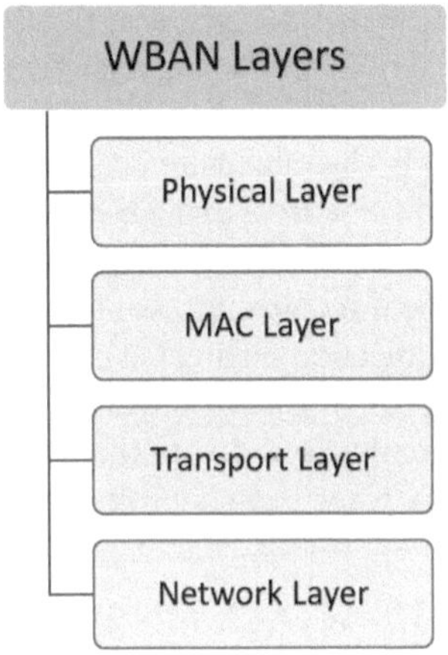

FIGURE 13.4 Layers in WBAN prone to attacks.

TABLE 13.1
Types of Attacks in WBAN

Types of Attacks	Attacking Factor
DoS	A noisy signal is transmitted to drop out the original signal in the communication channel.
Eavesdropping	Communications between sensor nodes and base stations are intercepted. An impersonation attack was launched by an attacker.
Man-in-the-middle (MitM)	An attacker acts as a relay or proxy between sensor node and the controller. Attackers manipulate the data without sender and receiver knowledge.
Insider attack	By introducing a physically compromised node, which leads to misrouting or dropping the data packet in the network.
Sniffing attack	The attacker retrieves the source address and destination address, protocol, and port information.
Physical attack or node-compromising attack	An attacker can alter the patient data by introducing a physical attack on the node using malicious code.
Sinkhole attack	Compromised node drops the neighboring node and redirects the data packet.
Masquerading attack	The attacker clones the device and injects false information into the patient database.
Spoofing and Sybil attack	A spoofing attack occurs when a malicious node poses as an authorized system object in order to cause network disruption and evade discovery.
Data falsification/Data modification	Attackers place the compromised node to send the modification data.
Replay attack	Attacker continuously resends the same data to confuse.

(DOS) at the MAC layer. In the network layer, a Sybil attack is presented that accepts single node as well as multiple node identifications. Another method is for a hacked node to fake or replay network nodes. It was the first to use selective forwarding node assaults. Nevertheless, WBANs encounter challenges when it comes to ensuring security because of their limited resources and adverse environmental circumstances. Therefore, ensuring security and privacy is a difficult task. It is a very challenging task to keep the patient's medical records accessible to only authorized persons over an unsecured wireless communication channel. Many studies have been carried out to overcome the security issues in WBAN. However, some issues are still dominant factors in the WBAN layer architecture due to their limited resources. A low-cost, lightweight security framework plays a crucial role in this network structure.

13.3 VARIOUS SECURITY ATTACKS IN WBAN

A WBAN connects all implanted, nonimplanted, and wearable human sensor devices in e-health applications in an efficient and dependable manner. Since health information transmitted over a WBAN may be exposed to unauthorized users or malicious adversaries, security services such as data collection, storage system transmission, and user access must be provided throughout the data stream. Data collection, transmission, and storage may be split into a three-layer network architecture.

The absence of adequate security measures might have negative impact. Hence, it is essential for WBANs to give precedence to strategies that ensure the security and privacy protection. To prevent threats, a WBAN must offer security services throughout the information life cycle of risks such as hacking, falsification, or data acquisition by a third party. In connectivity, the Internet of Things (IoT) can be used in health services. As a result, the attacks also include the Internet of Medical Things (IoMT). The present healthcare system integrates cutting-edge technology into medical practices. The implementation involves many platforms and prototypes for data gathering, data processing, and data transfer components.

Trust Management (TM) facilitates the establishment of a strong and secure platform for IoMT devices by implementing multiple security protocols inside the IoMT network. In general, if a device is unable to connect seamlessly with other devices inside a network, the intended objective of the communication process is not met, leading to dissatisfaction for one device in comparison to others. To address such issues, the Fuzzy-based TM method for defending Sybil attacks for the FTM-IoMT is presented. Through the establishment of IoMT infrastructures, the FTM-IoMT provides telemedicine services to persons utilizing e-health systems. The implementation functions as an intelligent tool for recognizing Sybil or fraudulent nodes inside the system.

13.4 WBAN'S SECURITY MECHANISM

Many security methods are used in WBAN to safeguard data from being accessed by unauthorized parties via insecure routes. Password authentication is one such basic approach discussed by Li et al. [13]. It is insufficient to rely just on one component to avoid infiltration since that aspect may be compromised. An ideal WSN/WBAN security protocol should offer both high security and low-cost calculation.

Systems are proposed using two or more components, such as biometrics, smart cards, mobile phones, and passwords, to increase the security level of authentication [14]. Consequently, additional security measures are implemented to improve authentication. A smart card is an essential part of two-factor authentication in WSNs and WBANs, in addition to a password [13].

Xu Li et al. [9] presented two schemes for authentication utilizing hash functions and smart cards, as well as mobile users. It does not include symmetric encryption or decryption. The established concept is divided into three phases: the initial phase, the first phase, and the second phase. The hash function in the proposed scheme has been changed [15]. The authors described a method for anonymous mutual authorization and key agreement based on fundamental cryptographic primitives such as the hash function and the XOR operation. To protect the distribution of a cipher key as well as inter-WBAN communications, Poon et al. [16] designed a biometric approach that takes use of a natural human body characteristic. Wang et al. [17] proposed a linked biometric-based security architecture that makes advantage of biometric traits shared by sensor nodes positioned throughout the body. Encryption is done using a person's biometric key. This employs a selective encryption method that uses less computing resources by just encrypting the essential data and leaving the rest of the biometric data unencrypted. Furthermore, when static ECG characteristics are used for encryption, key exchange is not necessary. Mana et al. [18] presented an efficient and energy-efficient trust key management system for WBANs capable of producing symmetric secret keys for both encryption and decryption, as well as a secure key distribution method. A human body's ECG signal is used to produce the keys, protecting both privacy and energy.

Sampangi et al. [19] developed the KEMESIS (Key Management and Encryption for Securing Inter-Sensor Communication) and IAMKeys (Independent and Adaptive Management of Keys) systems, which eliminate the requirement for key exchange. A key management mechanism known as KEMESIS is used to secure inter-WBAN communication. IAMKeys is a versatile and self-contained key management system that enhances WBAN security. By generating a random key for encrypting data frames at both communicating ends individually, these techniques eliminated the necessity for key exchange in the network [19].

Baqai et al. [20] investigated challenges like authentication and sensor recognition, energy-efficient packet transfer and reception with interference rejection, and others. This protocol offers a practical, low-cost solution in terms of computing and storage expenses. Kumar et al. [21] used genetic algorithm to develop guidelines for data storage and transmission security, as well as to create a more random, sophisticated, and surprising shared key for data security in WBAN.

Sammoud et al. [22] developed and enhanced a biometric-based key setup mechanism for WBAN that uses less energy. This protocol developed and communicated a symmetric key among two WBAN nodes using ECG data. To create identical sequences and minimize variation between two observed ECG signals, the error-correcting algorithm Bose–Chaudhuri–Hocquenghem (BCH) is utilized. The morphing function is used to remove the association between the symmetric keys employed and the ECG signal in order to safeguard user anonymity and privacy. The third node serves as a secure key distribution point between two WBAN nodes. The third node's selection has a substantial influence on resource utilization and the crucial setup procedure.

Irum et al. [23] developed a hybrid key management system for both tier 1 and tier 2 WBAN communications, which included both key preloading and auto-generation. Due to the difficulty of linear time, only one key is preloaded in tier 1 communication when the personal server (PS) is compromised, and the other keys are produced automatically from the ECG, but tier 2 communication employs a preloading-based approach that increases privacy by removing the key exchange phase.

He et al. [24] investigated WBAN anonymous authentication techniques and found them to be problematic for e-healthcare applications. The authors showed an impersonation threat and found that Liu et al.'s [25] anonymous authentication approach is subject to impersonation attacks. As a result, they established the anonymous authentication system, taking into consideration the various security requirements for WBANs. Three entities comprise the network concept: the WBAN client, the network manager, and the application provider. The anonymous authentication technique consisted of three steps: system initialization, client registration with Network Manager, and client authentication with Application Provider. Network Manager is a trustworthy third party that generated the Application Provider's system settings and public/private keys and keeps them secure and safe. After registering with Network Manager, WBAN client received its secret key represented by Application Provider. As a result, Application Provider and WBAN client may authenticate each other. Because the Application Provider was intended to be installed on hospital property, it was easily exposed to physical engineering attacks such as firmware modification and tampering with memory data [25]. The possibility of impersonation can be mitigated if data for authentication purposes is stored in the Network Manager's database rather than the Application Provider's database, as recommended by anonymous authentication.

According to the National Institute of Standards and Technology (NIST), Advanced Encryption Standard (AES) model has a high data rate and speed with low energy. However, it suffers from management key exchange in the network. According to Hercigonja et al. [26], asymmetric cryptographic algorithms need high processing time and large memory. However, this cryptography method is more secure than symmetric methods. According to Priya et al. [27], using the elliptic curve cryptography (ECC) method alone required higher processing time and computing power.

Herba et al. [28] investigated the benefits of both asymmetric and symmetric encryption methods for secure transmission. To create the secure WBAN communication, they employed AES, Rivest–Shamir–Adleman (RSA), and Hash. In their model, AES is used to encrypt patient data, RSA to guarantee secure key exchange, and HASH to assure data integrity. However, while it provides high-level protection, it necessitates more energy and memory.

Basnet et al. [29] used hybrid cryptography to reduce energy consumption in WBAN while maintaining strong data encryption levels. They researched at how WBAN was affected by the hybrid AES/RSA scheme. The authors suggested two energy modes, energy-rich mode and energy-saving mode, and they preferred the ECC method over the RSA algorithm for energy savings. When operating in energy-saving mode, the sensors' residual energy is below the threshold value, whereas when operating in energy-rich mode, it exceeds the threshold. Based on the type of energy mode, the encryption scheme is selected. All three algorithms—AES, ECC, and HMAC—are used in energy-rich mode; in energy-saving mode, however, patient data encryption and integrity are handled by AES and HMAC.

One of the major challenges to the global acceptance of WBAN is the protection of patient data. Various security techniques for the protection of patient data have been published in the literature. Nonetheless, due to resource constraints, WBAN faces a number of obstacles. Behind these constraints, hybrid security approaches assist in establishing high security, resulting in improved network lifespan. In the following session, all WBAN hybrid key cryptography protocols are compared with respect to their objectives, benefits, drawbacks, techniques, and other aspects.

13.5 LIGHTWEIGHT CRYPTOGRAPHY ALGORITHMS FOR WBAN

The TWINE cipher is a recently developed 64-bit block cipher that boasts a lightweight design. The system supports keys of 80 bits and 128 bits. The TWINE encryption algorithm utilizes a type 2 generalized Feistel structure (GFS) in order to address its deficiencies. This design choice allows TWINE to be very efficient when implemented on (embedded) software platforms while also enabling a compact integration of both encryption and decryption functionalities. The absence of bit permutation, an expanded Feistel-based method, and the absence of a Galois-Field matrix are the primary components of the proposal. The system consists of a 4-bit S-box and a 4-bit XOR operation [30].

The PRESENT algorithm, as seen in the Substitution-Permutation network (SPN), consists of a total of 31 rounds. Each block is 64 bits in length, and the system supports two extra key lengths: 80 bits and 128 bits. The security protections applied are considered adequate for the low-security applications often seen in tag-based deployments [31].

A new type of block cipher with two separate sets of keys, each independently controlling distinct encryption and decryption functions, was proposed to enhance security and operational flexibility. The KATAN ciphers, especially KATAN32, KATAN48, and KATAN64, are included in the first series of cryptographic algorithms. Each of the three ciphers is capable of accepting keys that are 80 bits in length while also possessing distinct block sizes. Specifically, KATANn has a block size of n bits. These three block ciphers exhibit a high level of compactness and are able to satisfy the minimum size requirement, while providing sufficient security. The second set, consisting of KTANTAN32, KTANTAN48, and KTANTAN64, achieves lower block ciphers at the expense of more flexibility. The KTANTANn algorithm exhibits more compactness compared with the KATANn algorithm. However, it is important to note that KTANTANn is only applicable in scenarios when the device is started with a single key that remains unalterable during its operation [32]. Another technique was taken into account in the construction of the PRINT cipher family. The PRINT cipher-48 cryptography technique runs on 48-bit blocks and is based on a predefined 48-bit secret key. It enhances its security by including a secret algorithm variability that contributes an extra 32 bits of protection. By either lowering the degree of security, for example, to 64 bits, or increasing the block size to 96 bits, other trade-offs may be made [33].

A group of block ciphers known as the KLEIN cipher have a fixed block size of 64 bits and a variable key length that might be 64, 80, or 96 bits. The ciphers will be denoted as KLEIN64/80/96, based on their relative key lengths. The significance of the key length and block size in the trade-offs between security and speed is well acknowledged in the context of block ciphers. When evaluating the efficiency of low-resource implementations, it is important to take into account the impact of key registers and intermediate results on the overall footprint. Furthermore, within the realm of ubiquitous computing, it is improbable for data flows to exhibit high-speed throughput. Consequently, using a large block size or key length for data encryption and authentication may prove to be unnecessary. The vulnerability of a 64-bit key length in terms of security is a subject of concern when considering attack scenarios that include pre-computation and significant storage capacity [10].

The LED cipher key is distinguished by its portability. A 64-bit block cipher, which is the one in question, is distinguished by its capacity to handle data in units of 64 bits. It has two main examples, each of which utilizes keys of either 64 or 128 bits in length. The cipher state is organized in a notional grid of dimensions 4×4. The Galois Field $GF(2^4)$ is represented by each nibble, or element, inside the grid. The field multiplication in this context is governed by the polynomial $X^4 + X + 1$. The LED cipher employs the well-known SBox, which has been widely used in a variety of lightweight cryptographic algorithms. The fundamental constituent of an LED consists of a series of four identical circular components, devoid of any further elements. The fundamental entity, hereafter referred to as a "step," facilitates the establishment of security boundaries for the building in a straightforward manner [34].

Piccolo encryption technique operates on 64-bit blocks and uses either 80-bit or 128-bit keys. Its design is iterative and may be categorized as a type of generalized Feistel network. Piccolo-80 and Piccolo-128 are the identifiers for the 80-bit and 128-bit key modes, respectively. Both ciphers are composed of two major parts: a data processing component and a critical scheduling component. The differences between the two major modes are defined by differences in the number of rounds for the data processing component and the key scheduling component. Piccolo's security level is rated acceptable for combating known kinds of analysis, such as current related-key differential and MitM attacks (see Table 13.2) [35].

13.6 LIGHTWEIGHT CRYPTOGRAPHY AUTHENTICATION SCHEME

Z. U. Rehman et al. [36] developed a lightweight cryptographic authentication approach with three steps that allow anonymized key agreement and authenticity for wireless data transfer. The proposed authentication method illustrates its efficiency in defending against several known cyberattacks, such as base station compromise and sensor node impersonation attacks. Formal validation was performed using BAN logic, while informal simulation was performed using the Automated Validation of Internet Security Protocol and Applications (AVISPA) tool.

Zia Ur Rehman [37] presented an anonymous, hybrid authentication approach that used physiological signals in conjunction with a lightweight cryptographic mechanism to provide strong security against well-known vulnerabilities, including

TABLE 13.2
Comparison of Various Lightweight Cryptography Algorithms

S. No.	Types of Algorithm	Block Size (Bits)	Key Length (Bits)	No. of Rounds	Functional Module	Features
1	KLEIN	64	64/80/96	12	Key expansion module, sub key mixing, SBox, PBox	Variable key length, hardware-friendly model
2	PRINT	48/96	80/160	48	—	3-bit operation
3	KATAN	32/48	64/80/96	254	SBox, PBox, key generation	Efficient implementation in resource-constrained environment
4	TWINE	64	80/128/192	32/36	SPN structure, key expansion, SBox, linear layer, key mixing	Its small code size and minimal computational requirements
5	PRESENT	64	80/128	31	SBox, PBox	Low-power IoT devices, radio-frequency identification (RFID) tags, and constrained sensor networks
6	Low Energy Device (LED)	64	64/128	48	SBox, key generation, PBox	Minimize computational and memory requirements

key escrow, base station compromise, and session intractability. To provide formal proof of mutual authenticity and key agreement, the widely established BAN logic is used.

O. Delgado-Mohatar et al. [38] implemented a novel energy-efficient technique for key formation and authentication that is specifically designed for sensor networks. The only symmetric cryptographic primitives needed for the method are encryption algorithms and keyed-hash functions (HMAC). Ensuring message secrecy and authenticity is its primary objective. Furthermore, the protocol has session-key transport features that minimize the quantity and amount of messages that are exchanged.

Sivasangari et al. [39] proposed SEKEBAN (Secure and Efficient Key Exchange for Wireless Body Area Networks). The method described here assures security by using an ECG signal-based key setup. The cluster head selection, logistic chaotic RC7 algorithm, and modified present algorithm are used in the SEKBAN model. The use of ECG signals for key creation is justified by the need to provide security as well as by the distinctiveness and uniqueness of each individual key. Since an attacker is unable to predict the keys, cryptographic keys must be unpredictable and keys derived from unpredictable temporal signals offer excellent security. Electrodes are positioned on the body to record the electrical activity of the heart, a procedure known as ECG. A P wave and QRS waves make up the standard ECG heartbeat trace.

The Two-Party Lightweight Authentication Protocol (TLAP) for WBANs in healthcare applications is proposed by Lara et al. [40]. To minimize processing overhead, TLAP makes use of self-certified public keys based on symmetric key encryption, scalar point multiplication, ECC, lightweight operations XOR, and traditional hash functions.

The study by Shen et al. [41] suggests a two-layer communication paradigm for WBANs, with the first layer concentrating on personal digital assistant (PDA) and sensor authentication and the second layer concentrating on PDA and access point (AP) authentication. Various authentication mechanisms are created for every tier, accounting for the restricted resources of individual sensors and the PDA's processing power. While the second layer protocol is totally wireless and occasionally does not require human intervention, the first layer protocol permits artificial intervention. The creation of a group key and authentication among group members are two features of the paper's safe and reliable one-to-many authentication system for WBANs. Furthermore, a brand-new nonpairing certificate less anonymous authentication procedure is suggested; it is reliable and effective even in the face of assaults from it.

13.7 DISCUSSION AND FINDINGS

The chapter covered the significance of WBAN's security mechanism. It explored data integrity, authenticity, and confidentiality for WBAN along with several security mechanisms for various levels and communication model formats. In fact, it provides a brief overview of several physical and network layer threats, including Sybil and jammer attacks. There is also a summary of several security attacks and their contributing components. In particular, the IoMT concept for trust management system was considered. According to the survey mentioned earlier, there are instances in which a single authentication system would not be adequate to protect the data.

A multi-authentication structure was implemented in this situation, enhancing the security mechanism. For example, two authentication techniques using smart card systems and hash functions were proposed by Xu Li et al. [9]. In an alternative approach, where the random number generation method is used to strengthen security, key management systems play a crucial part in security.

Subsequently, in order to develop the security mechanisms in tier 1 and tier 2 communication, several researchers worked on the hybrid key management system. Key preloading and the auto-generation technique are part of it. The significance of symmetric key encryption and decryption over asymmetric key encryption is summed up in the second section. It began with an enhanced version of many lightweight encryption and decryption techniques as well as a generic symmetric AES approach for security. The security model in lightweight cryptography has been implemented and analyzed using a variety of parameters. Based on the key length, the number of rounds, kind of function to be implemented, and power consumption, a lightweight encryption method has been selected.

13.8 CONCLUSION AND FUTURE WORK

To secure the WBANs, various lightweight cryptography algorithms are proposed based on various key factors such as energy, computational cost, area resources, and computational speed. All the algorithms mentioned are combined with additional security models and new schemes framed to enhance their authentication, data availability, security, and so on. Because of their resource limits and availability, these lightweight algorithms produce the greatest results in WBAN. Many researchers have proposed various authentication systems for various tiers. Authentication is ensured at each layer using a combination of lightweight algorithms and extra authentication mechanisms such as biometrics.

Further research can be conducted to explore lightweight cryptographic mechanisms that can provide strong security against vulnerabilities in WBANs, such as key escrow and base station attacks. The development of energy-efficient security protocols specifically designed for WBANs should be a focus of future work, considering the significant energy availability restrictions in these networks. Investigating the use of advanced encryption techniques, such as ECC, for secure transmission in WBANs could be a potential area of future research. Exploring the use of hybrid authentication approaches that combine physiological signals with lightweight cryptographic mechanisms to enhance security in WBANs could be another avenue for future work

REFERENCES

[1] M. S. Hajar, M. O. Al-Kadri, and H. K. Kalutarage, "A survey on wireless body area networks: Architecture, security challenges and research opportunities," *Comput. Secur.*, vol. 104, p. 102211, May 2021, http://doi.org/10.1016/j.cose.2021.102211.

[2] P. C. Paul, J. Loane, G. Regan, and F. McCaffery, "Analysis of attacks and security requirements for wireless body area networks—A systematic literature review," in *Systems, Software and Services Process Improvement*, vol. 1060, A. Walker, R. V. O'Connor, and R. Messnarz, Eds., in Communications in Computer and Information Science, vol. 1060, Cham: Springer International Publishing, 2019, pp. 439–452. http://doi.org/10.1007/978-3-030-28005-5_34.

[3] F. Noor, T. A. Kordy, A. B. Alkhodre, O. Benrhouma, A. Nadeem, and A. Alzahrani, "Securing wireless body area network with efficient secure channel free and anonymous certificateless signcryption," *Wirel. Commun. Mob. Comput.*, vol. 2021, p. e5986469, Oct. 2021, http://doi.org/10.1155/2021/5986469.

[4] J. Y. Khan and M. R. Yuce, *Wireless Body Area Network (WBAN) for Medical Applications, InTechOpen*, 2010. London, UK.

[5] M. T. Arefin, M. H. Ali, and A. K. M. F. Haque, "Wireless body area network: An overview and various applications," *J. Comput. Commun.*, vol. 5, no. 7, Art. no. 7, May 2017, http://doi.org/10.4236/jcc.2017.57006.

[6] M. F. Shaik, M. M. Subashini, and N. Swathi, "Implementation of a ZigBee based network for WBAN," in *2021 7th International Conference on Advanced Computing and Communication Systems (ICACCS)*, IEEE, 2021, pp. 188–192.

[7] L. Filipe, F. Fdez-Riverola, N. Costa, and A. Pereira, "Wireless body area networks for healthcare applications: Protocol stack review," *Int. J. Distrib. Sens. Netw.*, vol. 11, no. 10, p. 213705, 2015.

[8] D. Asam, et al., "Challenges in wireless body area network," *Int. J. Adv. Comput. Sci. Appl.*, vol. 10, Jan. 2019, http://doi.org/10.14569/IJACSA.2019.0101147.

[9] M. R. K. Naik and P. Samundiswary, "Wireless body area network security issues—Survey," in *2016 International Conference on Control, Instrumentation, Communication and Computational Technologies (ICCICCT)*, IEEE, 2016, pp. 190–194.

[10] S. Zou, Y. Xu, H. Wang, Z. Li, S. Chen, and B. Hu, "A survey on secure wireless body area networks," *Secur. Commun. Netw.*, vol. 2017, 2017.

[11] S. Al-Janabi, I. Al-Shourbaji, M. Shojafar, and S. Shamshirband, "Survey of main challenges (security and privacy) in wireless body area networks for healthcare applications," *Egypt. Inform. J.*, vol. 18, no. 2, pp. 113–122, 2017.

[12] M. Asam, T. Jamal, A. Ajaz, Z. Haider, and S. A. Butt, "Security issues in WBANs," *ArXiv Prepr. ArXiv191104330*, 2019.

[13] X. Li, J. Niu, M. Karuppiah, S. Kumari, and F. Wu, "Secure and efficient two-factor user authentication scheme with user anonymity for network based e-health care applications," *J. Med. Syst.*, vol. 40, pp. 1–12, 2016.

[14] D. He, N. Kumar, J. Chen, C.-C. Lee, N. Chilamkurti, and S.-S. Yeo, "Robust anonymous authentication protocol for health-care applications using wireless medical sensor networks," *Multimed. Syst.*, vol. 21, pp. 49–60, 2015.

[15] C.-C. Lee, M.-S. Hwang, and I.-E. Liao, "Security enhancement on a new authentication scheme with anonymity for wireless environments," *IEEE Trans. Ind. Electron.*, vol. 53, no. 5, pp. 1683–1687, 2006.

[16] C. C. Poon, Y.-T. Zhang, and S.-D. Bao, "A novel biometrics method to secure wireless body area sensor networks for telemedicine and m-health," *IEEE Commun. Mag.*, vol. 44, no. 4, pp. 73–81, 2006.

[17] H. Wang, H. Fang, L. Xing, and M. Chen, "An integrated biometric-based security framework using wavelet-domain HMM in wireless body area networks (WBAN)," in *2011 IEEE International Conference on Communications (ICC)*, IEEE, 2011, pp. 1–5.

[18] M. Mana, M. Feham, and B. A. Bensaber, "Trust key management scheme for wireless body area networks," *Int. J. Netw. Secur.*, vol. 12, no. 2, pp. 75–83, 2011.

[19] R. V. Sampangi, S. Dey, S. R. Urs, and S. Sampalli, "A security suite for wireless body area networks," *ArXiv Prepr. ArXiv12022171*, 2012.

[20] A. Baqai, F. A. Umrani, and B. S. Chowdhry, "A novel protocol with patient and node identification for optical WBAN with inherent security and interference rejection," *Wirel. Pers. Commun.*, vol. 95, no. 4, pp. 4211–4224, 2017.

[21] P. Kumar and A. Sharma, "Data security using genetic algorithm in wireless body area network," *Int. J. Adv. Stud. Sci. Res.*, vol. 3, no. 9, 2018.

[22] A. Sammoud, M. A. Chalouf, O. Hamdi, N. Montavont, and A. Bouallegue, "A new biometrics-based key establishment protocol in WBAN: Energy efficiency and security robustness analysis," *Comput. Secur.*, vol. 96, p. 101838, 2020.

[23] S. Irum, A. Ali, F. A. Khan, and H. Abbas, "A hybrid security mechanism for intra-WBAN and inter-WBAN communications," *Int. J. Distrib. Sens. Netw.*, vol. 9, no. 8, p. 842608, 2013.

[24] D. He, S. Zeadally, N. Kumar, and J.-H. Lee, "Anonymous authentication for wireless body area networks with provable security," *IEEE Syst. J.*, vol. 11, no. 4, pp. 2590–2601, 2016.

[25] J. Liu, Z. Zhang, X. Chen, and K. S. Kwak, "Certificateless remote anonymous authentication schemes for wireless body area networks," *IEEE Trans. Parallel Distrib. Syst.*, vol. 25, no. 2, pp. 332–342, 2013.

[26] Z. Hercigonja, "Comparative analysis of cryptographic algorithms," *Int. J. Digit. Technol. Econ.*, vol. 1, no. 2, pp. 127–134, 2016.

[27] C. L. Priya and U. S. Visalakshi, "Secure and efficient communication using ECC algorithm in wireless body area network," *Int. J. Eng. Sci.*, vol. 7, p. 10073, 2017.

[28] E. S. I. Harba, "Secure data encryption through a combination of AES, RSA and HMAC," *Eng. Technol. Appl. Sci. Res.*, vol. 7, no. 4, pp. 1781–1785, 2017.

[29] A. Basnet, A. Alsadoon, P. W. C. Prasad, O. H. Alsadoon, L. Pham, and A. Elchouemi, "A novel secure patient data transmission through wireless body area network: Health tele-monitoring," *Int. J. Commun. Netw. Inf. Secur.*, vol. 11, no. 1, pp. 93–104, 2019.

[30] T. Suzaki, K. Minematsu, S. Morioka, and E. Kobayashi, *TWINE: A Lightweight, Versatile Block Cipher*, Berlin, Heidelberg: Springer, 2011.

[31] U. Ambili, P. Rajasekar, and T. Parani, "FPGA implementation of lightweight present-GRP for RFID," *Int. J. Curr. Trends Eng. Res.*, vol. 3, no. 5, pp. 70–78, 2017.

[32] C. Cannière, O. Dunkelman, and M. Knežević, *Katan and Ktantan—A Family of Small and Efficient Hardware-Oriented Block Ciphers*, Berlin, Heidelberg: Springer, 2009, p. 288.

[33] L. Knudsen, G. Leander, A. Poschmann, and M. J. B. Robshaw, "PRINTcipher: A block cipher for IC-printing," in *Cryptographic Hardware and Embedded Systems, CHES 2010*, vol. 6225, S. Mangard and F.-X. Standaert, Eds., in Lecture Notes in Computer Science, vol. 6225, Berlin, Heidelberg: Springer Berlin Heidelberg, 2010, pp. 16–32, http://doi.org/10.1007/978-3-642-15031-9_2.

[34] J. Guo, T. Peyrin, A. Poschmann, and M. Robshaw, "The LED block cipher," in *Cryptographic Hardware and Embedded Systems–CHES 2011: 13th International Workshop, Nara, Japan, September 28–October1, 2011. Proceedings 13*, Springer, 2011, pp. 326–341.

[35] K. Shibutani, T. Isobe, H. Hiwatari, A. Mitsuda, T. Akishita, and T. Shirai, "Piccolo: An ultra-lightweight block cipher," in *Cryptographic Hardware and Embedded Systems—CHES 2011*, vol. 6917, B. Preneel and T. Takagi, Eds., in Lecture Notes in Computer Science, vol. 6917, Berlin, Heidelberg: Springer Berlin Heidelberg, 2011, pp. 342–357, http://doi.org/10.1007/978-3-642-23951-9_23.

[36] Z. U. Rehman, S. Altaf, and S. Iqbal, "An efficient lightweight key agreement and authentication scheme for WBAN," *IEEE Access*, vol. 8, pp. 175385–175397, 2020, http://doi.org/10.1109/ACCESS.2020.3026630.

[37] Z. U. Rehman, S. Altaf, S. Ahmad, S. Huda, A. M. Al-Shayea, and S. Iqbal, "An efficient, hybrid authentication using ECG and lightweight cryptographic scheme for WBAN," *IEEE Access*, vol. 9, pp. 133809–133819, 2021, http://doi.org/10.1109/ACCESS.2021.3115706.

[38] O. Delgado-Mohatar, A. Fúster-Sabater, and J. M. Sierra, "A light-weight authentication scheme for wireless sensor networks," *Ad Hoc Netw.*, vol. 9, no. 5, pp. 727–735, Jul. 2011, http://doi.org/10.1016/j.adhoc.2010.08.020.

[39] A. Sivasangari, P. Ajitha, and R. M. Gomathi, "Light weight security scheme in wireless body area sensor network using logistic chaotic scheme," *Int. J. Netw. Virtual Organ.*, vol. 22, no. 4, pp. 433–444, Jan. 2020, http://doi.org/10.1504/IJNVO.2020.107577.

[40] E. Lara, L. Aguilar, and J. A. García, "Lightweight authentication protocol using self-certified public keys for wireless body area networks in health-care applications," *IEEE Access*, vol. 9, pp. 79196–79213, 2021.

[41] J. Shen, S. Chang, J. Shen, Q. Liu, and X. Sun, "A lightweight multi-layer authentication protocol for wireless body area networks," *Future Gener. Comput. Syst.*, vol. 78, pp. 956–963, 2018.

Index

S

T

W